Remote Sensing of Aquatic Coastal Ecosystem Processes

Remote Sensing and Digital Image Processing

VOLUME 9

For other titles published in this series, go to
www.springer.com/series/6477

REMOTE SENSING OF AQUATIC COASTAL ECOSYSTEM PROCESSES

Science and Management Applications

edited by

LAURIE L. RICHARDSON

Florida International University, Miami, U.S.A.

and

ELLSWORTH F. LeDREW

University of Waterloo, ON, Canada

Including the WASI program
by Peter Gege, available at http://extras.springer.com

Springer

A C.I.P. Catalogue record for this book is available from the Library of Congress.

ISBN-10 94-017-8425-6
ISBN-13 978-94-017-8425-2
ISBN-10 1-4020-3968-9 (eBook)
ISBN-13 978-1-4020-3968-3 (eBook)

Published by Springer,
P.O. Box 17, 3300 AA Dordrecht, The Netherlands.

www.springer.com

Cover Image: 50 km Sea Surface Temperature Anomaly product for 31st July, 2002. NOAA. Also, see chapter 2, figure 2.

Printed on acid-free paper

Revised and corrected reprint 2010

Printed in the Netherlands.

Contents

Section II – Monitoring Applications

Chapter 6
INFRASTRUCTURE AND CAPABILITIES OF A NEAR REAL-TIME METEOROLOGICAL AND OCEANOGRAPHIC *IN SITU* INSTRUMENTED ARRAY, AND ITS ROLE IN MARINE ENVIRONMENTAL DECISION SUPPORT
JAMES C. HENDEE, ERIK STABENAU, LOUIS FLORIT, DEREK MANZELLO, AND CLARKE JEFFRIS

Chapter 7
AIRBORNE LASER ALTIMETRY FOR PREDICTIVE MODELING OF COASTAL STORM-SURGE FLOODING
TIM L. WEBSTER AND DONALD L. FORBES

Chapter 8
INTEGRATION OF NEW DATA TYPES WITH HISTORICAL ARCHIVES TO PROVIDE INSIGHT INTO COASTAL ECOSYSTEM CHANGE AND VARIABILITY
JENNIFER GEBELEIN

Section III – Management Applications

Chapter 9
OBSERVING COASTAL WATERS WITH SPACEBORNE SENSORS
BRIAN G. WHITEHOUSE AND DANIEL HUTT

Chapter 10
THE ROLE OF INTEGRATED INFORMATION ACQUISITION AND MANAGEMENT IN THE ANALYSIS OF COASTAL ECOSYSTEM CHANGE
STUART PHINN, KAREN JOYCE, PETER SCARTH, AND CHRIS ROELFSEMA

WASI Program and Manual available at http://extras.springer.com

LIST OF CONTRIBUTORS

ANDREAS ALBERT
DLR, Remote Sensing Technology Institute, P.O. Box 1116, 82230, Wessling, Germany

SERGE ANDRÉFOUËT
UR Coreus, Institut de Recherche pour le Développement, BP A5, 98848 Nouméa cedex, New Caledonia

FELIPE ARZAYUS
Coral Reef Watch, NOAA/NESDIS/ORA, E/RA31, SSMC1, Floor 5, 1335 East-West Highway, Silver Spring, MD 20910, USA

JOHN BROCK
U.S. Geological Survey, 600 4th Street South, St. Petersburg, FL 37701 USA

LAURETTA BURKE
Information Program, World Resources Institute, 10 G Street, NE, Washington, DC 20002 USA

LOUIS FLORIT
Atlantic Oceanographic and Meteorological Laboratory, National Oceanic and Atmospheric Administration, 4301 Rickenbacker Causeway, Miami, FL, 33149 USA

DONALD L. FORBES
Geological Survey of Canada, Bedford Institute of Oceanography, P.O. Box 1006, Dartmouth, Nova Scotia, Canada B2Y 4A2

JENNIFER GEBELEIN
Department of International Relations, Florida International University, Miami, FL, 33199 USA

PETER GEGE
DLR, Remote Sensing Technology Institute, P.O. Box 1116, 82230 Wessling, Germany

ROBERT HALLEY
U.S. Geological Survey, 600 4th Street South, St. Petersburg, FL 37701 USA

JAMES C. HENDEE
Atlantic Oceanographic and Meteorological Laboratory, National Oceanic and Atmospheric Administration, 4301 Rickenbacker Causeway ,Miami, FL, 33149 USA

DANIEL HUTT
Defence R&D Canada – Atlantic, P.O. Box 1012, Dartmouth, NS, B2Y 3Z7 Canada

CLARKE JEFFRIS
Atlantic Oceanographic and Meteorological Laboratory, National Oceanic and Atmospheric Administration, 4301 Rickenbacker Causeway, Miami, FL, 33149 USA

KAREN JOYCE
Centre for Remote Sensing & Spatial Information Science, School of Geography, Planning and Architecture, The University of Queensland, 4072, Australia

ELLSWORTH F. LEDREW
Department of Geography, University of Waterloo, Waterloo, ON, N2L 3G1, Canada

ALAN LIM
Department of Geography, University of Waterloo, Waterloo, ON, N2L 3G1, Canada

GAND LIU
Coral Reef Watch, NOAA/NESDIS/ORA, E/RA31, SSMC1, Floor 5, 1335 East-West Highway, Silver Spring, MD 20910 USA

DEREK MANZELLO
Atlantic Oceanographic and Meteorological Laboratory, National Oceanic and Atmospheric Administration, 4301 Rickenbacker Causeway, Miami, Florida 33149, USA

CANDACE M. NEWMAN
Department of Geography, University of Waterloo, Waterloo, ON, N2L 3G1, Canada

STUART PHINN
Centre for Remote Sensing & Spatial Information Science, School of Geography, Planning and Architecture, The University of Queensland, 4072, Australia

LAURIE L. RICHARDSON
Department of Biological Sciences, Florida International University, Miami, FL 33199 USA

JULIE A. ROBINSON
Earth Sciences and Image Analysis Laboratory, Johnson Space Center, 2101 NASA Parkway, Mail Code SA15, Houston, TX 7705, USA

CHRIS ROELFSEMA
Centre for Remote Sensing & Spatial Information Science, School of Geography, Planning and Architecture and Marine Botany Group, Centre for Marine Studies, The University of Queensland, 4072, Australia

JOHN SAPPER
Coral Reef Watch, NOAA/NESDIS/ORA, E/RA31, SSMC1, Floor 5, 1335 East-West Highway, Silver Spring, MD 20910 USA

PETER SCARTH
Centre for Remote Sensing & Spatial Information Science, School of Geography, Planning and Architecture, The University of Queensland, 4072, Australia

JOHN F. SCHALLES
Biology Department, Creighton University, Omaha, NE 68178 USA,

WILLIAM SKIRVING
Queensland Science and Engineering Consultants, PO Box 806, Aitkenvale, Queensland, Australia 4814 and Coral Reef Watch, NOAA/NESDIS/ORA, E/RA31, SSMC1, Floor 5, 1335 East-West Highway, Silver Spring, MD 20910 USA

ERIK STABENAU
Atlantic Oceanographic and Meteorological Laboratory, National Oceanic and Atmospheric Administration, 4301 Rickenbacker Causeway, Miami, Florida 33149 USA

ALAN E. STRONG
Coral Reef Watch, NOAA/NESDIS/ORA, E/RA31, SSMC1, Floor 5, 1335 East-West Highway, Silver Spring, MD 20910 USA

TIM L. WEBSTER
Applied Geomatics Research Group, Center of Geographic Sciences, Nova Scotia Community College, 50 Elliot Road, RR# 1 Lawrencetown, Nova Scotia, Canada B0S 1M0

BRIAN G. WHITEHOUSE
OEA Technologies Inc, 14 - 4 Westwood Blvd, Suite 393, Upper Tantallon, NS, B3Z 1H3 Canada

KIMBERLY YATES
U.S. Geological Survey, 600 4th Street South, St. Petersburg, FL 37701 USA

PREFACE

This book was produced at the invitation of our Publishing Editor at Springer, Petra van Steenbergen. We thank her for the invitation. It is one of a series of remote sensing books (Remote Sensing and Digital Image Processing), each of which addresses a specific topic within the remote sensing field.

Our contribution, unlike the other books in this series, is not aimed at remote sensing scientists. As such, there is very little optical theory in the book with the exception of one chapter which presents in detail the optical algorithms that form the basis of an interactive program (WASI, the "Water colour Simulator"). The program is offered at http://extras.springer.com, and is in association with a chapter by Peter Gege and Andreas Albert. In this case, technical details are provided for remote sensing scientists who are interested in the optical and theoretical bases for the program. Remote sensing and non-remote sensing scientists alike can manipulate this program to view the effects of various substances (pigments, sediment) on spectra. We thank these authors for this valuable contribution.

Our main target audience consists of aquatic scientists, and managers of coastal aquatic ecosystems. We are both involved in remote sensing at both the science and image processing levels – Laurie is an aquatic biologist, and became versed in remote sensing image processing techniques while employed at NASA's Ames Research Center, and Ells is a remote sensing scientist who actively carries out science-based field research in support of aquatic remote sensing. Additionally, Ells has been very involved in transferring remote sensing technology to managers in developing nations. Thus we offer a perspective which we hope can be shared by scientists and managers.

The book is divided into three sections: Science Applications, Monitoring Applications, and Management Applications. Each chapter of these sections was written with the appropriate target audience in mind. Thus managers are encouraged to review the final chapters and, if interested, can then delve into the more technical earlier chapters. The book was also written as a resource manual for those who would like a concise overview of specific sensors and their applicability for specific aspects of coastal ecosystems. Again, discussions of optics and sensor development theory are kept at a minimum.

We would like to thank our authors, who graciously accepted fitting into our specific concept for this book. We especially would like to thank Candace Newman, who produced the final figures for the book and Alan Lim who took on the laborious task of the index.

Each chapter in this book was reviewed by one or both of us, as well as anonymously by at least one of the chapter first authors.

Laurie L. Richardson
Ellsworth F. LeDrew

Chapter 1

REMOTE SENSING AND THE SCIENCE, MONITORING, AND MANAGEMENT OF AQUATIC COASTAL ECOSYSTEMS

LAURIE L. RICHARDSON[1] AND ELLSWORTH F. LEDREW[2]
[1]*Department of Biological Sciences, Florida International University, Miami, Florida 33199 USA*
[2]*Department of Geography, University of Waterloo, Waterloo, Ontario, N2L 3G1, Canada*

1. Introduction

Many books and scientific journals address the use of remote sensing in the context of its applicability to aquatic ecosystems. The most successful and widespread aquatic remote sensing connection to date has been remote sensing of the open oceans. This effort has been supported by many in-depth studies of the optics of "blue" or Case 1 waters. Tremendous strides in this field have led to global data bases in support of scientific models now used to study our Earth as a whole, and have led to integration of quantitative world-wide data sets into studies of such hemispheric and global scale problems as climate change. This success has been matched by the design and deployment of an ongoing series of space-borne remote sensors specific to the oceans. It has also led to the existence of global data sets and data products that are easily accessed and interpreted by managers.

A much less studied, but at least equally important, aquatic ecosystem that could benefit greatly from remote sensing technologies is the aquatic coastal zone. Such zones are of importance in terms of ecology and human populations. They are valuable in terms of biodiversity, as resources, and for their role in connectivity between terrestrial and aquatic habitats. Over 50% of human populations live in coastal zones.

The coastal zone has long been a target for science-based study, often in the context of the biology and ecology of this complex and interacting system. Algal and fish biologists, in particular, have intensively studied this zone. Scientific disciplines specific to coastal zones include study of estuaries, coral reefs, and the coastal zone as a nursery. Such studies are often interdisciplinary, relying on techniques in biology, chemistry, physical processes, and more recently molecular markers. Many of these studies involve coastal ecosystem dynamics and change.

Perturbations and long-term changes in coastal stability have ramifications at many levels, including impacts on fisheries, flooding of human populations and infrastructure, eutrophication, development of toxic algal blooms, etc. It is within the realm of aquatic ecosystem managers to be aware of and prepared to counteract or mitigate such perturbations. This task is daunting, largely as a result of the regional scale and dynamic nature of aquatic coastal zones. It also often demands interactions with aquatic scientists.

One of the most useful tools for both scientists and managers for the study of coastal zones is remote sensing. The benefits or remote sensing include synoptic, quantitative data sets that are regional (as well as local and global) in scale, and that can offer repeat sampling. In many cases archival remote sensing data are available that are invaluable in providing a history of the region. The disadvantages of remote sensing

L.L. Richardson and E.F. LeDrew (eds.), Remote Sensing of Aquatic Coastal Ecosystem Processes: Science and Management Applications, 1-7. © 2006 *Springer. Printed in the Netherlands.*

commonly include cost, lack of remote sensors tailored for coastal zones, and, perhaps most important, an underdeveloped line of communication between scientists, remote sensing experts, and managers. Despite these drawbacks, there are many current and ongoing examples of the successful use of remote sensing to support aquatic coastal ecosystem science, and to integrate remote sensing into the management of coastal zones. We hope to provide an overview of the integration of remote sensing and both science and management in this book.

2. Coastal zones

The apparent lag in coastal aquatic remote sensing as compared to open ocean remote sensing is, we believe, in large part due to the complexity of the coastal zone. This complexity is multifaceted, and includes both optical and biological complexity as well as an overlay of temporal and spatial dynamics. In terms of optical complexity, many factors are involved. These include the following:

- coastal zones are often shallow, thus bottom reflectance contributes to an optical water-leaving signal
- benthic communities of differing complexities and depths also contribute to optical water-leaving signals
- phytoplankton and suspended sediments are present in much higher concentrations in coastal zones as compared to "blue" offshore water
- spectral signatures of different types of phytoplankton and sediments are highly variable and can strongly affect spectral reflectance
- spectral signatures of phytoplankton and sediments interact in such a manner that specific algorithms must be derived that are tailored for coastal regions.

In terms of biological complexity, the following factors must be addressed:

- the presence of macroalgae, dense invertebrate communities (such as found on coral reefs) and other bio-optically active organisms can contribute to reflectance
- the effect of biota, both benthic and suspended, on traditional (Case 1) chlorophyll algorithms often results in erroneous results
- seasonal patterns of biomass and dominance by successive members of the biological community commonly occur.

Finally, temporal and spatial dynamics are much more relevant for remote sensing of aquatic coastal zones than for those of open oceans. Thus, transient phytoplankton blooms, runoff from storms, flooding, and resuspension of sediments due to storm events all must be considered within spatial scales that may not be resolved by existing satellite sensors.

3. Spectral signatures

We believe that one of the most promising features of coastal aquatic systems that can be exploited for both science and management applications of remote sensing is that of spectral signatures. Case 2 (coastal) waters, as opposed to Case 1 (oceanic) waters, are dominated by in-water features as opposed to scattering. This is particularly true of phytoplankton at the concentrations found in the coastal zone, which are typically orders of magnitude higher than those of Case 1 waters. An important ramification of this fact is that the widely used chlorophyll algorithms for ocean chlorophyll break down under high concentrations of phytoplankton and/or suspended sediment such as are typically found along the coast. Depending on the concentrations, types, and proportions of phytoplankton and suspended sediments, Case 1 chlorophyll algorithms can lead to either under- or over-estimation of chlorophyll concentration. This is the basis for the ongoing and costly reanalysis of Coastal Zone Color Scanner (CZCS) data, which were originally analyzed using algorithms designed for blue offshore waters. The spectral bands on the CZCS, in fact, were selected based on Case 1 chlorophyll algorithms.

The optical properties of Case 2 waters are very different from those of Case 1 waters. Spectral analysis of Case 2 waters can provide relatively much more information about the type(s) of both phytoplankton and sediment present in the water column, information which is useful to scientists and managers alike. This is possible due to the strong effects that these water constituents have on the water-leaving optical signals of coastal waters when they are present at high concentrations such as found near the coast. This subject is addressed in detail in Chapter 3 by John Schalles. In this chapter experimental manipulations of phytoplankton (type and concentration), sediments of different types, and combinations of phytoplankton and sediment were analyzed in terms of spectral reflectance and existing chlorophyll algorithms. This concept is further developed and illustrated in Chapter 4 by Peter Gege and Andreas Albert. These authors have also provided an interactive software program in which users can manipulate sediment and pigment constituents in a water column and observe the effects on spectra. Chapter 4 includes the optical theory and algorithms that support the interactive program.

In addition to phytoplankton and sediment signatures, many benthic communities exhibit distinctive spectral signatures detectable using remote sensing. One of the most successful coastal applications taking advantage of this is that of remote sensing of coral reefs, which is addressed in several chapters in this book (e.g. Chapter 11 by Newman et al.). Thus although the coastal zone is optically complex, this complexity in itself can be the basis for extending the potential of remote sensing beyond what is possible for Case 1 waters.

4. Science based connections between ecosystem processes and remote sensing

Many features of aquatic coastal ecosystems have strong optical signals. These include the various photoreactive, photoprotective, and light harvesting pigments of phytoplankton, submerged and emergent macroalgae, and coastal (terrestrial) plants. Many of these organisms also have unique pigment combinations as well as indicator fluorescence signals. Thus such populations can be monitored, measured, and identified using optical properties.

Use of optical features in coastal biology is widespread and implemented by both scientists and managers alike. The most common is optical (spectrophotometric) detection of chlorophyll to estimate phytoplankton concentration and thus monitor potential eutrophication as a proxy for the "health" of the aquatic ecosystem. This approach is now being expanded to detect pigment signatures specific to toxin producing algae. In each case, the fundamental basis for the signals (i.e. the pigments themselves) can be detected using remote sensing.

Most of the Case 1 studies on remote sensing of pigments (primarily chlorophyll *a*) have focused on optics and algorithm development. The parameter that is deemed most important is radiative transfer, which is a composite of backscattering, absorption, transmission, fluorescence, etc. As a result, much of this work results in publications that are dominated by optical modeling and radiative transfer equations. This approach is necessary because most water-leaving reflectance spectra in Case 1 waters are quite similar.

In contrast, the water-leaving reflectance of Case 2 waters is often highly variable and can be detected with the eye alone. Thus, in a different approach, there is a body of work on Case 2 algal pigment remote sensing that foregoes preliminary radiative transfer modeling and instead focuses directly on spectral signatures. Differences in the spectra themselves then become the basis for algorithm development.

Coastal phytoplankton are often dynamic in terms of phytoplankton population composition and quantity of cells, both of which can dramatically affect spectral reflectance. Thus detection of specific spectral patterns, or signatures, can often detect a specific type of algal bloom. The connection between pigments and algal type is one of the most promising and applicable examples of the potential for bridging coastal ecosystem processes and remote sensing.

As mentioned above, one of the benefits of remote sensing for studies of aquatic coastal zones is the ability of remote sensing to provide synoptic data sets at different scales. This is accomplished by the availability of sensors with different spatial resolutions. An enhancement of this capability is the fact that many sensors offer different spectral, as well as spatial, resolution. The result is that synoptic remote sensed data can be attained for a given coastal zone that can detect and assess many different aspects of that particular coastal aquatic ecosystem. Therefore, in addition to discriminating between phytoplankton types, remote sensing can allow for habitat mapping, coastal shoreline anomalies, and change detection.

Remote sensing can also detect certain physical properties that are directly or indirectly crucial to aquatic ecosystem processes. One of the most important of these is water surface temperature. Temperature is an important factor in the physiological functioning and health of organisms. It is also a major factor in controlling population dynamics of many aquatic organisms. The combination of optical signals that can detect, identify, and quantify different types of aquatic organisms, along with the capability to detect an important and regulatory factor such as temperature, leads to the potential for remote sensing as a powerful tool to study and quantify aquatic ecosystems at the physiologically functional level.

Current research is aimed at using remote sensing to directly scale up aquatic ecosystem studies at the process level. An example of such an effort is presented by John Brock and colleagues in Chapter 5. This research group is using a suite of remote sensors with different spatial and spectral resolution as well as different remotely sensed factors to measure and extrapolate carbon biogeochemical processes on a coral reef to the regional scale. Their program includes remotely sensed mapping of the reef itself (geomorphology), remote sensing based detection of different habitats/groups of organisms (biotopes), and scaling up of in situ experimental measurements, based on the remote sensing data, to examine reef "metabolism". While this chapter is particularly innovative and beyond the scope of most monitoring and managing programs, it is included as an example of a feasible and existing remote sensing application. Another example of this type is found in Chapter 6 by Jim Hendee et al., who are moving towards remote sensing of coral health by remote measurement of coral photosynthetic efficiency.

5. Monitoring coastal zones, habitats, and ecosystems

Although many aquatic monitoring programs incorporate remote sensing, often the degree of incorporation is prohibited by cost and a well-founded perception that remote sensing requires expertise in a highly specialized field. Despite this, the success of remote sensing in monitoring can only suggest that this approach will become increasingly common in the future. Buoys with monitoring sensors are increasing in number, as are their data capabilities. Such buoys are now used to, for example, measure and estimate chlorophyll and turbidity in the water column - an assessment of water quality. Some innovative programs, particularly in Australia and Europe, include fully integrated optical remote sensing (satellite and aircraft) data input for the early detection of harmful algal blooms. A specific case study is presented in Chapter 10 in which Stuart Phinn et al. discuss the use of remote sensing monitoring to detect toxic blooms of the cyanobacterium *Lyngbya majuscula*. In this particular case, the use of remote sensing in monitoring is also integrated into management.

Satellite platforms that support remote sensing now provide global coverage. Currently remote sensing capabilities and data streams exist that could be used much more extensively in monitoring programs. Additionally, historical data are available which can potentially be a source of baseline data, an increasingly important concept in monitoring aquatic coastal environments. An overview, discussion of data sets, and example of use of historical data are given in Chapter 8 by Jennifer Gebelein.

One global ecosystem that is currently the focus of study on a world-wide basis, and that makes extensive use of satellite data for mapping and monitoring, is the coral reef ecosystem. Many of the chapters in this book include some aspect of remote sensing of coral reefs. These include mapping on a global basis, which is being conducted as the first step of global monitoring of coral reefs; remote sensing based studies of coral reef health; use of remote sensing to scale up coral reef process level studies; and integration of historical data in coral reef/land interaction studies. Rather than compile these efforts into one section on remote sensing of coral reefs, we have placed individual studies within the framework of remote sensing and science, monitoring, and management as a recurrent thread throughout the book.

It is our hope that this book will alleviate some of the hesitation of those involved in monitoring of aquatic coastal zones and ecosystems by providing a non-technical overview of this field as well as a compilation of remote sensing resources.

6. Management of coastal zones, habitats, and ecosystems

Even more so than within monitoring programs, remote sensing is an underutilized tool that would greatly benefit management programs that are responsible for, or include, coastal ecosystems. However, managers are often more reluctant than those involved in monitoring to introduce remote sensing. This is, again, based on the view that expertise if required, and is enhanced by the fact that many managers, as opposed to monitoring personnel, do not have a science background. In cases where underdeveloped nations are involved, the problem is pronounced.

We have included in this book particular attention to bridging the gap between managers a nd rem ote sensing. Chapte r 9 by Brian Whitehouse a nd Daniel Hutt presents an overview, aimed at managers, that discusses the benefits and potential of remote sensing for management of coastal aquatic ecosystems, as well as a discussion of cost and financial issues. Chapter 11 by Candace Newman et al. extends this discussion to specifically address technology transfer of remote sensing to underdeveloped countries, including local communities. These concepts are further extended to policy making in Chapter 10 by Stuart Phinn and colleagues.

One of the most important and encompassing rotes of managers is integration of data within and between the diverse areas of social, natural resource, economic, and environmental frameworks. Incorporation of remote sensing at the management level also includes data integration, but with a different aim – that of not only extracting and integrating data from various sensors, but transferring this knowledge to the wider community. Such topics are addressed, and examples of successful efforts are provided, in Chapter 12 by Julie Robinson et al.

7. Integrating remote sensing, science, monitoring, and management

The integration of science and remote sensing is now a reality. One of the best examples of success is the use of remote sensing to predict coral bleaching, as presented by William Skirving and colleagues in Chapter 2. The science-based data that have proven the connection between physiological thermal stress and coral bleaching are now fully integrated into automated remote sensing data analyses that allow real-time world-wide coral bleaching "alerts". The accuracy of these alerts in predicting bleaching is extremely high. Thus in this case scientists have provided a quantitative link between an aquatic ecosystem process and a factor (sea surface temperature) detectable at global scales using remote sensing. Managers now routinely access real-time satellite-derived predictive data that is directly relevant to the health of the ecosystem they are managing – coral reefs.

This approach has been extended using another type of remote sensing instrument package – permanently moored buoys that support an array of different sensors that measure factors of importance to aquatic scientists, in particular biologists, and managers. A state of the art system is described here in Chapter 6 by Jim Hendee and colleagues. In addition to an overview of the system and examples of how these data can support science applications, a guideline for design and deployment of the system is provided for managers.

Another example is the use of remote sensing in support of management of coastal flooding. In Chapter 7, Tim Webster and Donald Forbes present a detailed case study in which remote sensing is being used to mitigate the effects of coastal flooding based on the results of light detection and ranging (LIDAR). Thus in this case, as opposed to those discussed above, purely physical features (details of varying coastal elevations) are integrated with historical data bases to predict flooding effects. City managers are actively integrating these remote sensing data with planning.

8. Summary

All of the chapters in this book are meant to serve as resources for both scientists and managers. In addition to the specific examples of remote sensing in science, monitoring, and management briefly summarized above, the chapters by lead authors Jennifer Gebelein, Brian Whitehouse, Stuart Phinn, Candace Newman, and Julie Robinson include overviews of data bases, comparisons of sensor capabilities, and the existence of archived data sets. These are provided in potential support of scientists and managers who would like to integrate remote sensing into individual projects. We hope that this resource will further inspire incorporation of remote sensing into the study and management of aquatic coastal and shallow-water ecosystems and processes.

Section I

Science Applications

Chapter 2

EXTREME EVENTS AND PERTURBATIONS OF COASTAL ECOSYSTEMS

Sea Surface Temperature Change and Coral Bleaching

WILLIAM SKIRVING[1,2], ALAN E. STRONG[2], GAND LIU[2], FELIPE ARZAYUS[2], CHUNYING LIU[2]AND JOHN SAPPER[2]

[1]*Queensland Science and Engineering Consultants, PO Box 806, Aitkenvale, Queensland, Australia 4814 William.Skirving@noaa.gov*

[2] *Coral Reef Watch, NOAA/NESDIS/ORA, E/RA31, SSMC1, Floor 5, 1335 East-West Highway, Silver Spring, MD 20910-3226, USA*

1. Introduction

Remote Sensing of Aquatic Coastal Ecosystem Processes presents many examples of remote sensing tools which could be used in an operational sense for the benefit of management of various aspects of coastal ecosystems. This chapter will present the World's first global operational satellite products designed specifically to help coral reef managers map and monitor anomalous sea surface temperatures (SST) and hence better understand and predict mass coral bleaching. These products are possibly the only global suite of operational satellite products currently being used for the management of any marine ecosystem.

Coral bleaching occurs when there is widespread loss of pigment from coral, mainly due to the expulsion of symbiotic algae (Yonge and Nicholls, 1931). The algae are usually expelled in times of stress, often caused by sea surface temperatures which are higher than the coral colony's tolerance level. This may be as little as 1 to 2°C above the mean monthly summer values (Berkelmans and Willis, 1999; Reaser et al., 2000).

A number of publications in the early 1990s reported that links had been observed between significant bleaching and anomalously warm water several weeks earlier (e.g. Bermuda: Cook et al. 1990; Indonesia: Brown and Suharsono, 1990; Jamaica: Gates, 1990; Andaman Sea: Brown et al., 1996). About the same time a number of journal articles proposed a mechanism that linked sea surface temperature with mass coral bleaching (e.g. Lesser et al., 1990; Glynn and D'Croz, 1990). Coles and Jokiel (1977) and Jokiel and Coles (1990) proposed the existence of a universal critical threshold temperature for bleaching. They proposed that regardless of location, the threshold could be defined as a 1°C increase over the mean local summer maximum temperature. As an extension of these ideas, Goreau and Hayes (1994) produced maps of "ocean hot spots" using monthly global ocean temperature anomaly maps from NOAA Climate Diagnostic Bulletins (monthly maps derived mostly from *in situ* data with a small satellite contribution). These "ocean hot spots" identified areas whose SSTs exceeded long term averages by more than 1°C. Bleaching seemed to be occurring within the boundaries of the "ocean hot spots" for the warm season in each region.

One of us (Dr. Alan Strong-NOAA/NESDIS) recognized that there was an opportunity to derive an automated satellite product based on this relationship. As a result, the HotSpot product was born, which served as the basis of the NOAA Coral Reef Watch Program. HotSpot was first presented as an experimental product in 1997 (Strong et al., 1997). This proved to be somewhat fortuitist given that we now know

L.L. Richardson and E.F. LeDrew (eds.), Remote Sensing of Aquatic Coastal Ecosystem Processes: Science and Management Applications, 11-25. © 2006 *Springer. Printed in the Netherlands.*

that 1998 turned out to be the most significant year for mass coral bleaching in known history.

The HotSpot product gained widespread recognition when, in early 1998, Dr. Strong announced that the Great Barrier Reef (NE Australia) was bleaching before either the Great Barrier Reef Marine Park Authority (GBRMPA) or the Australian Institute of Marine Science (AIMS) knew about it. Out of the realization that this product was potentially a very powerful satellite tool for coral reef management, a formal agreement was signed between NOAA, AIMS and the GBRMPA to work together on the development of new and improved satellite products for monitoring coral reef health. As a result of this agreement, and a series of workshops, the HotSpot product suite has grown and the understanding of the mechanisms and management actions associated with coral bleaching have vastly improved.

This chapter will describe the operational HotSpot product suite and each of its derivations.

2. Sea Surface Temperature Data

NOAA has been producing sea surface temperatures from satellite data since 1972. Monitoring SST from earth-orbiting infrared radiometers has had a wide impact on oceanographic science. Currently, one of the principal sources of infrared data for SST measurement is the Advanced Very High Resolution Radiometer (AVHRR), which has been carried on NOAA's Polar Orbiting Environmental Satellites (POES) since 1978. It is a broad-band, four or five channel (depending on the model) scanner, with sensors in the visible, near-infrared, and thermal infrared portions of the electromagnetic spectrum (http://www.ngdc.noaa.gov/seg/globsys/avhrr.shtml). The POES satellite system offers the advantage of daily global coverage, by making near-polar orbits roughly 14.1 times every 24 hours. *In situ* SSTs, from buoys (drifting and moored) are used operationally for calibration purposes to maintain accuracy, removing any biases, and compiling statistics with time (McClain, 1985; Strong, 1991; and Strong, et al., 2000).

The composite AVHRR-SST products have a resolution of 50km and are produced twice-weekly in near real-time. On Tuesdays data from the previous three days (Saturday through Monday) are used, and on Saturdays data from the previous four days (Tuesday through Friday). Since the AVHRR is limited to a temporal resolution of six hours (when combining data from both operational satellites), it is not possible to accurately characterize the diurnal variation in SST caused by the cyclic heating of the sea surface as the sun changes it's angle during the day and its absence during the night. While some of NOAA's operational SST products use blended day and night retrievals, nighttime-only observations are used in many SST products in an effort to minimize the effects of diurnal variation. Nighttime SST products compare most favorably with *in-situ* buoy SSTs at 1 meter depths (Montgomery and Strong, 1995). An example of the global nighttime composite AVHRR-SST product can be seen in Figure 1.

The nighttime AVHRR-SST products were primarily developed for NOAA's Coral Reef Watch (CRW) Program for both monitoring and assessment of coral bleaching. CRW's other satellite monitoring and assessment products include SST anomalies, coral bleaching HotSpots, Degree Heating Weeks (DHW) , Tropical Coral Bleaching Indices, and SST time series, including on-line animations. The development and production of these CRW products takes place within the NOAA National Environmental Satellite, Data, and Information Service (NESDIS). This NESDIS team is comprised of

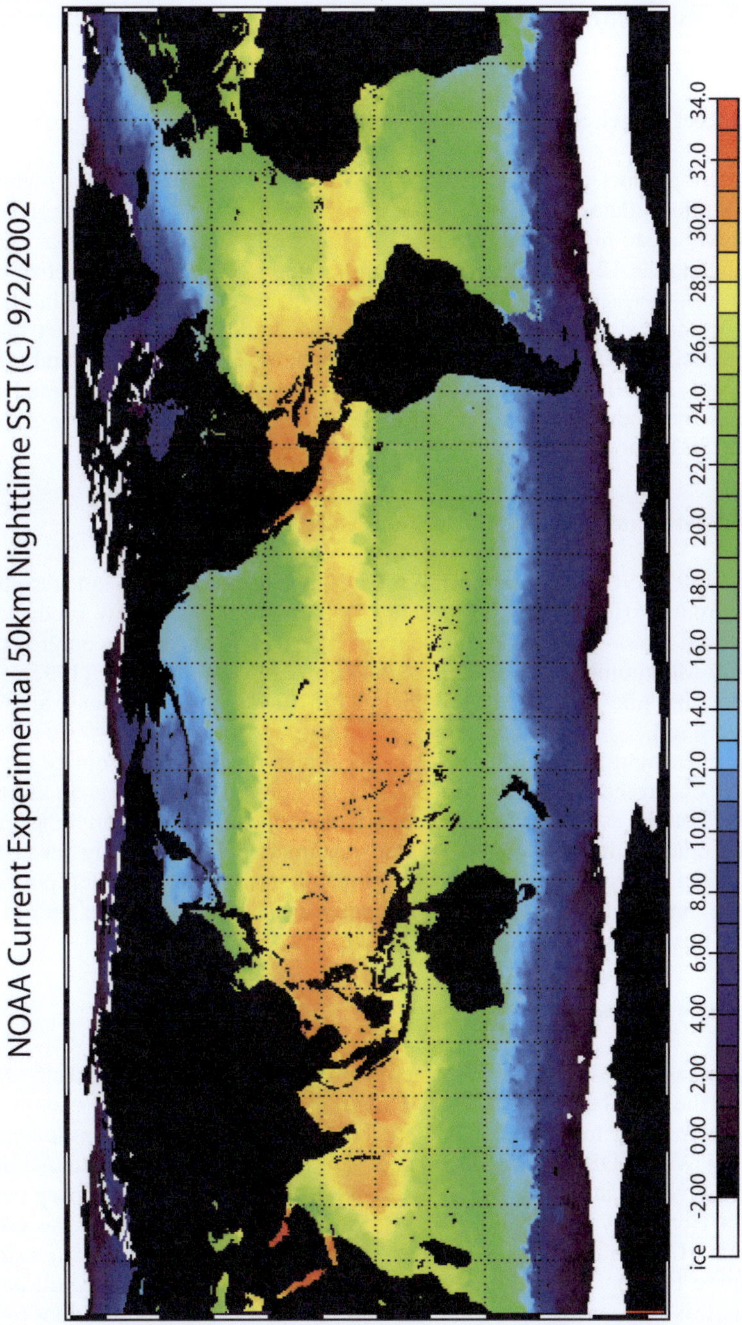

Figure 1. NOAA 50km Nighttime SST product for 2nd September, 2002.

scientists from the Marine Applications Science Team (MAST) in the Oceanic Research and Applications Division (ORAD) of the Office of Research and Applications (ORA) and from the Product Systems Branch (PSB) of the Information Processing Division (IPS) of the Office of Satellite Data Processing and Distribution (OSDPD).

3. The Coral Reef Watch SST Anomaly Product

CRW's first product consisted of a simple SST anomaly based on a nighttime AVHRR climatology data set from 1985 to 1993. The climatology was derived by averaging weekly mean nighttime SSTs into monthly means over the nine year period, and then producing the 12 monthly means from all nine years of data. The actual daily SST anomaly was derived after the monthly means had been interpolated into daily means. This product supports wide ranging uses and is the preferred product for the NESDIS' National Climate Data Center (NCDC) when observing the onset and effects of El Nino and La Nina (http://www.ncdc.noaa.gov/oa/climate/elnino/elnino.html). The SST Anomaly product became officially operational in 2001 (see Figure 2 for an example of the SST Anomaly product).

4. The Coral Reef Watch Bleaching HotSpot Product

Early in 1997, NESDIS began developing global satellite 50-km resolution experimental products (initially SST Bleaching HotSpots, and then DHW products) as indices of thermal stress-related coral bleaching. These products are the outgrowth of earlier work by Montgomery and Strong (1994) and Gleeson and Strong (1995). The HotSpot product became officially operational in 2001 (see Figure 3 for an example of the HotSpot product).

The coral bleaching HotSpot is not a typical climatological SST anomaly. It is a measure of the occurrence of the hottest SST for a region and as such is an anomaly that is not based on the average of all SST, but on the climatological mean temperature of the climatologically hottest month (i.e. the maximum of the monthly mean SST climatology, often referred to as the MMM climatology). Since the HotSpot is an anomaly based on the maximum of the monthly mean SST, negative values are meaningless in this context; therefore only positive values are displayed.

5. The Coral Reef Watch DHW Product

HotSpot values provide a measure of the intensity of sea surface thermal stress, but do not measure the cumulative effects of that thermal stress on a biological system such as a coral reef. In order to monitor this cumulative effect, a thermal stress index, the Degree Heating Week (DHW), was developed. DHW represents the accumulation of HotSpots for a given location over a rolling 12-week time period. Preliminary results indicate that a HotSpot value of less than one degree is insufficient to cause visible stress on corals. Consequently, only HotSpot values $\geq 1°C$ are accumulated (i.e. if we have consecutive HotSpot values of 1.0, 2.0, 0.8 and 1.2, the DHW value will be 4.2 because 0.8 is less than one and therefore not used). One DHW is equivalent to one week of HotSpot levels staying at 1°C or one ½ week of HotSpot levels at 2°C, and so forth. For an example of the DHW product, see Figure 4.

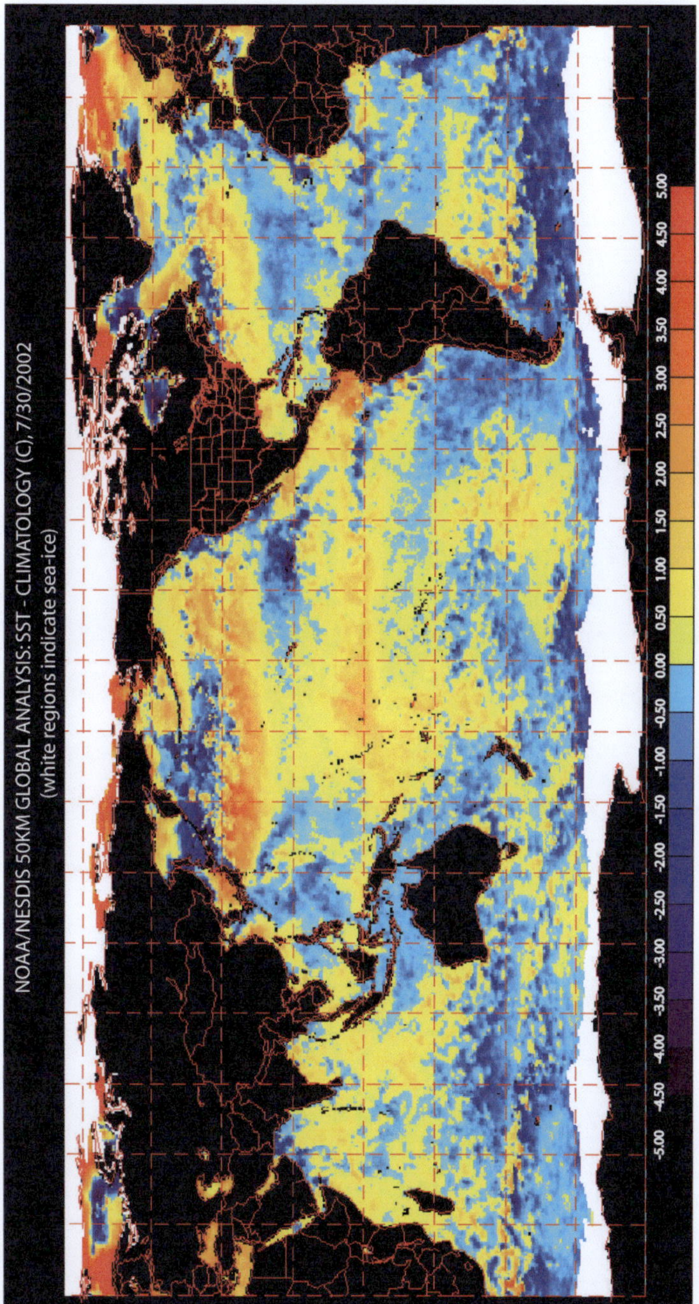

Figure 2. NOAA 50km SST Anomaly product for 31st July, 2002.

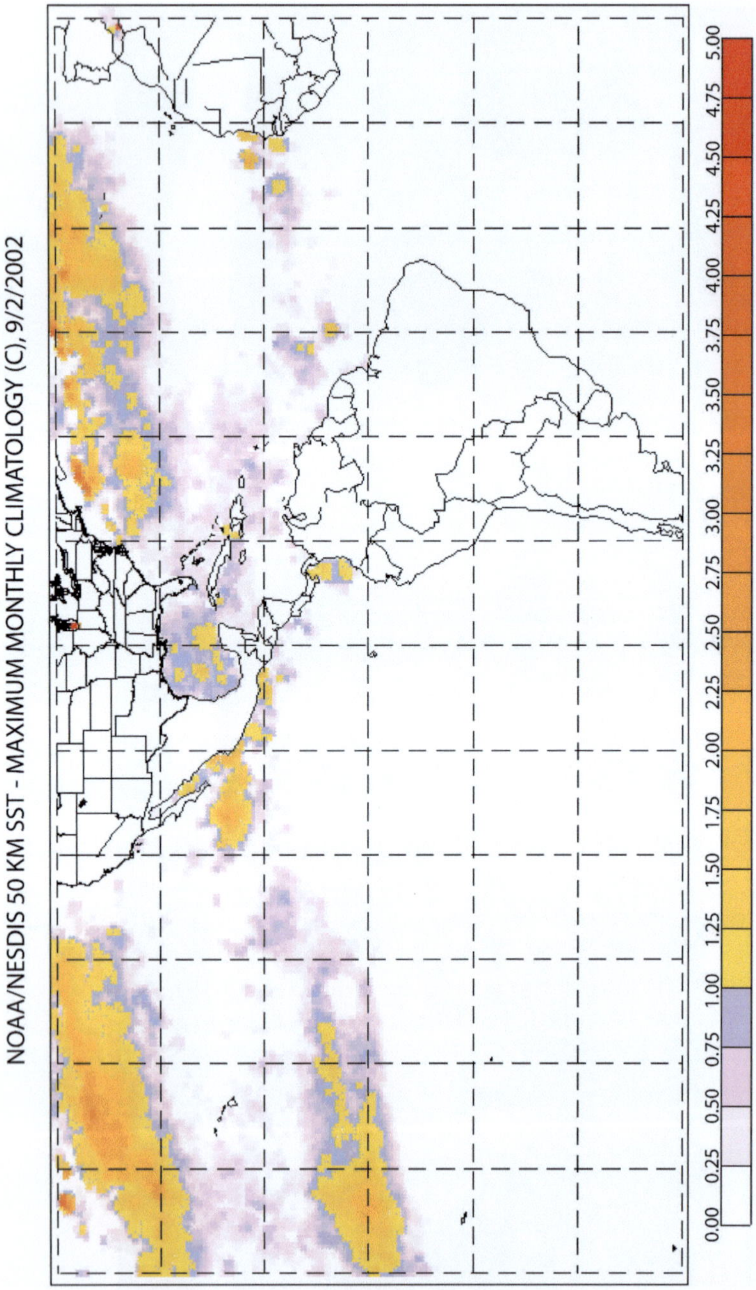

Figure 3. NOAA 50km HotSpot product for 2nd September, 2002.

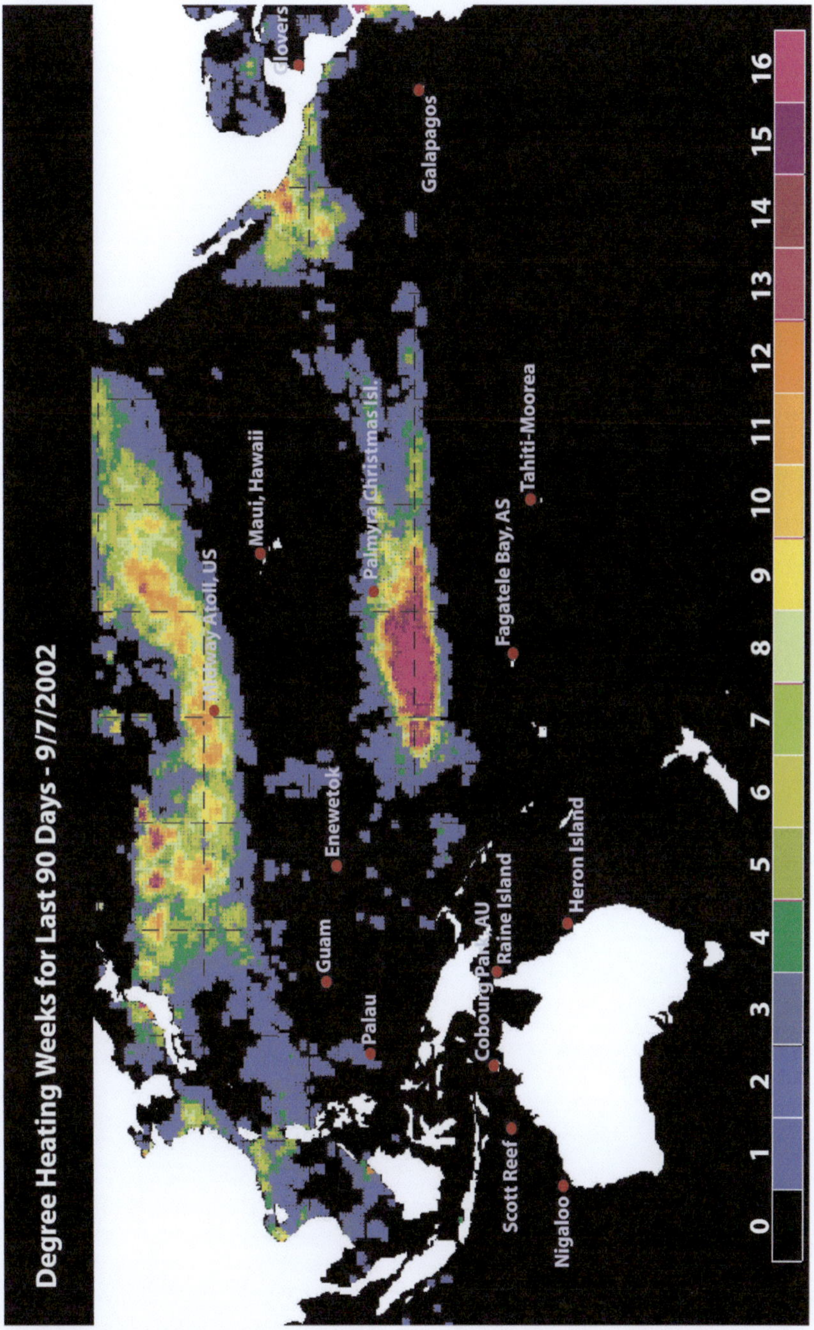

Figure 4. NOAA 50km DHW product for 7th September, 2002.

Figure 5. Geographic distribution of coral bleaching warning responses (dots).

Field observations (most of which are subjective measurements presented as informal reports) with coincident satellite data are only available for a limited number of years. These observations consistently indicate that there is a correlation with bleached corals and a threshold DHW value of 4. By the time DHW values reach 8, widespread bleaching is likely and some mortality can be expected. CRW has applied these DHW values for the last two years when generating satellite bleaching warnings and alerts (Liu et al., 2003).

All of the products discussed in this chapter have been used successfully in monitoring major coral bleaching episodes around the globe (e.g. Goreau et al., 2000; Wellington et al., 2001). Since 2000, when the DHW product was being used in an experimental mode (it became operational along with the HotSpot product in 2001), CRW has issued over 100 bleaching warnings. On 46 occasions field reports on the status of the corals were received and on every occasion bleaching was confirmed. This is not a comprehensive test of the DHW product but is a very good indication that it is at the very least a very good conservative measure of coral bleaching. Figure 5 is a plot of the geographic distribution of these reports and demonstrates that the success of the DHW product is widespread, i.e. not restricted to a regional level.

The data products discussed in this chapter are available as HotSpot and DHW charts (produced by CRW twice weekly in near-real time from the composite nighttime AVHRR SST products). These products, along with descriptions of the methodologies, are Web-accessible at http://orbit-net.nesdis.noaa.gov/orad/coral_bleaching_index.html.

6. Coral Reef Watch in Action

6.1 CASE STUDY: MIDWAY ATOLL

During July of 2002 CRW team members noticed a SST anomaly building in the central North Pacific Ocean (Figure 3). This anomaly grew until temperatures at Midway Island on 1st August passed the threshold for coral bleaching, where they remained. Based on the accumulated heat stress (recorded in the DHW product), CRW issued a bleaching warning on 7th August. There was, however, no one on site to confirm the existence of bleaching. Continued monitoring revealed that SSTs in this area remained above the threshold for more than a month, and that SSTs did not drop back below the threshold until 7th September (Figure 6).

At this time, the Hawaii Laboratories of NOAA's National Marine Fisheries Service (NMFS) were due to lead a multi-agency scientific cruise to the North West Hawaii Islands (NWHI) as part of their 2002 NWHI Reef Assessment and Monitoring Program. This cruise was scheduled almost one year prior to this bleaching event as part of their on-going monitoring efforts. The cruise was conducted between September 8th and October 7th, 2002. During the cruise, surveys were conducted at Midway Atoll on September 20th, 21st and 25th. Although the CRW bleaching report did not instigate the cruise, the cruise plan was altered to focus more attention on conducting surveys in shallow backreef habitats based on the information that the field team gained from the HotSpot products combined with NMFS *in situ* buoy data from the region.

During the field work at Midway Atoll, which took place two weeks after the thermal stress abated, the survey team found extensive fresh bleaching. In most cases, the corals were bleached white with no algal cover (indicative of recent bleaching). In a

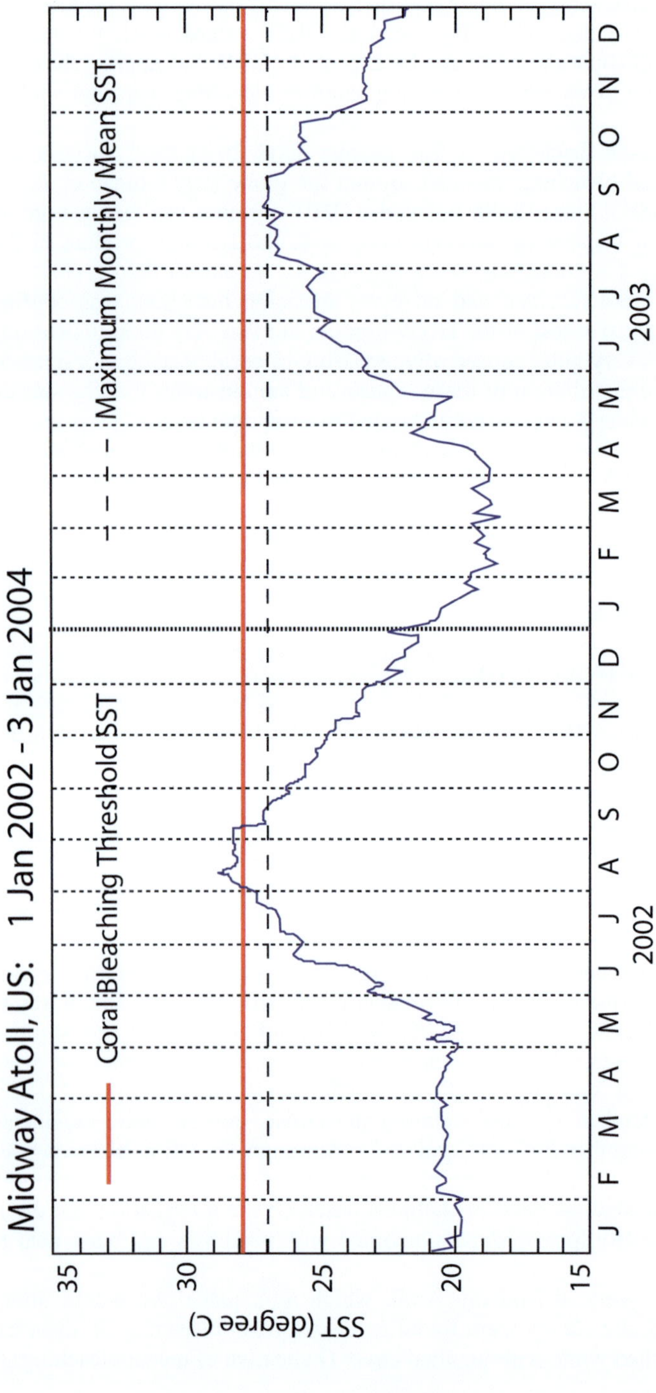

Figure 6. SST plot for Midway Atoll, 1st Jan. 2002 to 3rd Jan 2004. The dashed line is the value of the MMM climatology for this site and the red line is the coral bleaching threshold.

few instances, recent algae colonization was observed. The cause of the bleaching could not be confirmed as related to the SST anomaly,

The data from this cruise revealed that the bleaching was a NWHI event, and was not simply restricted to Midway Atoll. The most severe bleaching was observed at Pearl & Hermes, Midway, and Kure Atolls. Additional bleaching was also observed at Lisianski and French Frigate Shoals.

7. Bleaching Warnings for Managers

In an attempt to assist managers and non-physical scientists with the use of the CRW product suite, a summary page, called the "Indices Page", has been developed (http://www.osdpd.noaa.gov/PSB/EPS/CB_indices/coral_bleaching_indices.html). This website allows the user to zoom in on 24 individual sites (Figure 7). The initial page allows a quick look at the crucial numbers associated with the CRW products, including the names and locations of sites. For each site the current DHW value, the maximum (record) DHW value for that site since 1985, the current SST and the value of the MMM climatology for that site are provided (Figure 7). Together these values provide an index of potential bleaching.

By clicking on the location within each individual box (see Figure 7) the user is taken to a map showing the location of the pixel whose value is being reported for that location. At the top of this page the user has further links that will allow access to zoomed in images of HotSpots, DHWs, SST, and ocean winds. There are also links to a current time series plot of SST and another site that allows access to buoy locations and data. Of importance for managers, the Indices Page has a warning symbol that appears when any HotSpot value is greater than zero. This symbol increases in size and flashes according to the severity of the current heat stress.

8. The Future of Coral Reef Watch Products

There are a number of improvements planned for CRW products in the near future. These include the incorporation of as many relevant satellite-based environmental measurements as possible, such as wind, turbidity, light, waves, etc. Non-satellite data that supports the existing and proposed space-borne measurements would also be incorporated. It is possible that these may include information about tides, currents, bathymetry, etc.

Beyond the near future, CRW is working on ways of providing 1 to 4 week forecasts of coral bleaching, and improved data sets for climate modeling. For the latter we are using our now-casting knowledge and 20 years of satellite data to help understand the outputs from current Global Climate Models (GCM) and their relationship to thermal stress of corals (and hence bleaching). Ultimately this will assist in the design of "coral friendly" GCMs.

ATLANTIC OCEAN Peak Season (N) Jul-Sep, (9) Jan-Mar	PACIFIC OCEAN Peak Season (N) Jul-Sep, (9) Jan-Mar		INDIAN OCEAN Peak Season (N) Jul-Sep, (9) Jan-Mar
BERMUDA 32.0 N, 64.5 W 12WK ACCUM TODAY 0.0 MAX 12 WK* 7.5(1998) CURRENT SST (C) 19.7 CLIMATOLOGY** 27.2	**MIDWAY ATOLL, US** 28.5 N, 177.5 W 12WK ACCUM TODAY 0.0 MAX 12 WK* 4.6(1987) CURRENT SST (C) 20.0 CLIMATOLOGY** 26.9	**GUAM** 13.0 N, 144.5 E 12WK ACCUM TODAY 0.0 MAX 12 WK* 5.1(1999) CURRENT SST (C) 28.3 CLIMATOLOGY** 29.5	**OMAN-MUSCAT** 24.0 N, 58.0 E 12WK ACCUM TODAY 0.0 MAX 12 WK* 14.7(1996) CURRENT SST (C) 26.6 CLIMATOLOGY** 30.3
SOMBRERO REEF, FL 25.0 N, 81.4 W 12WK ACCUM TODAY 0.0 MAX 12 WK* 6.0(1998) CURRENT SST (C) 23.9 CLIMATOLOGY** 29.3	**OAHU-MAUI, HI** 21.0 N, 158.0 W 12WK ACCUM TODAY 0.0 MAX 12 WK* 7.6(1996) CURRENT SST (C) 25.0 CLIMATOLOGY** 27.0	**ENEWETOK** 11.5 N, 162.0 E 12WK ACCUM TODAY 0.0 MAX 12 WK* 2.5(2001) CURRENT SST (C) 27.9 CLIMATOLOGY** 29.1	**MALDIVES-MALE** 4.0 N, 73.0 E 12WK ACCUM TODAY 0.0 MAX 12 WK* 10.5(1998) CURRENT SST (C) 30.2 CLIMATOLOGY** 29.9
BAHAMAS, LEE STOCKING IS 23.5 N, 76.5 W 12WK ACCUM TODAY 0.0 MAX 12 WK* 7.8(1998) CURRENT SST (C) 25.5 CLIMATOLOGY** 29.3	**PALMYRA, ISL** 6.0 N, 162.0 W 12WK ACCUM TODAY 0.0 MAX 12 WK* 6.0(1994) CURRENT SST (C) 28.0 CLIMATOLOGY** 28.7	**PALAU** 7.5 N, 134.0 E 12WK ACCUM TODAY 0.0 MAX 12 WK* 10.4(1998) CURRENT SST (C) 28.8 CLIMATOLOGY** 29.5	**SEYCHELLES-MAHE** 5.0 S, 56.0 E 12WK ACCUM TODAY 0.0 MAX 12 WK* 4.5(1998) CURRENT SST (C) 29.5 CLIMATOLOGY** 29.5
PUERTO RICO 18.0 N, 67.5 W 12WK ACCUM TODAY 0.0 MAX 12 WK* 6.8(1998) CURRENT SST (C) 27.1 CLIMATOLOGY** 28.5	**GALAPAGOS** 1.0 S, 90.0 W 12WK ACCUM TODAY 0.0 MAX 12 WK* 34.4(1998) CURRENT SST (C) 26.7 CLIMATOLOGY** 26.5	**DAVIES REEF** 19.0 S, 147.5 E 12WK ACCUM TODAY 2.0 MAX 12 WK* 4.8(1987) CURRENT SST (C) 26.8 CLIMATOLOGY** 28.5	**COBOURG PARK** 10.5 S, 132.5 E 12WK ACCUM TODAY 0.0 MAX 12 WK* 4.8(1987) CURRENT SST (C) 29.6 CLIMATOLOGY** 29.9
VIRGIN ISLANDS, US 18.0 N, 65.0 W 12WK ACCUM TODAY 0.0 MAX 12 WK* 7.5(1998) CURRENT SST (C) 26.8 CLIMATOLOGY** 28.5	**AMERICAN SAMOA-OFU** 14.0 S, 170.0 W 12WK ACCUM TODAY 0.0 MAX 12 WK* 7.4(1991) CURRENT SST (C) 29.8 CLIMATOLOGY** 29.3	**HERON ISLAND** 23.5 S, 152.0 E 12WK ACCUM TODAY 3.5 MAX 12 WK* 6.0(1987) CURRENT SST (C) 25.1 CLIMATOLOGY** 27.3	**SCOTT REEF** 14.0 S, 122.0 E 12WK ACCUM TODAY 0.0 MAX 12 WK* 13.3(1998) CURRENT SST (C) 29.4 CLIMATOLOGY** 30.1
GLOVERS, BELIZE 16.5 N, 88.0 W 12WK ACCUM TODAY 0.0 MAX 12 WK* 9.7(1998) CURRENT SST (C) 28.2 CLIMATOLOGY** 28.9	**TAHITI-MOOREA** 17.5 S, 150.5 W 12WK ACCUM TODAY 0.0 MAX 12 WK* 7.3(1991) CURRENT SST (C) 28.9 CLIMATOLOGY** 28.8	**FIJI-BEQA** 18.5 S, 178.5 E 12WK ACCUM TODAY 4.2 MAX 12 WK* 6.9(2000) CURRENT SST (C) 28.7 CLIMATOLOGY** 28.1	**NINGALOO, AU** 21.5 S, 114.0 E 12WK ACCUM TODAY 0.0 MAX 12 WK* 11.3(1999) CURRENT SST (C) 28.3 CLIMATOLOGY** 28.2

*** DERIVED BASED ON DHWs FROM 1985 ~ 2001**
**** ALSO KNOWN AS THE MAXIMUM MONTLY MEAN SST**

Figure 7. The Coral Reef Watch Indices web page.

9. References

Berkelmans, R. and B.L. Willis. 1999. Seasonal and local spatial patterns in the upper thermal limits of corals on the inshore Central Great Barrier Reef. Coral Reefs, 18:219-228.

Coles, S.L. and P.L. Jokiel, 1977. Effects of temperature on photosynthesis and respiration in hermatypic corals. Marine Biology, 43:209-216.

Gleeson, M.W. and A.E. Strong. 1995. Applying MCSST to coral reef bleaching. Advances in Space Research, 16(10):151-154.

Glynn, P.W. and L.D. D'Croz. 1990. Experimental evidence for high temperature stress as the cause of El Niño-coincident coral mortality. Coral Reefs, 8:181-191.

Goreau, T. J., and R.L. Hayes. 1994. Coral bleaching and Aocean hot spots.@ AMBIO, 23:176-180.

Jokiel, P.L. and S.L. Coles. 1990. Response of Hawaiian and other Indo-Pacific reef corals to elevated temperature. Coral Reefs, 8:155-162.

Lesser, M.P., W.R. Stochaj, D.W. Tapley, and J.M. Shiek 1990. Bleaching in coral reef anthozoans: effects of irradiance, ultraviolet radiation, and temperature on the activities of protective enzymes against active oxygen. Coral Reefs, 8:225-232.

Liu, G., W. Skirving, and A.E. Strong. 2003. Remote Sensing of Sea Surface Temperatures During the 2002 (Great) Barrier Reef Coral Bleaching Event, EOS, 84(15):137,141.

McClain, E.P. W.G. Pichel, and C.C. Walton. 1985: Comparative performance of AVHRR-based multi channel sea surface temperatures. Journal of Geophysical Research. 90:11,587.

Montgomery, R.S. and A.E. Strong. 1995. Coral bleaching threatens oceans, life. Eos, Transactions, American Geophysical Union, 75:145-147.

Reaser, J.K., R. Pomerance, and P.O. Thomas 2000. Coral bleaching and global climate change: Scientific findings and policy recommendations. Conservation Biology, 14,:1500-1511.

Strong, A.E., C.S. Barrientos, C. Duda, and J. Sapper. 1997. Improved satellite techniques for monitoring coral reef bleaching. Proceedings of the 8th International Coral Reef Symposium, Panama City, Panama, 1:1495-1498.

Strong, A.E. 1991. Sea surface temperature signals from space. In Encyclopedia of Earth System Science, Ed. W.A. Nierenberg, Vol. 4, Academic Press, San Diego, CA, pp 69-80.

Strong, A.E., E. Kearns, and K.K. Gjovig. 2000. Sea surface temperature signals from satellites - an update. Geophysical Research Letters, 27(11):1667-1670.

Winter, T.M. 1995. The development of an operational global ocean climatology through the use of remotely sensed sea surface temperature. Trident Scholar Project Report No. 235, U.S. Naval Academy, 67pp.

Yonge, C.M. and A.G. Nicholls. 1931. Studies on the physiology of the zooxanthellae, Science Report, Great Barrier Reef Expedition, 1928-1929, 1:135-176.

10. *Appendix*

SST Field Generation Algorithm

Algorithm Selection Procedure

For each target processed (approximately 60,000 targets per orbit, 14 orbits per day), the first step is to check the quality of the target and select one or more processing algorithms.

The SST is retrieved principally using either the daytime or nighttime AVHRR-only algorithm; however, as a fail-safe provision, it can be retrieved using one of the HIRS-only algorithms. A simultaneous parallel-test mode allows comparison of results from a new algorithm or modified threshold with the result of the operational algorithm. Use of the Target Rejection Decision Table allows test or implementation of an algorithm for a selected portion of the global ocean. Note: A target is an 11 by 11 array of AVHRR 4km Global Area Coverage (GAC) fields of view. These targets are centered on the High resolution Infrared Sounder (HIRS) fields-of-view (a sensor that flies on the same satellite as the AVHRR). Thus, the GAC targets overlap by 4 or 5 fields-of-view depending on the collocation.

1. Check Quality Control Flags and Calibration Consistency

Target data from the orbital processing program (MUT) contains quality control (QC) flags from the level 1b data base. Some of these flags, if set, are fatal for target processing. If a fatal QC flag is set, processing of the target is terminated. A count of the number of QC errors is accumulated by blocks of 500 scan lines to allow bad sections of data to be pinpointed for study. The magnitude and consistency of AVHRR and HIRS calibrated coefficients are also monitored, and targets with erroneous calibration data are rejected.

2. Perform Gross Land Test

In the gross land test, if AVHRR data are available, check the low-resolution land/sea tag value at the nearest ½ ^{0} latitude/longitude intersection for the following positions within the target:

   a. If satellite zenith angle < 26 ^{0} , check the four target corners

b. If satellite zenith angle >26 ^{0} but <50 ^{0}, check the four corners and the center
c. If satellite zenith angle >50^0 , check the four corners, center, and the two points equidistant from the center and the outside edge on the long axis of the target.

If all coordinates checked are sea, process the target. If all are land, do not process. If some coordinates are coastal interface with high resolution tags available, process. If any are coastal interface with no high resolution tags available, do not process. (High resolution tags are available globally now).
If no AVHRR data are available, check the HIRS center latitude/longitude against the low resolution land/sea tag for the nearest ½ ^{0} latitude/longitude intersection. Unless the tag is sea, reject the target.
3. Select Algorithms
Data input is used to select from 1 to 12 different algorithms to be used on each target. These are:

(1) AVHRR-only day operational	(7) AVHRR-only day test
(2) AVHRR-only night operational	(8) AVHRR-only night test
(3) HIRS-only day operational	(9) HIRS-only day test
(4) HIRS-only night operational	(10) HIRS-only night test
(5) AVHRR & HIRS day operational	(11) AVHRR & HIRS day test
(6) AVHRR & HIRS night operational	(12) AVHRR & HIRS night test

Note: Only algorithms 1, 2, 7 and 8 have been used in the last 15 years.
4. Determine whether it is Day or Night
First the solar zenith angle is checked. If it is less than 75^0, then it is daytime. If it is greater than 75^0, then check the center spot of one of the GAC Visible Channels (or the HIRS visible channel value if GAC is not available). If the visible channel value (currently channel 2) is greater than or equal to a reflectance threshold (currently 1%), then do not process. If it is less than the threshold, it is nighttime.
5. Call Retrieval Algorithm
Each selected algorithm will be used if it matches the actual time of the day (i.e., day or night) and if the Target Rejection Decision Table indicates that this algorithm is to be used for the geographical location of the target. The algorithm is then called to make one or more SST observations. Then the next algorithm is selected, and so on until the input list of desired algorithms is exhausted.

AVHRR-Only Nighttime Algorithm Summary
Note: All threshold values reflect current NOAA-14 settings as of 2/22/99.

1. Nighttime Gross Cloud Test T_4 > 268.16 for 30 or more target elements
2. Unit Array Selection
   Find warmest channel 4 elements in the target and construct four 2x2 unit arrays containing the warmest elements.
3. IR Uniformity Test
   Test each of the four unit arrays until one passes both of the following tests:
   $Cmax_4$ - $Cmin_4$ # 2
   where $cmax_1$ & $cmin_1$ are the maximum and minimum counts for channel 1
4. Average Unit Array Counts
   Find the average count for each of five channels for the unit array that passes the IR uniformity Test.
5. Calibrate the unit array averages.
6. Channel 4 Test T_4 ∃ 270^0
7. **(Not currently used)** 3.7-11 IR Cloud Test $T_3 = -25.0089766 + 1.0916 (T_4)$ $|T_3 . T_3|$ # 3^0 C
8. Channel 4-Channel 5 Test $T_4 - T_5$ # 4.0^0
9. **(Not currently used)** 4-5 IR Cloud Test $T_4 = -11.488987 + 1.0439 (T_5)$ $|T_d . T_3|$ # 1^0 C
10. Currently using T_5-T_3 option in the following (not using T_4-T_3)
    Nighttime Low Stratus Test or $T_5 . T_3$ Low Stratus Test (currently this is used)
    $|T_4 . T_3|$ # 0.7^0 C $T_5 . T_3$ # 0.0^0 C
11. Calculate SST with 3 Equations (SST in ^{0} C)
    Split, Dual, & Triple Window Equations
    Currently the equations used for NOAA-14 are the NLSST split-window, the MCSST dual-window, and the NLSST triple-window equations.
12. Unreasonable SST Test -2^0 C # T_{sst} # 35^0 C where T_{sst} is currently (2/22/99) the NLSST triple window equation.
13. SST Intercomparison Test [Max (T_{sst1}, T_{sst2}, T_{sst3}) - MIN (T_{sst1}, T_{sst2}, T_{sst3})] # 2.0^0C
    The equations used to compute the three SSTs are those noted in 11.
    Where MAX (A, B, C) is the maximum of A, B, C
    MIN (A, B, C) is the minimum of A, B, C
14. Compute Earth Location of Center of Unit Array
15. Climatology Test $|T_{sst}$ - CLIMO$| < 10^0$C

Where CLIMO is the climatological SST for the nearest 1^0 latitude/longitude intersection

16. HIRS Cloud Test Using HIRS and Field and Climo Temps (refer to last page)
17. Store SST in SST Initial Storage File

Note:

AVHRR Channel	Wavelength (Φm)
1	0.550- 0.90
2	0.725- 1.10
3	3.550- 3.93
4	10.500-11.50
5	11.500-12.50

Use of HIRS data to detect cloud contamination

HIRS algorithm

$D=3.5 - 0.2333 (H_8 - H_7) + 0.038446*SATFLD + 1.612 (SEC\ \theta - 1)$

Where: H_8 - HIRS Channel 8 Temperature
H_7 - HIRS Channel 7 Temperature

HIRS cloud test (Night and Day relaxed visible cloud mode)

IF $D > D_1$ Test is failed

Field test

If $| MCSST - A.SAT\ FLD - (1- A)\ CLIMO | > D_2$ Test if failed

A= input coefficient = 0.666

D_1 AND D_2 - Threshold values are read in from data files

Current settings: Daytime (only used in WRMSPT mode) D1=0.5 D2=3.0
Nighttime D1=0.5 D2=2.5

This is a combined test. Both tests musts be failed for rejection (HIRS and field test).

Chapter 3

OPTICAL REMOTE SENSING TECHNIQUES TO ESTIMATE PHYTOPLANKTON CHLOROPHYLL *a* CONCENTRATIONS IN COASTAL WATERS WITH VARYING SUSPENDED MATTER AND CDOM CONCENTRATIONS

JOHN F. SCHALLES
Biology Department, Creighton University, Omaha, NE 68178, USA, jfsaqua@creighton.edu

1. Introduction

"Roughly 98% of the world's oceans and coastal waters fall into the Case 1 category, and almost all bio-optical research has been directed toward these phytoplankton-dominated waters. However, near-shore and estuarine Case 2 waters are disproportionately important to human interests such as recreation, fisheries, and military operations. It is therefore likely that Case 2 waters will receive increasing attention in coming years." – Curtis Mobley, 1994. Light and Water: Radiative Transfer in Natural Waters

The estimation of chlorophyll concentrations is one of the most scientifically relevant and commonly used applications of remote sensing to aquatic coastal systems. This, however, is a far from trivial task because of the complex composition and distribution of optically active constituents in many coastal waters. In the decade since the publication of Mobley's "Light and Water", the hydrologic optics of turbid Case 2 waters (Morel and Prieur, 1977) have been vigorously examined. The advantages and challenges of remote estimation of chlorophyll concentrations in Case 2 waters are now relatively well defined, although operational monitoring schemes are not fully mature and await further sensor and algorithm improvements (International Ocean-Color Coordinating Group, 2000). This chapter will survey this important component of coastal remote sensing. There is much to be gained from examining the commonalities in the optical environments of inland and coastal water, and I will draw primarily from empirical studies that span these diverse and optically complex environments. This chapter also considers the transferability of the well developed, satellite remote sensing approaches for chlorophyll assessments of open ocean, Case 1 waters (O'Reilly et al., 1998) to Case 2 waters.

The terminology and rationale for Case 1 and Case 2 water classifications were established by Morel and Prieur (1977) in their seminal work on the bio-optical basis for ocean color variations. The optics and emergent color signals of Case 1 waters are largely dominated by: 1) living phytoplankton cells; 2) organic tripton (detritus) particles from death and decay of phytoplankton and the grazing products of zooplankton; and 3) the dissolved organic matter produced by phytoplankton metabolism, and decay of organic tripton (modified from Gordon and Morel, 1983). Case 2 waters contain Case 1 constituents plus materials introduced from outside the water column, which also effect optical properties. These include: 4) turbulent resuspension of bottom particles in shallow areas; 5) inorganic and organic tripton from river drainages, glaciers, and/or wind transport; 6) terrigenous and littoral zone colored dissolved organic matter (CDOM); and 7) anthropogenic particulate and dissolved

L.L. Richardson and E.F. LeDrew (eds.), Remote Sensing of Aquatic Coastal Ecosystem Processes: Science and Management Applications, 27-79. © 2006 *Springer. Printed in the Netherlands.*

substances (Klemas and Polis, 1977; Gordon and Morel, 1983; Alberts and Filip, 1994; Bukata et al., 1995). Although Case 1 conditions normally prevail offshore, they may also occur: 1) in coastal and near shore areas in arid regions, where shelf geology is limited and deep water occurs just off the coastline; 2) during droughts resulting in highly reduced or no runoff; or 3) other instances where currents introduce offshore water in localized areas lacking significant terrestrial runoff. Obviously, Case 2 conditions intergrade spatially with Case 1 conditions in relation to distance from coastlines and river plumes. Physical dynamics and seasonality can result in state changes for a given location. The chlorophyll prediction algorithms for Case 1 and Case 2 waters are often different, and both sets may be required in coastal zones. Note that certain *in situ* water column processes (ex. phytoplankton blooms or amplified scattering by carbonates from coccolithophores or whiting events) can shift the optics of offshore waters away from true Case 1 conditions.

Most approaches for remote estimation of phytoplankton biomass are based on the absorption of sunlight by algal pigments in the presence of light scattering by algal cells and non-algal particles. This interplay of pigment-specific absorption and particle scattering produces graded responses in the emergent reflectance signals detectable by radiometer instruments. This chapter will examine the nature of these graded responses and their utilization in various quantitative and qualitative assessment schemes for spatial and temporal patterns of phytoplankton abundance.

Coastal, estuarine, and inland water habitats present serious challenges to the interpretation of diagnostic light signals because of the diversity of optically active constituents, which partially mask fundamental phytoplankton absorption and scattering relationships (Carder et al., 1989; Gallegos et al., 1990; Ritchie et al., 1994; Schalles et al., 1998a). Field spectroscopy measurements in turbid inland and coastal Case II waters have increased substantially in the past decade (Wang et al, 1996; Kahru and Mitchell, 1998; Dekker et al., 1997; Schalles et al., 1998a, 1998b; Gitelson et al., 2000; Cunningham et al., 2001; Lahet et al., 2001; Gons et al., 2002; Brando and Dekker, 2003; Dall 'Olmo et al.,2003). These and similar studies have led to the production of new chlorophyll prediction algorithms for Case 2 waters. Several Case 2 algorithms are operational for sensors on aircraft (Kallio et al., 2003) and satellites (Ruddick, 2004).

The chlorophyll algorithms for Case I waters (Gordon and Morel, 1983; O'Reilly et al., 1998) are universally based on a simple interaction of phytoplankton density with water - as cell densities increase, chlorophyll and carotenoid absorptions increasingly dominate at blue wavelengths and generally cause decreased reflectance, whereas cell scattering increasingly dominates at mid-green wavelengths and causes increased reflectance (see detailed explanations below). A simple blue to green ratio has a robust and sensitive relationship to chlorophyll *a* concentrations in Case I waters. The relationship becomes less sensitive at higher chlorophyll levels (above ~ 30 μg/l chl *a*). The relationship is also sensitive to changes in the ratios of chlorophyll *a* (hereafter, chl *a*) to total algal pigment and total carotenoids (Aiken et al., 1995). Furthermore, this simple relationship can be highly compromised by the effects of CDOM and/or tripton particles in turbid, Case II estuarine and near shore waters and even waters well offshore in regions of large river plumes. For example, the influence of the Orinico plume can extend across the entire central Caribbean (Muller-Karger et al., 1989). CDOM material in the Orinoco plume caused significant overestimation of chlorophyll using Coastal Zone Color Scanner (CZCS) data. In many estuaries, CDOM and tripton dominate the optical processes of water column components (Bowers et al., 2003). Cole and Cloern (1987) estimated that phytoplankton accounted for only 5% of light attenuation in South San Francisco Bay. Biogenous calcite can substantially increase water reflectance, particularly from resuspension of carbonate sediments during stormy periods and the mass appearance of coccoliths during the declining stages of coccolithoporid blooms (Brown and Yoder, 1994) and extensive "whiting" conditions in inland lakes and coastal and offshore waters.

2. Chlorophyll as an Integrative Bioindicator

Chl *a* is the dominant light harvesting pigment and is universally present in eukaryotic algae and the Cyanophyceae (cyanobacteria or "bluegreen algae") (Rowan, 1989). The only potential water column phototrophs lacking chl *a* are certain photosynthetic bacteria such as Purple or Green bacteria, which contain bacteriochlorophylls with different absorption maxima, and phototrophic Archaebacteria containing bacteriorhodopsin as the light-harvesting pigment.

Chl *a* is commonly measured in water quality monitoring programs for coastal and inland waters (Jordan et al., 1991; Morrow et al., 2000;, Cracknell et al., 2001;, Casazza et al., 2003), in surveillance programs for harmful algal blooms (Munday and Zubkoff; Paerl, 1988; Richardson, 1996; Kahru and Mitchell, 1998;, Pettersson et al., 2000), and in ecological studies of phytoplankton biomass and productivity (Cole and Cloern, 1987; Talling, 1993;, Gallegos and Jordan, 2002; Lefevre et al., 2003). The contribution of phytoplankton derived primary production on continental margins (shelves, slopes, and rises) to both global ocean production and particulate organic carbon export to the deep oceans has been underexamined, but appears to be highly significant (Walsh, 1991; Jahnke, 1996).

It is important to keep in perspective that remote measurements of *in situ* phytoplankton using passive solar reflectance are generally incapable of isolating the chlorophyll signal from other cell components and other optically active compounds (OACs) in the water column. Even the chlorophyll fluorescence signals detected by active LIDAR systems can be affected by other fluorescing materials, red reabsorption by chl *a* and optical processes of the extracellular environment (Doerffer, 1993; Pozdnyakov et al., 2002). Chl *a* is packaged with other pigments within the refractive and reflective contents of algal cell walls, membranes, and other cytoplasm constituents. An extensive analysis (n ~ 5600 stations) of ocean pigment data found an average total pigment to chl *a* ratio of 2.164, with a range of 1.876 to 2.876 across cruises and ocean provinces (Aiken et al., 1995).

Uniform sphere models of cells are unable to reasonably describe scattering in marine phytoplankton cells (Quinby-Hunt et al., 1989). A three layered model using a high refraction index for the outer layer (cell wall component) and low indices for middle (chloroplasts) and inner (cytoplasm) layers performed well in simulating a mixed particle size distribution from the central Pacific gyre (Kitchen and Zaneveld, 1992). Non-extractive, remote sensing schemes to detect and monitor phytoplankton chlorophyll must contend with the complex scattering and absorbing features of this total cell package. In most natural assemblages, cell populations of mixed species with variable shapes and sizes (submicron to tens of microns or more) and differing cell wall constituents and pigment suites occur in spatially heterogeneous patterns. Other photosynthetic and photoprotective pigments have overlapping spectral absorption with chl *a*, especially at the blue, Soret band of chlorophyll (Figure 1). An example of an aggregate absorption spectrum (chl *a* plus accessory chlorophylls and carotenoids) is compared to an isolated chl *a* spectrum in Figure 2. The data in Figure 2 were produced using the specific absorption coefficients of Bidigare et al. (1990; see Figure 1) and a global ocean average of the proportions of each of the five classes of pigments from Aiken et al. (1995). Note that the data in Figure 2 are scaled to a chl *a* concentration of 1 μg/l. Between about 460 and 570 nm, accessory pigments account for most absorption. Even at 440 nm, accessory pigments account for 52% of the total absorption. However, above 600 nm chl *a* absorption is dominant, and accounts for nearly all absorption above 650 nm. Most notable is the *in vivo* red absorption maximum of chl *a* at 674 nm (Figure 2). Chl *a* pigment absorption is present but barely detectable in the range of about 698 to 730 nm. In this spectral range, specific absorption coefficients decrease from 0.0010 to 0.0001 $m^2 * mg^{-1}$ (Bidigare et al., 1990).

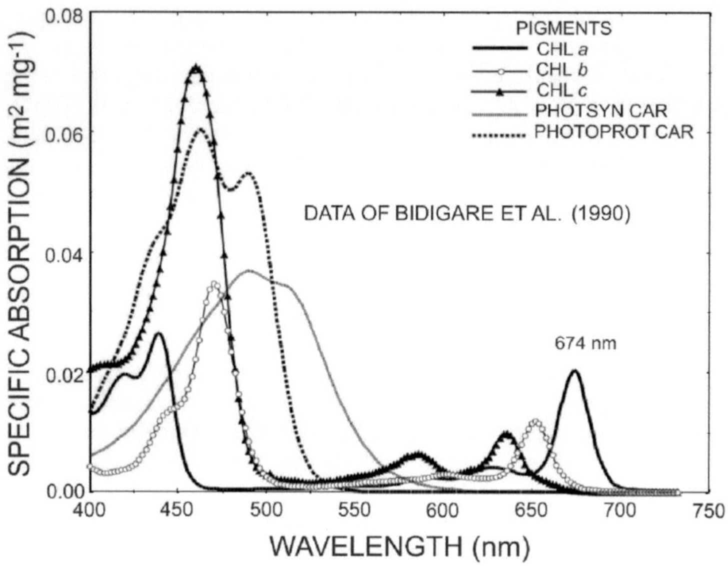

Figure 1. Calculated *in situ*, weight specific absorption coefficients for the major phytoplankton pigments: chl *a, b, c* = chlorophylls *a, b,* and *c*, Psyn Car = photosynthetically active carotenoids, Pprot Car = photoprotective carotenoids. Note red absorption maximum for chl *a* at 674 nm.

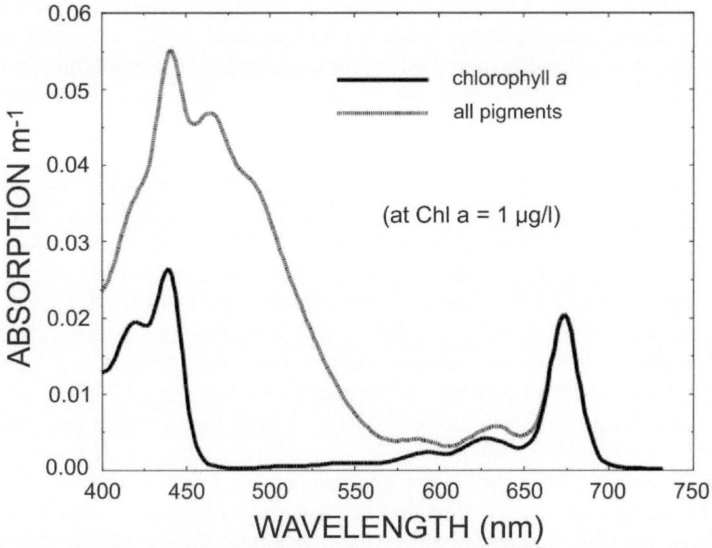

Figure 2. Comparison of *in situ* chl *a* absorption (at a chl *a* concentration of 1 μg/l) versus total pigment absorption (with all chlorophylls and carotenoids) based on the specific absorption coefficients for five pigment classes (Figure 1) and global ocean averages for the proportions of these five classes (see text).

Chlorophyll distribution may have significant vertical variation, particularly when stable stratification develops in coastal and estuarine waters (Paerl, 1988) or with the characteristic deep chlorophyll maximum of oligotrophic ocean waters (Lefevre et al., 2003). Optical remote sensing above the water surface generally cannot resolve deep chlorophyll layers and other vertical heterogeneity patterns (Andre, 1992). A further complication, rarely considered, arises when multiple wavelength algorithms use bands representing differing optical depths. For example, in a turbid eutrophic water column, blue and red reflectance signals attenuate much more rapidly (in both the down and upwelling fields), and the composite upwelling blue and red signals return from shallower depths than the green signals.

Chlorophyll is a comparatively efficient and effective, albeit indirect, measure of phytoplankton biomass. There are three primary analytical approaches for laboratory analyses of chl *a* concentration in field samples: spectrophotometry, fluorometry, and high-performance liquid chromatography (American Public Health Association, 1998). Most analyses begin with filtration of the chlorophyll-containing pigment fraction. Pigments are then extracted with an organic solvent (acetone, methanol, or less commonly ethanol or diethyl ether (Rowan, 1989). Freezing, grinding, and other techniques to physically disrupt algal cells generally improve the extraction pigment yield. Note that these extraction schemes also bring accessory pigments (except for the water soluble phycobiliproteins) into solution. Following clarification of the extract with filtration or centrifugation, the pigment content is assessed using pigment specific absorption bands in a spectrophotometer, via fluorescence using proper excitation and emission wavelength settings, or through chromatographic column separation (usually HPLC) together with in-line optical detection of column fractions. HPLC is now widely used to simultaneously identify and quantify the entire suite of extractable pigment fractions. These techniques provide both direct estimation of pigment in samples and calibration information for *in situ* measures.

In situ chl *a* is currently assessed using fluorometry (submersible or shipboard probes or airborne LIDAR systems; see Yoder et al., 1992), wavelength specific path attenuation in transmissometers (Morrow et al., 2000), and measurement of water column reflectance from close range, airborne (Kallio et al., 2003), or satellite sensors (Brando and Dekker, 2003). When data are collected from high altitude aircraft and satellite sensors, the signal from water column reflectance is usually far lower in magnitude than the atmospheric scatter. Removal of this "atmospheric effect" through correction techniques is very important for standardizing data and normalization to reflectance (Bukata et al., 1995; Gordon, 1997; Brando and Dekker, 2003).

Antoine et al. (1996) used CZCS satellite data to estimate an average chl *a* value of 0.19 μg/l for the world's oceans. These workers further calculated that 55.8% of the world's oceans (between 50°S to 50°N) averaged less than 0.1 μg/l chl *a*, 41.8% were between 0.1 and 1 μg/l, and only 2.4% exceeded 1 μg/l, levels which they classified as oligotrophic, mesotrophic, and eutrophic, respectively. In the low phytoplankton densities of the open oceans, productivity is comparably low. On average, the quantum yield in the oceans is about one third that of terrestrial ecosystems, with only one CO_2 molecule fixed per 1500 incident photons (Falkowski and Raven, 1997).

Most coastal waters exceed the 0.19 μg/l chl *a* average, with more productive upwelling and offshore river plume areas commonly reaching the 1-10 range and an additional order of magnitude greater in highly eutrophic and higher latitude sites. Overall, the global range of blue water, coastal, and estuarine phytoplankton chlorophyll is nearly 5 orders of magnitude (0.01 to 1000 μg/l, Figure 3). The relative impact of chl *a* and accessory pigments on water column optics is thus highly variable and renders problematic the development of a single, robust algorithm to estimate pigment content by remote sensing.

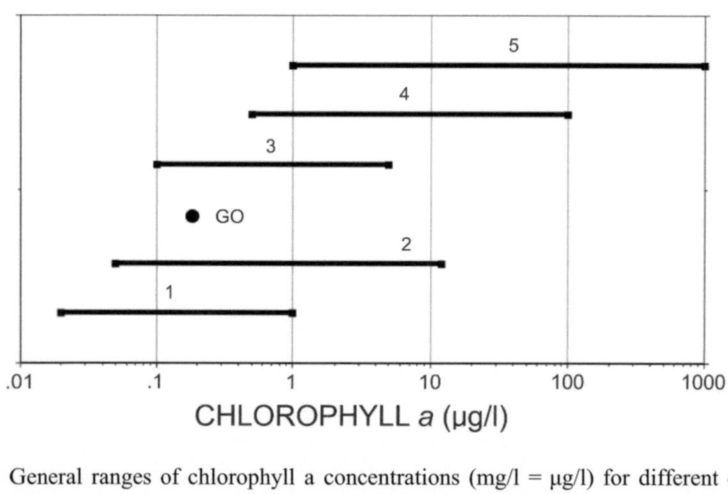

Figure 3. General ranges of chlorophyll a concentrations (mg/l = μg/l) for different ocean and coastal provinces: 1 - Sargasso Sea, Equatorial Pacific, Caribbean, 2 - California Current, 3 - Estuaries and Coastal Waters, 4 - North Atlantic, 5 - harmful algal blooms (Munday and Zubkoff, 1981; Carder and Steward, 1985; Kahru and Mitchell, 1998). GO = global ocean average of 0.19 (from Antoine et al., 1996).

Control of chl *a* levels and phytoplankton production are complex phenomena involving the interactions of physical, chemical, and biological processes (Jumars, 1993). Chlorophyll levels are generally controlled by nutrient and light availability, ratios of limiting resources, water column residence times and structural stability, water temperatures, and grazing pressure. Estuaries and other coastal embayments often have greater chlorophyll concentrations than adjacent offshore areas. However, offshore upwelling and plume activity can generate elevated pigment levels and increase spatial heterogeneity. Estuarine zones commonly have complex spatial patterns of biogeochemical constituents related to wind and tidal driven circulation patterns, longitudinal mixing gradients and point source inputs, and anthropogenic disturbances (Jassby et al., 1997). These spatial variability patterns are a major challenge in the design of estuarine sampling strategies.

In Georgia river transects extending from nontidal river reaches to offshore, we found a consistent pattern of increased chlorophyll within the estuarine mixing zone compared to upriver and offshore stations (Figure 4; also see Schalles et al., 1998a; Alberts et al., 2004). In these Georgia habitats, the mid and lower salinity reaches of the estuaries often have upper single to double digit chlorophyll levels and offshore shelf areas are typically low to mid single digit levels. The lower reaches of these estuaries also have elevated total and inorganic seston levels (Alberts et al., 2004). Upstream nutrient inputs, increased water residence times, nutrient sequestration and recycling in tidal wetlands and benthos, complex circulation patterns, ebb tide export of littoral zone phytoplankton and epibenthic algae, and wind-driven sediment resuspension all contribute to greater productivity and biomass in many estuaries.

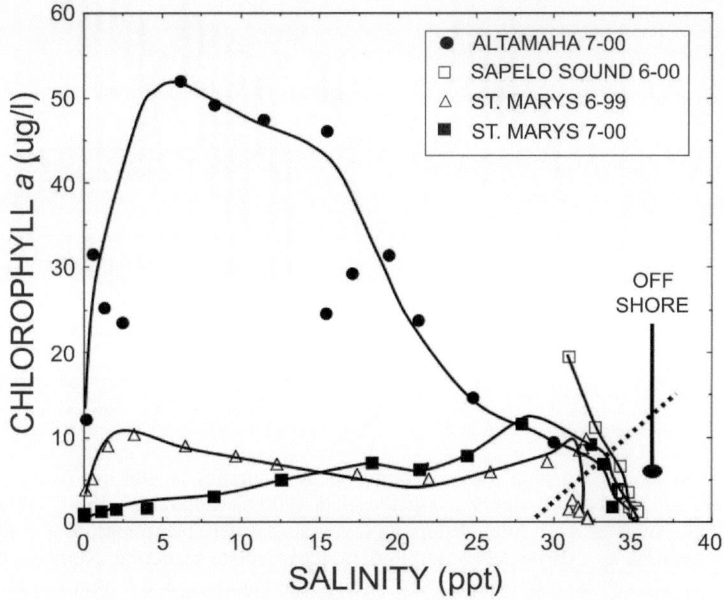

Figure 4. Chlorophyll *a* (not corrected for phaeophytin) versus salinity patterns for three Georgia estuaries and adjacent offshore areas (dashed, slanted line demarcates offshore stations). Data from two transects are shown for the St. Marys River.

3. Inherent Optical Properties and Constituent Absorption and Scattering

Chlorophyll remote sensing is based on detectable patterns within apparent optical properties. Radiative transfer equations describe the interactions between the inherent optical properties, or IOPs (e.g. absorption and scattering coefficients) and apparent optical properties, or AOPs (e.g. downwelling scalar irradiance attenuation, irradiance reflectance) for a given time, location, and atmospheric condition. Barnard et al. (1998) used a pair of *in situ* profiling ac-9 transmissometers (9 wavelengths; WET Labs, Inc), one with prefiltration of seawater, to operationally differentiate IOPs in a wide variety of ocean, coastal, and estuarine conditions. Their IOP data for 488 nm are summarized in Figure 5. The IOP measurements typically varied by one to two orders of magnitude for a particular location and cruise, and the total range for individual IOPs was two to four orders of magnitude. Not surprising, the equatorial Pacific had the lowest overall absorption and scattering values and the two estuarine sites (Chesapeake Bay and East Sound, Washington) had the greatest values. The coastal Case 2 waters generally had component absorption and scattering values above 0.1 m^{-1} whereas the clearest Case 1 waters had values between 0.0001 and 0.1. Considerable coherence existed between measurements at different wavelengths.

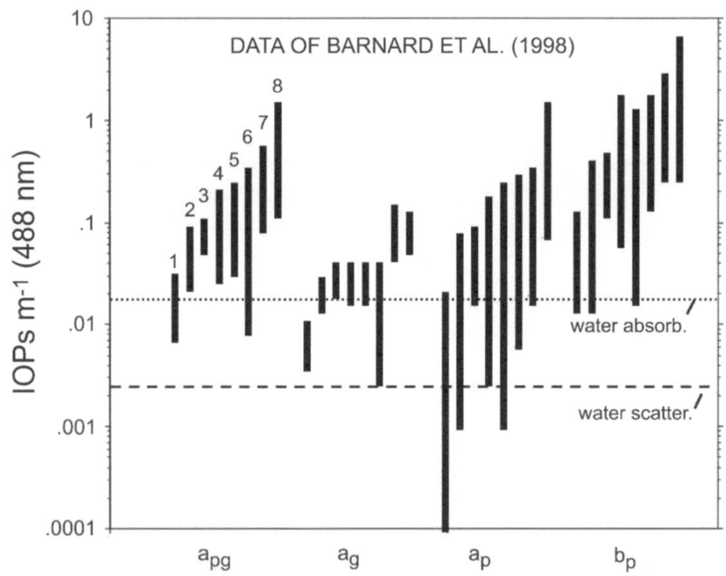

Figure 5. Summary of the ranges of inherent optical properties at 488 nm (from Table 2 in Barnard et al., 1998) for eight cruises and greater than 1900 measurements from 77 total profiles. Notation for coefficients: a_{pg} = total absorption (excluding water absorption), a_g = chromophoric dissolved absorption, a_p = particulate absorption, b_p = particulate scattering. Numbers above bars (shown only for a_{pg} values) refer to cruises: 1 - Equatorial Pacific, 2 and 5 - Gulf of California, 3 and 4 - North Atlantic Shelf, 6 - Near coastal California, 7 - Chesapeake Bay, 8 - East Sound, Washington. Absorption and scattering values for pure seawater (Mobley, 1994) are placed as reference lines.

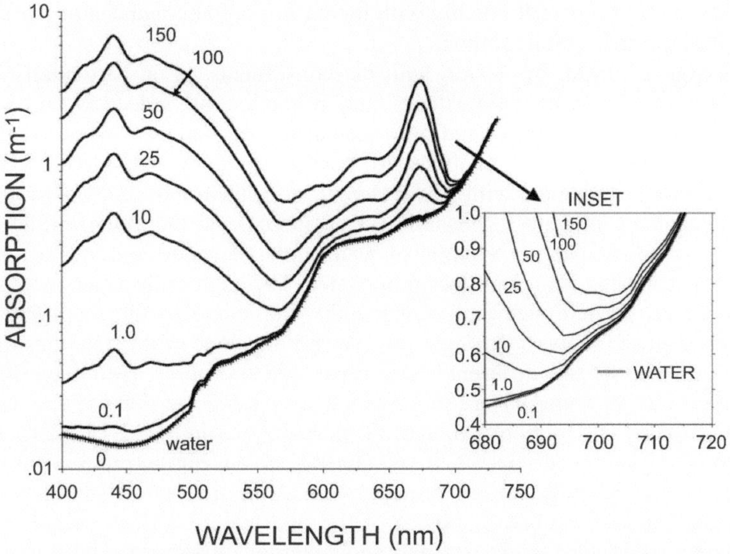

Figure 6. Semi-logarithmic plot of combined absorption coefficients for water and a model phytoplankton pigment suite at different concentrations of chlorophyll *a*. Pigment coefficients are based on the product of pigment specific absorption coefficients (m^2 mg^{-1}) and the average concentrations of chlorophyll and carotenoid fractions (µg/l) measured on four Sargasso Sea cruises (in Bidigare et al., – see text). Water absorption coefficients for pure seawater are from Smith and Baker, 1991. The inset shows the critical region near 700 nm where a localized peak of phytoplankton reflectance occurs at the point of minimum combined absorption by water and pigment.

The interaction of pigment and water absorption (Figure 6) provides important insights for the interpretation of minima and inflexion points in spectral reflectance curves. The pigment absorption data in Figure 6 assume the following fractional proportions, referenced to chl *a* (average values are from four Sargasso Sea cruises presented in Table 2 in Bidigare et al., 1990): chl *b* = 0.266, chl *c* = 0.083, photosynthetic carotenoids = 0.450, and photoprotective pigments = 0.236. At chl *a* = 0.1 and 1.0 µg/l, total pigment absorption adds comparatively little to water absorption above 500 nm. Hence, algorithms utilizing information from the red absorption peak will have little sensitivity for chl *a* values below 1, whereas pigment absorption has maximal separation from water in the Soret band region near 440 nm (Figure 6). A distinctive absorption minimum, corresponding to the green reflectance peak, occurs near 565 nm in this example. Note that a second local minimum occurs in the region of 680 - 710 nm (see inset in Figure 6). This minimum represents a region between stronger pigment absorption in the red Q band at adjacent lower wavelengths and much stronger water absorption at adjacent higher wavelengths. Note also that the wavelength position of this minimum shifts to higher wavelengths with higher pigment concentrations. This behavior has important bearing on the behavior of a corresponding peak in reflectance spectra in this same region produced by a combination of pigment fluorescence and phytoplankton cell scattering. Although many authors attribute this peak to chl *a* fluorescence (see discussion below), the wavelength position of chl *a* fluorescence should not shift notably with increasing pigment concentration, compared to the shift in the absorption minima.

Scattering of light by water and its constituents is a particularly complex phenomenon involving both reflectance and refraction of molecules and particulate matter (Kirk, 1994). Pure water and pure seawater (water and dissolved inorganic ions at a salinity of about 35 parts per thousand) have a strong, inverse relationship between wavelength and scatterance, with a wavelength dependence of $\lambda^{-4.32}$ (Morel, 1974). However, seawater ions are responsible for an increase in the random fluctuations of the number of molecules per volume of water, and seawater consequently has about 30% greater scattering (b_w) than does pure water. The angular distribution of scattering is dependent on particle size and wavelength. The shape of the scattering angle to volume scattering function in strongly inverse and the slope of this relationship appears relatively constant across different water types and turbidities (Mobley, 1994). At a temperature of 20°C, water has a refractive index of 1.3433 at 400 nm and an index of 1.3289 at 768 nm. The refractive indices of inorganic particles (~1.15-1.20) are greater than those of organic matter (~1.03-1.04) (Jerlov, 1976). Furthermore, algal cells with hard outer walls and soft, watery protoplasm may scatter more light than predicted based on the separate refractive indices of the two materials (Zaneveld et al., 1974). The complex refraction and external and internal reflectances of algal cells is apparently responsible for the larger scattering values per unit weight observed with phytoplankton compared to tripton (Dekker, 1993). In general, the scattering coefficients of turbid waters have less wavelength sensitivity than in pure water. An inverse relationship between wavelength and scattering also occurs in turbid waters, and spectral dependency (slope) is generally stronger for algae than for inorganic particles (Dekker, 1993).

Increased scattering results in an increase in the effective path lengths that light travels within water columns (Kirk, 1994). Increased scattering shifts the angular distributions of light and increases the probability that photons will re-emerge from the water column as a reflectance signal. One important consequence of increased scattering is a decrease in optical depth and reduction in the photic zone volume (Bukata et al., 1995). Also, in a water column with more diffuse light and longer effective path lengths, the probability of photon absorption increases. As discussed below, increasing particle densities thus amplify pigment absorption related differences in the magnitudes of the peak and trough reflectance used in chl *a* algorithms. In more productive and turbid waters, the "black pixel" assumption of negligible photon emergence at infrared wavelengths becomes invalid due to increased scattering and can lead to significant errors in chlorophyll retrieval (Siegel et al., 2000). Relaxation of this assumption can significantly improve chl *a* retrieval from satellite imagery in waters with chl $a > 2$ μg/l. Indeed, the NIR spectral region often yields the best remote sensing relationships between seston concentrations and water reflectance (Schiebe et al., 1992; Kirk, 1994).

4. Instrumentation, Calibration, and Biases

The dynamic range of radiometric measurement, spectral resolution, and number and position of bands are critical sensor design elements that greatly affect the operational abilities of instruments to detect chl *a* signals in different optical regimes. Field spectroradiometers and aircraft and satellite sensors have wide variations in the number of channels (bands) and bandwidths, foreoptics and viewing geometries, and detector array configurations. Chl *a* absorption of red light occurs in a rather narrow band. Optimal detection of the chlorophyll red absorption peak in laboratory spectrophotometers requires a spectral bandwidth of 2 nm or less (Rowan, 1989). The blue, Soret band of chl *a* has a comparably narrow absorption peak. Broader bandwidth not only reduces the ability to resolve these peaks, but also allows greater overlap with the absorption regions of accessory chlorophylls (especially chl *b*) and other pigments. These wavelength resolution issues may have similar importance in measurements with field radiometers. However, signal-to-noise constraints in field spectroradiometers and remote platform sensors may necessitate broader bandwidth (Dekker, 1993).

Wavelength selection for pigment detection is affected by a well known wavelength shift (roughly 5-10 nm) for the wavelength of maximum pigment absorption in extracted versus *in vivo* conditions. For example, chl *a* and *b* in 90% acetone have maximum absorption values at 664 nm and 647 nm (consensus averages from Rowan, 1989) whereas Bidigare et al. (1990) estimated the maximum red absorption *in vivo* at the longer wavelengths of 674 nm and 652 nm (see Figure 1). Also, the wavelengths of maximum red absorption for chl *a* can vary by about 5 nm, depending on the extractant and pH conditions (Rowan, 1989). Chl *a* has a prominent left "shoulder" of absorption associated with the blue, Soret band and a secondary red absorption peak broadly centered at 628 nm (Figure 1). Ratios of *in vivo*, weight specific absorption coefficients for the blue and red primary peaks and the blue and red secondary peaks are about 1.3 and 4.95, respectively (Bidigare et al., 1990). Global ocean comparisons reveal that chl *a*, total carotenoids, chl *c* and chl *b* account for about 47%, 43%, 8%, and 4% of total phytoplankton pigment, although a range of 35% to 53% for chl *a* was reported for different bio-optical provinces (Aiken et al., 1995).

Maintaining proper instrument calibration is an important but sometimes challenging issue. Wavelength to channel registration and radiometric energy calibration may not be stable over time (Evans and Gordon, 1994; Starks et al., 1995). Instruments should be checked periodically for gradual drift or erratic changes that may be related to operating temperature or to degradation or damage to the optical components or their light path alignments. Wavelength specific lasers, emission lines from lamps, and narrow pass filters are commonly used to establish a regression fit between known wavelengths and the instrument channels with maximum response to these signals (Starks et al., 1995). Radiometric responses of individual channels are calibrated against well characterized lamp emission responses. Proper radiometric calibration can be more difficult when foreoptics and/or fiber optic guides deliver light to the optical slit entrance of the instrument. Researchers performing close range measurements with spectroradiometers usually convert their water spectra measurements to fractional or percent reflectance, a normalization procedure in which upwelling and downwelling estimates are ratioed for each channel of an instrument or pair of instruments.

Bottom substrate reflectance contributes to and can cause bias in the composite upwelling irrandiance and reflectance signals in optically shallow waters (Akleson and Klemas, 1986; Maritorena et al., 1994; Lee et al., 1998; Odhe and Sigel, 2001; Albert and Mobley, 2003), and may interfere with interpretation of water column signals and chlorophyll retrieval. Rundquist et al. (1995) measured reflectance from black and white panels suspended at 10 cm depth intervals between 2 and 82 cm during dilution of an algal bloom in an outdoor tank mesocosm. With the black panel at 82 cm, a ratio of the NIR peak reflectance to chlorophyll absorption band reflectance near 674 nm (Mittenzwey et al., 1992) had a linear relationship to chl *a* over a range of 30 to 336 µg/l (r^2 = -0.98, n = 15). The white panel had an amplifying effect on spectral reflectance, and also amplified the differences between spectral peaks and troughs in a manner similar to suspended clay additions (see below). In the experiment, the NIR/red ratio increased about 40-50% with a white, reflective bottom and optically shallow conditions. Because of wavelength dependent light attenuation patterns, signal return from bottom substrates may occur at some wavelengths and not others. In general, chl *a* indices at higher wavelengths will be less biased than indices using lower wavelengths because of greatly increased light attenuation by water at red and especially NIR wavelengths. Boss and Zaneveld (2003) demonstrated that interactions of bottom substrate and overlaying water may strongly affect the IOPs of the water column, with greatest effects on absorption and/or scattering in near bottom waters.

Water surface effects may also cause substantial changes in light fields and remote sensing reflectance (Kirk, 1994). Light reflectance at the air/water interface is strongly affected by incident solar angle and by wave activity. Workers generally seek to avoid direct solar beam reflectance sun glint (glitter) effects by: 1) reducing the instrument's field of view; 2) directing the view angle away from sun's incident angle; and 3) restricting the time of day for measurements to avoid high or extremely low zenith angles. During strong winds, wave turbulence can lead to surface foam and entrained air bubbles in the upper several meters of water. Bubble backscattering has minimal effect on downwelling irradiance, but can increase upwelling radiance and reflectance five fold in the upper several meters (Mobley, 1994).

5. Examples of Water Reflectance Spectra

A series of water reflectance spectra measured close range by my research group using a pair of USB2000 spectroradiometers (Ocean Optics, Inc.) are compared in Figure 7. The spectra were selected to illustrate water color reflectance signals in a broad range of water types. The ranges of OACs at these locations were quite diverse (C = chlorophyll *a*, range = 0.03 - 272 μg/l; S = total seston - i.e. suspended solids, range = 1 - 106 mg/l dry weight; A = CDOM absorption at 440 nm, range = 0.18 - 12.3 m^{-1}). The chl *a* values were not corrected for phaeophytins, and thus represent a combined pigment set. These measurements represent a four orders of magnitude range in chl *a*. Reflectance spectra were measured as downwelling solar irradiance in air and upward radiance (25° field of view) at nadir angle just below the surface, and were computed at each wavelength as the ratio of the upward radiance just below the surface, $L_u(0-)$, to the downward irradiance measured in air, E_d (0+).

The Caribbean spectra, measured in deep water (about 600 m) off the north coast of Roatan Island, Honduras demonstrates an ultraoligotrophic condition (chl *a* = 0.03 μg/l) with high blue reflectance (about 8.5% at 400 nm), strongly decreased reflectance with increasing wavelength, and virtually no measurable signal above 600 nm. This lack of return signal corresponds to the sharp increase in water absorption coefficients (0.08 m^{-1} at 580 nm versus 0.3 m^{-1} at 620 nm; see Figure 6). Conversely, the scattering coefficient for water is only 0.0014 m^{-1} at 600 nm as compared to 0.0076 m^{-1} at 400 nm (Smith and Baker, 1981). Water color at this station was blue-violet. The spectra resembled modeled and observed blue water spectra in the Sargasso Sea and Crater Lake, Oregon (Morel and Prieur, 1977) and exemplifies locations where the clear seawater end member dominates reflectance. The tropical lagoon spectra, also measured at Roatan Island, represented a water condition with 20 times greater chl *a*, and 4-5 times greater seston and CDOM concentrations. A broad green maximum with peak reflectance between 530 and 570 nm was present. Blue reflectance was sharply reduced compared to the offshore station. Reflectance was detectable to about 725 nm.

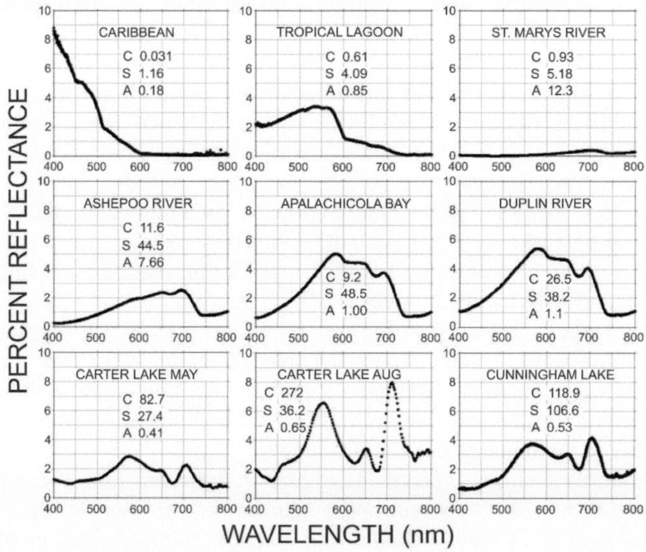

Figure 7. Water reflectance spectra measured by the author at various coastal, estuarine, and inland locations representing a broad range of optically active constituents. Caribbean and tropical lagoon sites - deep offshore waters and Man-O-War lagoon at Roatan Island, Honduras, C.A.. All other sites in the U.S.A.: St. Marys River - upstream station on Georgia/Florida border; Ashepoo River - mid river station in coastal South Carolina; Apalachicola Bay - western Florida; Duplin River - a tidal river at Sapelo Island, Georgia; ; Carter Lake - Missouri River floodplain lake on Nebraska/Iowa border; Cunningham Lake - flood control and recreation reservoir in Omaha, Nebraska. Note: C = µg/l of chl *a*, not corrected for phaeophytin; S = mg/l of total seston; A = m^{-1} of CDOM absorption at 440 nm.

The St. Marys River, at the eastern border of Florida and Georgia, typically has very high CDOM levels (Alberts and Filip, 1994). Spectra were acquired at a river station during a period with an extremely high CDOM level (ABS_{440} nm = 12.3 m^{-1}), very low chl *a*, and moderate total seston. Reflectance was well below 1% (Figure 7). A subdued reflectance peak (about 0.45%) was centered near 700 nm and no green peak was present. In this spectrum CDOM absorption (see below) suppressed scattering return at all wavelengths, but most notably in the blue and green regions. The scattering return was largely due to ultrafine clay-detritus aggregates (~ 16 µm diameter) that occur in these coastal plain rivers (Carlough, 1994). This curve is a good example of CDOM end member dominance. The Ashepoo River in South Carolina also had high CDOM concentrations and much higher chlorophyll loads and total seston loads than the St. Marys River. This spectrum was measured at a station with ABS_{440} (a measure of CDOM concentration) of 7.66, chl *a* of 11.6 µg/l, and total seston of 44.5 mg/l and had maximum reflectance of about 2.5% from 650 to 700 nm, with a modest pigment absorption trough near 670 nm. Reflectance remained below 1% between 400 and 500 nm and increased to 1.5% at 550 nm. The particulate load created notable NIR reflectance (about 1% between 735 and 800 nm) compared to the St. Marys River example (discussed below).

The Apalachicola Bay (Florida), and Duplin River (Sapelo Island, Georgia) spectra represent relatively turbid (total seston concentrations of 48.5 and 38.2 mg/l), moderately productive ecosystems (chl *a* concentrations of 9.2 and 26.5 μg/l) with about an order of magnitude lower CDOM than the previous two river examples (Figure 7). Both spectra had maximum reflectance (about 5% and 5.5%) near 580 nm and a similar NIR reflectance (about 1%) to the Ashepoo River spectrum. In both spectra, an inflexion near 600 nm and plateau extending from about 600 to 650 nm were caused by the interplay of seston scattering and the sharply higher water absorption in this spectral region (see Figure 6). In contrast to the St. Marys and Ashepoo examples, these spectra have noticeably larger chl *a* absorption features near 670 nm, with the larger chlorophyll load in the Duplin example causing a larger red trough feature and a more distinctive localized maximum near 690 nm. These local peaks represent the spectral position of minimum combined pigment and water absorption (Figure 6, see inset). Blue reflectance, although very low compared to the Caribbean blue water example, was greater than the St. Marys and Ashepoo examples. The higher blue and green reflectances in the Apalachicola and Duplin examples were presumably due to the much lower CDOM levels.

The final three example spectra in Figure 7 represent highly eutrophic sites in Nebraska. Carter Lake is a hypereutrophic lake on the Missouri River floodplain near Omaha (Schalles et al., 1998b; Schalles and Yacobi, 2000). A spring (May) example, during a bloom of the diatom *Synedra* sp. with a chl *a* concentration of 82.7 μg/l, had two spectral maxima near 574 nm and 702 nm, and relatively low (about 1%) and level reflectance between 400 and 500 nm (Figure 7). The upward wavelength shift of the NIR maximum (peak) above 700 nm is consistent with the shift in the absorption minimum with increased chl *a* (see inset, Figure 6). A pronounced chl *a* minimum was centered at 673 nm, with a prominent shoulder between 625 and 650 nm produced by both chls *a* and *c* (see Figure 1).

The Carter Lake August spectrum reveals extreme dominance by the filamentous cyanobacterium *Anabaena* sp., with a chl *a* concentration of 272 μg/l. The cyanobacterial pigment phycocyanin produced a distinctive reflectance minimum, centered at 625 nm (Dekker et al, 1996; Schalles and Yacobi, 2000). Phycocyanin concentration was estimated at 450 μg/l, yet the magnitude of its absorption trough was not as pronounced as the chl *a* trough. Phycocyanin absorption per unit of pigment is only 20% of the absorption per unit of chl *a* (Rowan, 1989). The position of the green peak (reflectance about 6.6%) was reduced by 20 nm, to 553 nm, compared to the May spectra (Figure 7). Carotenoid absorption (see Figure 1) "erodes" the left shoulder of the green peak and shifts its maximum to higher wavelengths. In response to the overall bloom dynamics, total carotenoids doubled from about 45 to 90 μg/l between May and August (Schalles and Yacobi, 2000). However, in the cyanobacteria bloom period phycocyanin has an even more dominant optical effect by eroding the right shoulder and shifting the green peak to lower wavelengths. A significant inverse correlation existed between phycocyanin concentration and green peak position ($r = -0.912$) in Carter Lake and resulted in regular, seasonal patterns of green peak position shifts during four years of observations (Schalles and Yacobi, 2000). Note, also, that the simultaneous erosions of the left and right shoulders of the green peak by carotenoids and phyocynanin may supress the overall height of this peak. At this high level of chl *a*, the blue, Soret band absorption (Figures 1 and 6) produced a distinctive minimum near 440 nm and the red absorption induced minimum (about 2% at 675 nm) was sharply lower than the adjacent NIR maximum (8% at 712 nm). The upward spectral shift of the NIR peak with increased chl *a* has been widely observed (Gitelson, 1992; Gitelson et al., 1999) and is explained by an upward shift in the position of minimum combined absorption by pigment and water with increased chlorophyll concentrations (Figure 6). The comparative heights of the NIR peak and the chl *a* red trough are used in several chlorophyll algorithms for Case 2 waters (Dekker, 1993; Ammenberg, 2000; Gitelson et al., 2000, Gons et al., 2002; see below).

Cunningham Lake is a eutrophic, clay-rich, turbid reservoir near Omaha within an agricultural watershed. A late summer example with high chl *a* and total seston concentrations of 118.9 μg/l and 106.6 mg/l is shown in the last panel of Figure 7. Phytoplankton composition was quite diverse, with abundant cynaobacteria and dinoflagellates as well as diatoms and chlorophytes. Green and NIR maxima occurred at wavelengths similar to the May Carter Lake example, but with higher reflectance. Reflectance at the NIR peak exceeded green peak reflectance (4.1% versus 3.7%) in contrast to green peak dominance in the first Carter Lake example, with an overall pattern more similar to the August spectrum. The dinoflagellate carotenoid pigment peridinin has significant activity in the mid-500 nm region and suppresses the green peak magnitude (Gitelson et al., 1999). Additionally, combined pigment and water absorption in the green peak region is more affected by increased pigment loads than the NIR peak position (Figure 6).

6. Tank Mesocosm Studies

6.1 PHYTOPLANKTON DENSITY AND COMPOSITION AND THEIR EFFECTS ON REFLECTANCE

Experiments with manipulations of OACs in indoor and outdoor tanks (ex. Krijgsman, 1994; Schalles et al., 1997; Schalles et al., 2001) have provided calibration data for pigment estimation algorithms, empirical evidence for interactions between phytoplankton and other constituents, and tests of bio-optical models. The results of spectral reflectance measurements from a tank mesocosm experiment at the former National Agricultural Water Quality Laboratory (USDA-Agricultural Research Service) in Durant, Oklahoma, illustrate the optical interactions of a graded series of phytoplankton densities with water. In this experiment (Figure 8, from Schalles et al., 1997), a mixed culture of green algae was grown to bloom state in an 80 m^3 outdoor tank. The tank depth was 3.2 m. *Chlorella* sp. accounted for 92% of cell density and *Scenedesmus* sp. for an additional 7% after bloom development. The upper 80% of the culture tank's water was transferred to a separate tank (dilution tank) to minimize settled detrital particles and brought to full volume with clear water. In the experiment, the bloom water was sequentially diluted with clear water from two additional tanks and volume was balanced by pumping the phytoplankton tank water to a clearwater, enrichment tank. Complete tank mixing was accomplished in 15-20 minutes with a pair of pumps drawing water from the bottom center, which was reinjected tangentially at the tank wall just below surface and at mid depth. For load balance, the enrichment tank was allowed to overflow to a drain. In this manner, phytoplankton cell densities were manipulated in a graded manner, achieving a chl *a* range of 0.4 - 62 μg/l. Initial cell density in the bloom water tank prior to dilution was 3.2×10^5 cells/ml.

A hinge point near 510 nm was observed in the resultant family of spectra (Figure 8). At wavelengths below 510 nm, reflectance decreased with increasing cell density - direct evidence that pigment absorption was dominant. Note that peak reflectance was about 1.5% at 400 nm, which is much lower than the 8.8% value for the Caribbean station in Figure 7 in the absence of deep water scattering. The chl *a* concentration was more than an order of magnitude greater in the tank compared to the Caribbean blue water example, and the black bottom of this tank, at a depth of 3.2 m, was visible in the clear water tank. At all wavelengths above 510 nm, reflectance increased with increasing cell density, although the magnitude of increase varied as a function of wavelength dependent absorption behaviors. Bukata et al. (1995) demonstrated similar results when their multi-component optical model for natural waters was parameterized with no CDOM present and no, or very low, amounts of suspended matter (tripton). With no CDOM or tripton present, their spectral pivot point occurred at 497 nm, the wavelength for which the ratio of $(b_b)_{chl} / a_{chl} = (b_b)_w / a_w$ (where b_b = backscatter, a = absorption, chl = an average phytoplankton condition, and w = pure water). Their modeled hinge point shifted to 528 nm with the addition of 0.1 mg/l tripton and to much higher wavelengths with higher tripton concentrations. Their model is also sensitive to the optical characteristics of the phytoplankton parameterization.

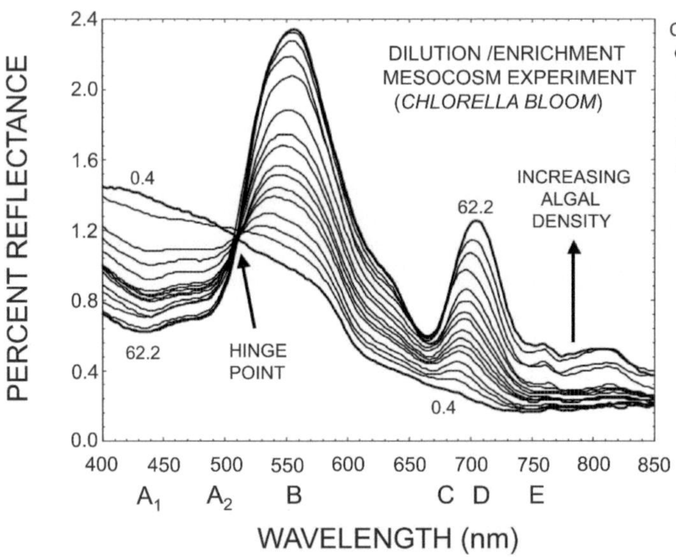

Figure 8. Graded series of reflectance spectra for different chl *a* levels (0.4 to 62.2 μg/l) for a dilution/enrichment scheme experiment (Schalles et al., 1997) involving two, 3.2 m deep mesocosm tanks. Note: letters along the X axis designate wavelength regions used for many chlorophyll algorithms (see text for explanation).

In the experiment in Figure 8, green reflectance increased from approximately 1 to nearly 2.4% as cell density increased. A noticeable peak near 550 nm became evident at chl *a* concentrations above 4 μg/l. The wavelength position of the green peak increased about 6 nm with increasing cell densities, presumably from erosion of the left shoulder of the peak by carotenoid absorption. A slight shoulder developed at about 620 nm, in the region of the secondary red peak of chl *b* (Figure 1). A well defined minimum (trough), centered near 670 nm and associated with the Q band of chl *a*, developed with increasing cell density (Figures 8, 9), as did the adjacent reflectance maximum (peak) near 700 nm. As seen in Figure 9, the position of maximum reflectance for this peak increased about 20 nm (from 685 to 705 nm), coincident with increasing pigment concentration and a shift in the position of minimum combined absorption by pigment and water (Figure 6).

A key observation apparent in Figures 8 and 9 is the appearance and intensification of peaks versus troughs with increasing cell densities. Increased scattering occurs at all wavelengths with increased cell densities. This scattering decrease with increasing wavelength is essentially monotonic (Mobley, 1994), and thus cannot account for the peaks and troughs coinciding with phytoplankton communities of different densities. Reflectance "troughs" develop at spectral regions of stronger pigment and/or water absorption and the "peaks" at regions of absorption minima (Figure 6). Therefore, phytoplankton density controls reflectance at these peak and trough positions, and are the primary signals available for remote sensing detection of chl *a* and, potentially, other pigments.

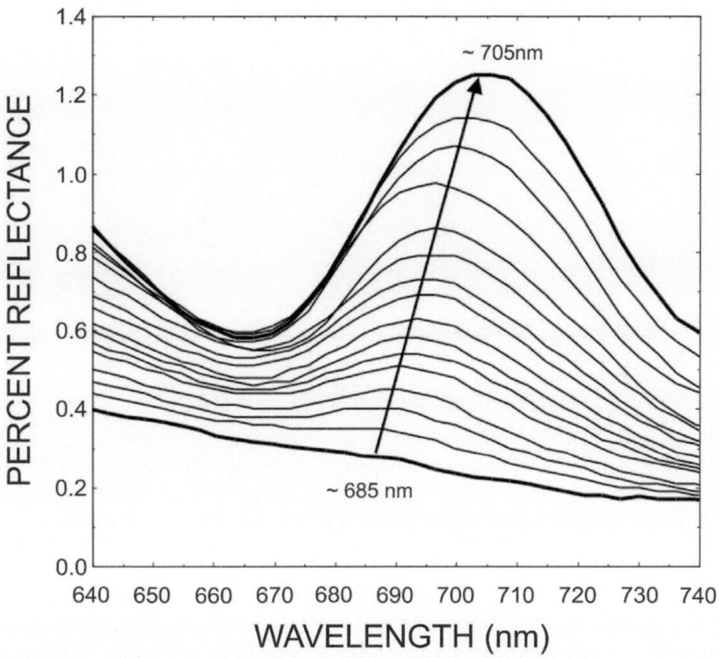

Figure 9. Magnification of the spectral region 640 to 740 nm from Figure 8, illustrating the increase in magnitude and the shift in wavelength position (arrow) of the NIR peak near 700 nm in a series of reflectance spectra with chl *a* levels of 0.4 to 62.2 μg/l.

The graded series of reflectance spectra (Figures 8 and 9) illustrate the basis for the predictive algorithms used in optical remote sensing of chl *a*. In ocean color schemes for Case 1 waters, bands in the blue and green regions (Figure 8, points A_1 - 443 nm, A_2 - 490 and/or 510 nm, and B - 550 or 555 nm) and others are commonly used (Gordon and Morel, 1983; O'Reilly et al., 1998). Oligotrophic, blue water conditions have an inverse relationship of reflectance and wavelength (Figure 7) due to increasing absorption by water with increasing wavelength (Figure 6) and blue dominated, wavelength specific scattering efficiencies (Smith and Baker, 1981). Modest increases in phytoplankton density, accompanied by increased pigment absorption (Figure 8), result in strongly decreased blue reflectance (tropical lagoon example in Figure 7 and point A_1 in Figure 8), while scattering outcompetes absorption in the green peak region and results in increasing reflectance (points B_1 and B_2 in Figure 8). The ratio of 443 to 555 nm (Figure 10) is strongly and inversely correlated and has been used to estimate chl *a*. The relationship, however, is nonlinear and becomes increasingly less sensitive with increasing pigment (especially above 10 μg/l).

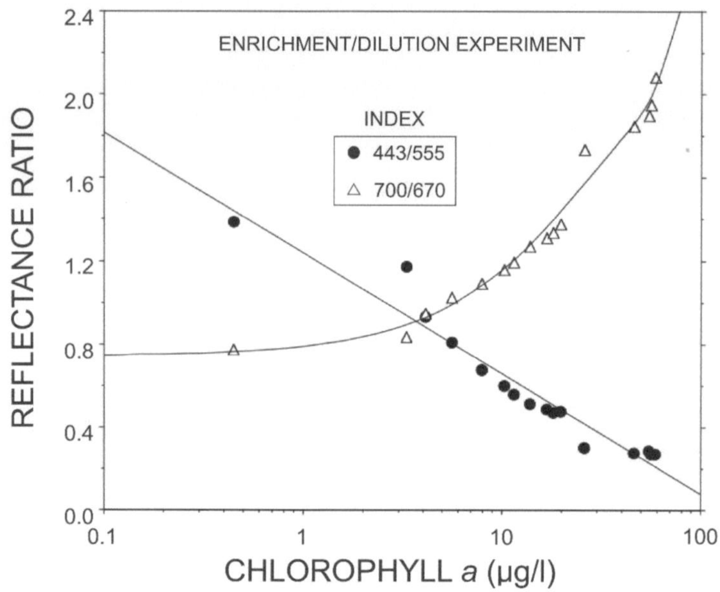

Figure 10. Comparison of two common chlorophyll algorithms: (1) ratio of 453/555 nm for Case 1 waters and (2) ratio of 700/670 nm for turbid and/or eutrophic Case 2 waters. Data are from the mesocosm dilution and enrichment experiment in Figures 8 and 9.

Algorithms based on Q band, red absorption and the NIR reflectance peak feature are favored in situations with high chl *a* and/or Case 2 optical conditions (Dekker et al., 1996; Gower et al., 1999; Ammenberg, 2000; Gitelson et al., 2000; Gons et al., 2002). A band ratio using the NIR peak near 700 nm and the red Q band may largely isolate the chl *a* signal from other pigment and CDOM absorption activity. The NIR feature builds much more rapidly than the Q band region with increasing cell densities (Figures 8 and 9), producing an exploitable response. A ratio of 700 nm to 670 nm (Gitelson et al., 2000), applied to our tank mesocosm data, produced a strong relationship to chl *a*, with good sensitivity above 10 μg/l (Figure 10).

Different phytoplankton taxa often have differentiable colors, i.e. are members of the "green algae", "golden brown algae", etc., with these variations clearly related to spectral differences in absorption by accessory pigments (Sathyendranath et al., 1987; Rowan, 1989; Richardson, 1996). Thus ecologically different oceanic provinces, with differing proportions of dominant phytoplankton groups, may have differing absorption and scattering properties that affect remote sensing bands and the band ratios for chlorophyll estimates. These differing bio-optical properties may partially explain the poorer performance of ocean color chlorophyll algorithms in waters at polar latitudes (Sathyendranath et al., 2001). In Figure 11 the reflectance characteristics of phytoplankton from four taxomic divisions are compared, with each containing chl *a* concentrations of about 60 μg/l. In this figure, two spectra were measured in our Durant, Oklahoma mesocosm experiments (*Navicula* and *Chlorella*) and two were measured from lakes with nearly monotypic assemblages (*Anabaena* and *Peridinium*). *Navicula* is a pinnate diatom, *Chlorella* is a small, spherical non-filamentous chlorophyte, *Anabaena* is a filamentous cynabacterium, and *Peridinium* is a large, oval dinoflagellate. Large quantitative differences exist in the four spectra. Higher reflectivity of the silica frustule in *Navicula* probably accounts for the overall stronger reflectance, including the NIR region beyond the region of pigment activity. In addition to prominent Soret and Q band chl *a* minima, the diatom spectrum had conspicuous shoulders near 480 nm and 630 nm, coincident with carotenoid and chl *c* absorption maxima (Figure 1). The *Anabaena* spectrum was measured in Carter Lake (compare to Carter Lake examples in Figure 7) in early July, 1996, as the bloom was developing. The NIR peak was nearly as large as the green peak, with the green peak position about 15 nm to the left of the diatom and dinoflagellate positions. Additionally, a significant phycocyanin trough (620 nm) was present. The *Peridinium* curve, from Lake Kinneret, Israel (Gitelson et al., 1999), had a narrower green peak, probably due to strong accessory pigment activity on each side. This curve was taken from a series of spectra, acquired on the same date in different bloom densities, which demonstrated that green peak reflectance decreased with chlorophyll concentrations above 10 μg/l (Gitelson et al., 1999). In contrast, the NIR peak increased with increasing cell and chl *a* levels. Green peak suppression was probably due to the broad absorption profile of peridinin (Richardson, 1996), a pigment unique to dinoflagellates. Peridinin also suppresses the green reflectance peak in zooxanthellae symbionts of corals (Meyers et al., 1999). The *Chlorella* spectrum, taken from the tank experiment discussed above (Figure 8), had the lowest overall reflectance, and a relatively broad and symmetrical green peak.

Using the data shown in Figure 11, the Case 1 water, blue to green ratio for chl *a* (443 nm to 555 nm) was computed for each taxon's spectrum, and compared to the Case 2 water ratio of 670 nm to 700 nm (Figure 12). Note that the denominator and numerator of the Case 2 ratio is reversed from the last case discussed (Figure 10) to allow a more direct comparison with the Case 1 ratio. In this comparison, the ocean color ratio was more variable (coefficients of variation for the Case 1 and Case 2 ratios were 34.0% and 8.7%, respectively). The Case 2 band ratios (chl *a* estimate) using 670 and 700 nm were more consistent when applied to the four spectral data sets, compared to the Case 1 ratio (433 to 555). In the latter case, the ratio differed (for almost the same chl *a* concentration – see Figure 11) for all four taxa, and by almost a factor of 3 for *Chlorella* (62 μg/l chl *a*) vs. *Peridinium* (61 μg/l chl *a*). The higher variability in the spectral responses for the 443 to 555 ratio was likely due to the greater impact of accessory pigments, especially carotenoids, at these wavelengths.

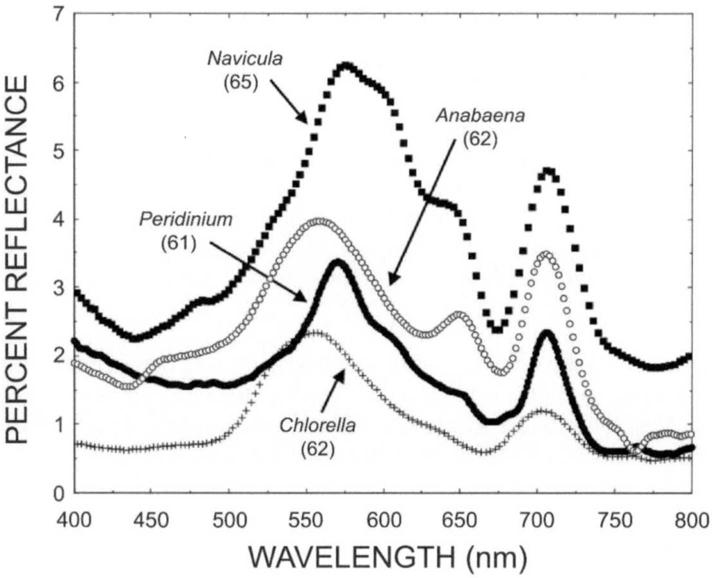

Figure 11. Comparison of reflectance spectra for algal taxa representing four different divisions. The measured spectra are examples for nearly identical chlorophyll concentration (chl *a* concentrations, in µg/l, are given in parentheses).

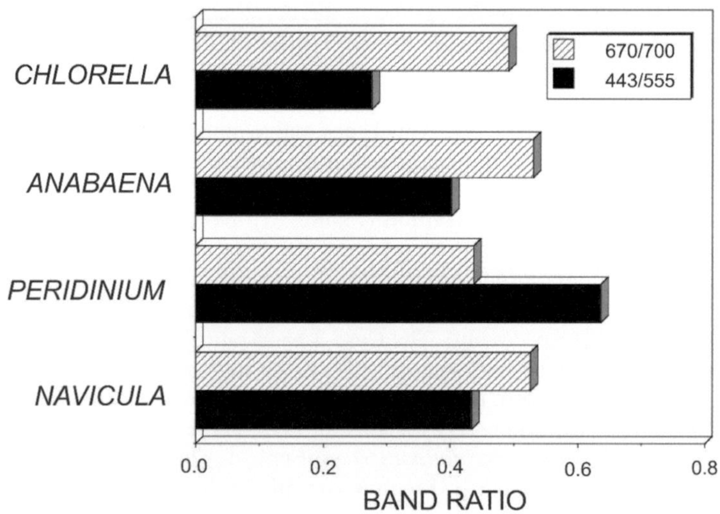

Figure 12. Comparison of two band ratios for chl *a* estimation for the reflectance spectra of four different algal taxa with nearly identical (measured) chl *a* concentrations (see Figure 11).

6.2 CLAY SUSPENSIONS AND THEIR EFFECT ON REFLECTANCE

In general, inorganic tripton particles such as clays and carbonate minerals increase the albedo of water in a density dependent manner, due to strong scattering efficiency and comparatively low absorption. In contrast, absorption by organic tripton often resembles CDOM, with a strongly inverse correlation between wavelength and concentration (Dekker, 1993). In an experiment with the Durant, Oklahoma tank mescosms (Schalles et al., 1997), a suspension of white, kaolin clay was added in seven steps to clear water (Figure 13). Added clay concentrations ranged from 1.9 to 39.3 mg/l. Median particle size of the suspended clay was 0.79 μm (the 1st and 3rd quartiles were 0.66 and 0.95 μm). As the clay was added, the water turned a milky, white chalk color, with increased brightness. Reflectance increased in a stepwise manner, with the magnitude of reflectance above 500 nm inversely correlated with the spectral pattern of water absorption (Figure 6). The position of peak reflectance shifted from approximately 455 nm at 1.9 mg/l clay to 540 nm at 39.3 mg/l clay. At the highest clay level, peak reflectance was a remarkable 52.5%, notably higher than values reported for many other sediments in water (Curran and Novo, 1988; Han and Rundquist, 1994). Reflectance decreased sharply near 600 nm for all clay concentrations, in response to the sharp increase in water absorption (see Figure 6). Consistent with other findings, the reflectance response versus clay level was non-linear, becoming less sensitive at higher clay loads. Substantial NIR reflectance was found at the higher clay loads. For example, at 750 nm the reflectance was 2.7% and 7.1% for clay concentrations of 18.2 and 39.3 mg/l. The reflectances in this experiment probably approach the upper limits of scattering and return signals in water. Note, however, that clay turbid rivers and lakes may have tripton concentrations well in excess of 100 mg/l, and the highest concentrations may approach 1000 mg/l (Schiebe et al., 1992).

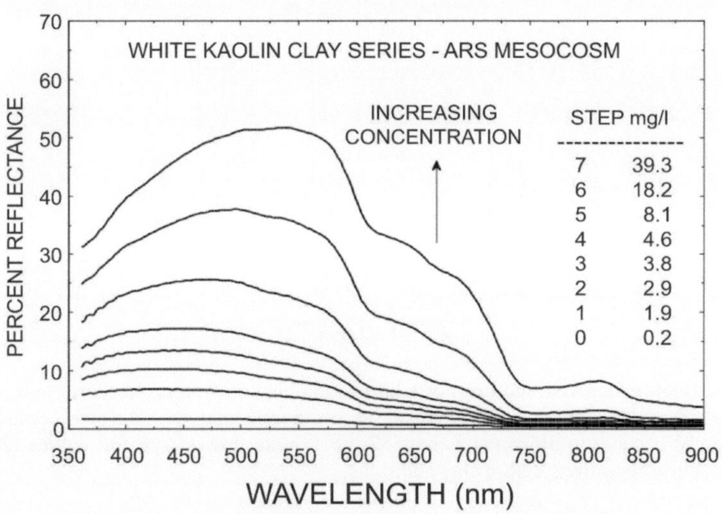

Figure 13. Effect of stepwise additions of white kaolin clay suspension on water reflectance. Clay was added to an 80 m^3 tank mesocosm with clear water (lowest curve), which had a seston concentration of 0.2 mg/l (the initial, pre-clay condition).

6.3 CLAY INTERACTIONS WITH PHYTOPLANKTON

In a separate mesocosm experiment, algal bloom water from a single, fertilized experimental tank was divided and added to two clean, empty 80 m^3 tanks (Schalles et al., 1997). One third of the bloom water was added to the first tank and two thirds to the second tank, taking care not to draw settled materials from the bottom of the fertilized culture tank. Clear, dechlorinated water was then used to fill the remaining volumes of the first and second experimental tanks. In this manner, algal suspensions were established with chl *a* concentrations of 31 and 57 μg/l. Both tanks had identical communities of algae, with the same age and size distributions, and the only difference in the starting conditions of the two ponds was cell density. In this experiment, nearly identical additions of the same white, kaolin clay used in the experiment shown in Figure 13 were made to both tanks in a stepwise manner. Clay was added in 6 increments, and the final increment achieved a concentration of 72 mg/l dry weight in both tanks.

Prior to clay additions, the tank with the higher chl *a* concentration had higher reflectance at all wavelengths, as seen in the bottom spectra with initial seston concentrations (Figure 14). The increase in reflectance with higher algal density was greater for both the green peak (4.4% in the high density versus 2.6% in the lower density tank) and NIR peak locations (2.7% versus 1.4%). Clay additions increased reflectance at all wavelengths, in a dose dependent manner. The white clay had little overall affect on the wavelength positions of peaks and troughs. Interestingly, at the higher clay levels, reflectance was greater in the lower algal density tank (a reversal from the starting condition with only phytoplankton present - see below for explanation).

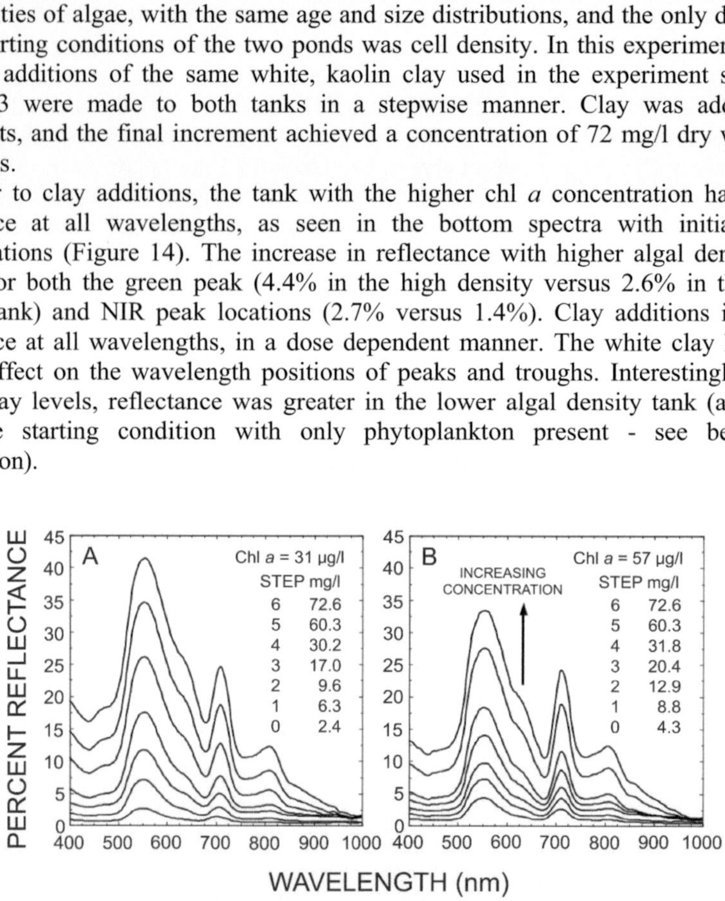

Figure 14. Effect of stepwise additions of white kaolin clay on water reflectance in replicate tank mesocosms with algal suspensions identical in taxonomic and particle size composition but differing in chl *a* concentrations (31 versus 57 μg/l). Note: lowest spectral curves (2.4 and 4.3 mg/l) are baseline conditions before clay additions.

The impact of clay additions on the 443 to 555 nm and 670 to 700 nm ratios for the data depicted in Figure 14 is summarized in Figure 15. The 443 to 555 ratio before clay additions was 0.252 in the low density tank and 0.198 in the high density tank. Overall, clay additions resulted in higher ratios. The ratio increased 55.2% in the low density and 69.2% in the high density tanks. An increase in ratio values will result in underestimation of the chl *a* concentrations. Additionally, the ratios for the two tanks and chl *a* concentrations converged at intermediate clay levels, which is also problematic for algal quantification. These data agree with a model developed by Bowers et al. (1996) that predicted decreased slope and sensitivity in the chlorophyll relationship to the blue to green ratio with increasing mineral suspended solids (inorganic tripton).

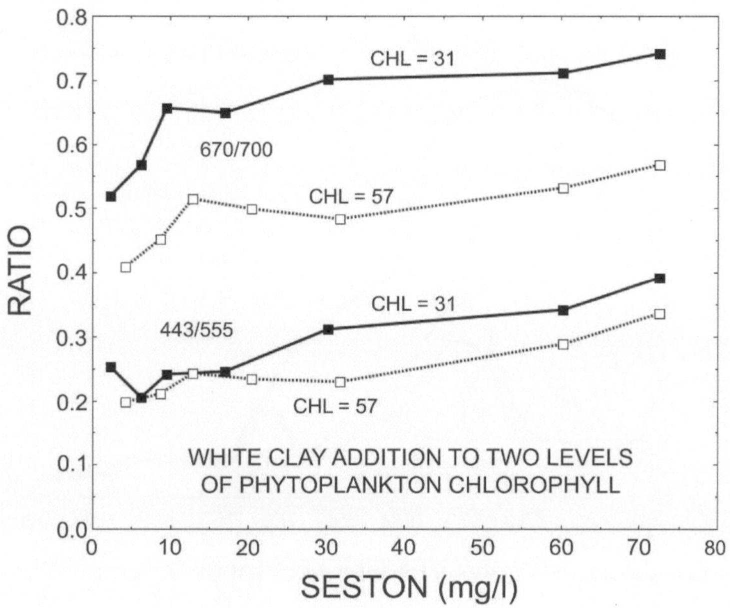

Figure 15. Effect of stepwise white clay additions to two tanks on two band ratios used for chl *a* estimation. The tanks contained identical algal communities at two different concentrations (31 versus 57 μg/l chl *a*; see Figure 14).

The 670 to 700 nm ratio in Figure 15 also increased with increased clay concentrations, which would again lead to underestimation of algal chlorophyll. However, the increases were more modest than those for the 443 to 555 ratio: 42.8% and 38.9% for the low and high density tanks, respectively. In contrast to the 443 to 555 ratio data, the response of the 670 to 700 ratio for the two algal concentrations was similar when clay was added in that the curves remained separate and mostly parallel (Figure 15).

At step four in the white clay addition to algae experiments (Figure 14), the total suspended solids (i.e. seston) concentrations were similar (17.0 mg/l and 20.4 mg/l) in the low and high algal density tanks. The 6th step in the clay addition to clear water experiment (Figure 13) resulted in a similar total suspended solids concentration (18.2 mg/l). The spectral curves for these three conditions (chl a = 0, 31, and 57 μg/l), each with similar total seston concentrations, are compared in Figure 16. Note the convergence of all three spectra at wavelengths above 725 nm, which corresponds to a region with virtually no algal pigment absorption (see Figures 1 and 6). The tank with the higher chl *a* (57 μg/l) had the lowest reflectance in the active pigment region, while reflectance in the region of pigment absorption was about 25-30% higher in the low algal concentration tank (Figure 16). In contrast, the tank with no algae had substantially higher reflectance, particularly in the spectral regions corresponding to the greatest overall pigment activity (see Figure 6). At these pigment concentrations, pigment absorption is nearly two orders of magnitude greater than water absorption at wavelengths below about 500 nm. This disparity decreases rapidly with increasing wavelength (Figures 6, 16). The comparability of scattering activity in the three tanks was confirmed by the close agreement in reflectance values above 725 nm (Figure 16).

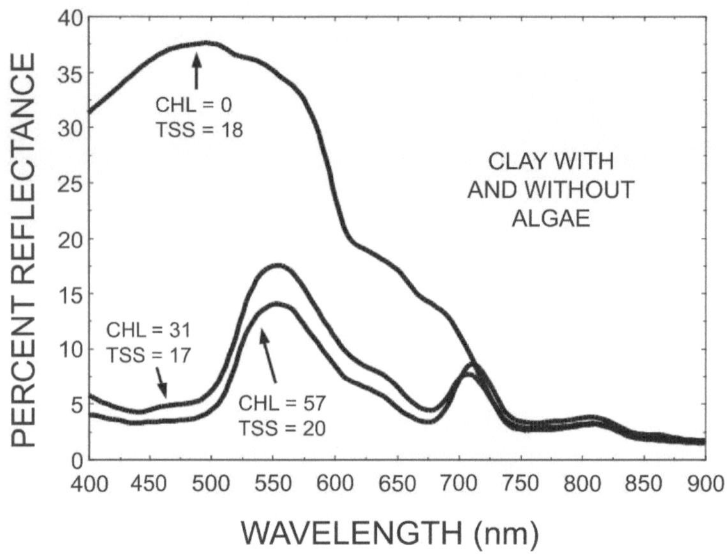

Figure 16. Comparison of three tank mesocosms with comparable suspensions of white clay (17 - 20 mg/l) and no phytoplankton chlorophyll versus chlorophyll concentrations of 31 and 57 μg/l.

Additional experiments were conducted in tank mesocosms using different clays and phytoplankton conditions (Schalles et al., 2001). In one example, six stepwise additions of an Oklahoma red clay suspension were made to a tank containing a mixed chlorophyte phytoplankton bloom with 53 μg/l chl *a* (Figure 17). The final step resulted in a clay concentration of 55.4 mg/l. Similar to the white clay additions, reflectance again increased at all wavelengths. With red clay, however, the position of the green peak shifted 45 nm (from 546 to 591 nm), and the NIR peak shifted 4 nm (from 700 to 704 nm). We noted an observable color shift from green to reddish brown. In the red clay experiment, maximum reflectance reached 24.8% at 591 nm, which is less than the maximum values of 41.4% and 33.4% in the previous experiment in the low and high chlorophyll tanks with comparable white clay concentrations of approximately 57 mg/l (Figure 14).

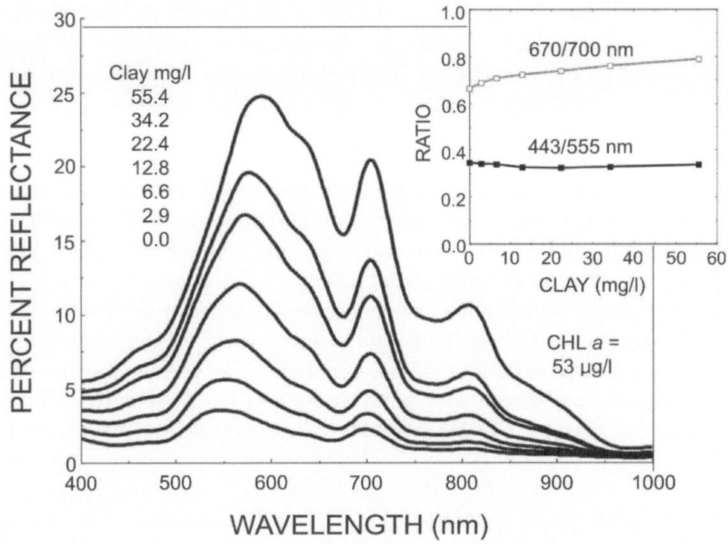

Figure 17. Reflectance spectra for a tank mesocosm experiment (Schalles et al., 2003) in which a red clay suspension was added stepwise to a tank containing a phytoplankton bloom of mixed Chlorophytes. Resulting clay concentrations are shown. Inset shows the effect of clay additions on band ratios for two chl *a* algorithms.

The red clay additions resulted in a much smaller change in the 443 to 555 nm ratio (-2.7%) and a moderate change (+ 19.2%) in the 670 to 700 nm ratio (see inset, Figure 17) when compared to white clay additions (Figure 15). The height of the NIR peak above a reference baseline from 675 to 750 nm increased from 0.9 to 6.2 percent reflectance (689% change). This NIR peak height calculation is an effective algorithm for chl *a* estimation in many Case 2 inland waters (Gitelson et al., 2000). In 17 separate mesocosm experiments with clay/algae interactions (Schalles et al., 2001), clay concentrations of 50 mg/l caused average changes of 9.4% in the 670 to 700 nm ratio and 325% in the NIR peak height above baseline. The clay additions caused a large shift to higher wavelengths in the green peak position. For clay additions of 50 mg/l, the average shift was +17.2 nm for the position of the green peak and +2.4 nm for the NIR peak.

7. CDOM Optics and Interference with Chlorophyll Algorithms

The term CDOM is an acronym for either colored (= coloured) or chromophoric dissolved organic matter. Both versions of the acronym refer to organic matter which strongly absorbs light in the ultraviolet and blue regions and less intensively at higher wavelengths. Chromophoric is a general term for molecular structures that absorb light. These light absorbing organic materials, largely composed of humic substances, are also variously named gilven (pale yellow color), gelbstoff, yellow substance, yellow organic acids, and humolimnic acid (Kirk, 1994). Their light absorbing properties are especially related to the presence of large molecular weight, aromatic humic and fulvic acids compounds. The chemical and light absorbing properties of CDOM are well investigated (Carder et al., 1989; Alberts and Filip, 1994; Schwarz et al., 2002). An important, but sometimes overlooked, contribution of CDOM to spectral signatures is CDOM fluorescence. The CDOM emission wavelength is dependent on the excitation wavelength (Mobley, 1994), while CDOM fluorescence yield (a ratio of photons emitted to photons absorbed) is generally wavelength independent and varies from about 0.005 to 0.015.

CDOM substances largely originate from the decomposition of plant materials and are especially abundant in the discharges of streams and rivers with soft water, acidic chemistries and riparian zones with abundant plant biomass and productivity. Salt marsh and mangrove communities are also important sources of CDOM in the lower reaches of estuaries and other coastal habitats. CDOM concentrations are generally much lower in offshore waters and the open oceans (Carder et al., 1989) and originate from highly diluted inputs from land and from *in situ* phytoplankton production. CDOM dominates the water column optics in some estuaries and is responsible for the dark-stained, "blackwater" appearance of rivers such as the Rio Negro in Brazil and the Suwannee River in Florida. CDOM abundance is reported in either units of optical absorption of sample filtrate (ex. optical density or absorption units at a fixed wavelength in the ultraviolet or blue regions) or as carbon based, organic matter estimates (ex. dissolved organic carbon in mg/l), usually from oxidation of fixed carbon to carbon dioxide in carbon analyzer instruments.

At wavelengths greater than 300 nm, absorption by CDOM is inversely related to wavelength and is described by an exponential decay curve (Figure 18; see also Bricaud et al., 1981; Carder, 1989; Dekker, 1993; Schwarz et al., 2002). The magnitude of absorption at a single wavelength is highly correlated with CDOM concentration. However, the optical activity per unit carbon can vary in space and time and the slope of the semilogarithmic equation describing the absorption to wavelength relationship varies accordingly (Jerlov, 1976; Kirk, 1994). The average CDOM absorption from 28 estuarine and near shore stations on the Georgia coast was calculated in order to obtain a representative spectral curve (Figure 19). The average ABS_{440} value for these 28 stations was 2.12 m^{-1}. The average curve was then adjusted upward by one order of magnitude and downward by one and two orders of magnitude to create a family of curves. In turn, these curves were compared to the absorption spectrum of water (Figure 19). The curve for ABS_{440} = 21.2 exceeded water absorption values at all wavelengths below about 750 nm whereas the ABS_{440} = 0.021 curve exceeded water absorption only at wavelengths below 470 nm. Although the affect of CDOM absorption is strongest at lower wavelengths, this material can reduce reflectance signals at all visible wavelengths and into the lower NIR. When absorption of CDOM and water are combined (see inset, Figure 19), the water column can become highly absorptive across visible and NIR wavelengths. Note that the minimum combined absorption of CDOM and water shifts strongly to higher wavelengths with increasing CDOM concentrations (inset, Figure 19). At $ABS_{440} \sim 2\ m^{-1}$, the minimum absorption is a rather broad region between 580 and 680 nm.

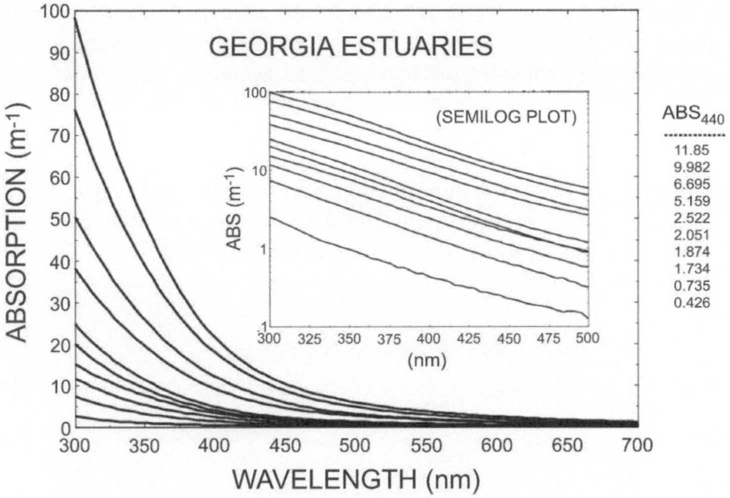

Figure 18. Absorption spectra (300 - 700 nm) for dissolved organic matter (CDOM) in representative Georgia estuaries and offshore waters. Inset shows semilogarithmic plot of the same set of spectra. The respective ABS440 values (m^{-1}) for each of these spectra are shown to the right of the figure.

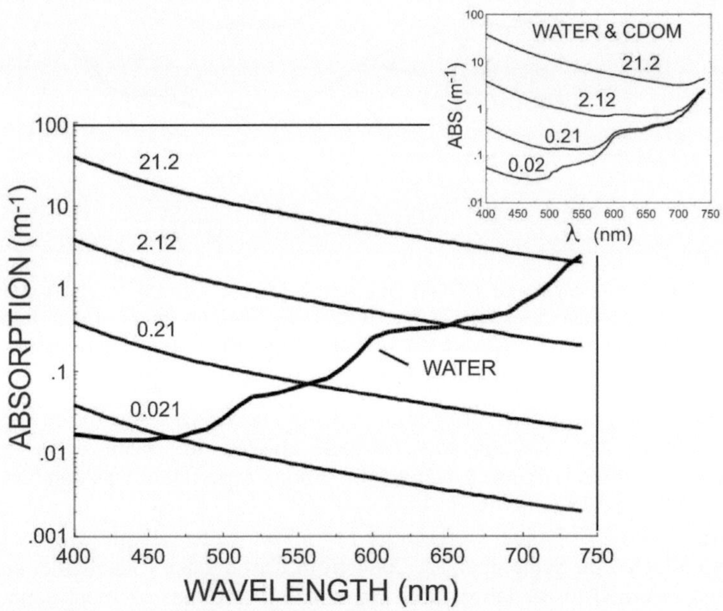

Figure 19. Modeled absorption spectra based on average absorption of Georgia sites (2.12 m^{-1} at 440 nm curve). The other 3 curves represent order of magnitude scaling. In turn, the family of CDOM absorption curves are compared to pure seawater. The inset shows the combined absorptions of CDOM and water for these four CDOM levels.

The combined water and CDOM absorption spectra shown in Figure 19 are compared in Figure 20 with selected total pigment absorption spectra from Figure 6. At chl *a* < 1 μg/l, pigment absorption activity is generally lower than the range of combined CDOM and water absorption typically encountered in coastal waters. At wavelengths below 550 nm and chl *a* > 1 μg/l, substantial overlap occurs between these different conditions. Only very high pigment levels in the red region result in higher absorption than the combined absorption of CDOM and water. When ABS_{440} (CDOM) exceeds 10 m^{-1}, the combined CDOM and water absorption exceeds the chl *a* = 100 μg/l absorption at all visible (PAR) wavelengths (Figure 20).

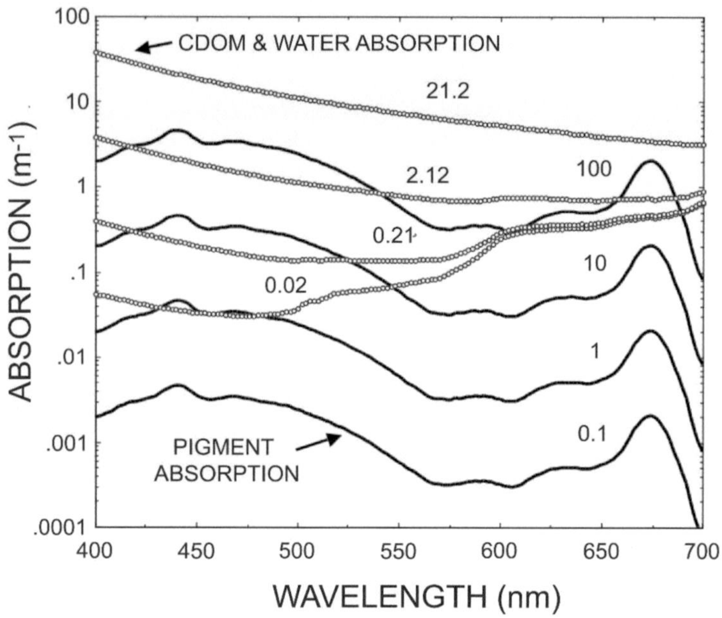

Figure 20. Combined water and CDOM absorption at four CDOM levels (see Figure 19) compared to algal pigment absorption at four levels (see Figure 6, pigment data from Bidigare et al., 1990). CDOM and chl *a* levels are shown for their respective spectral absorption curves.

The shapes and magnitude of combined water and CDOM absorption spectra have important implications for spectral features utilized in ocean color and other chlorophyll algorithms for Case 2 waters. An indoor experiment was conducted with a small (16.5 L) tank and involved stepwise additions of purified humic acid (as a sodium salt, Aldrich Chemical Corporation) to a lake water sample (Figure 21). Lake water was collected on September 12, 2004 from Carter Lake (Schalles et al., 1998b), immediately returned to the laboratory, and added to the measurement container. The lake water was dominated by the filamentous cyanobacterium *Anabaena*. Field Secchi transparency was 24 cm, and chl *a* was 116.3 μg/l. All tank reflectance measurements were completed within 2.5 hrs of sample collection. The ABS_{440} (CDOM) value for the lake water was 0.7 m^{-1}. Eighteen stepwise additions of humic acids brought the final CDOM absorption to 31.6 m^{-1}. Altogether, 1.328 g of humic acid (powder) was used, which achieved a final concentration of 80.24 mg/l. Tank water depth was 26 cm (bottom not visible), and the tank was illuminated by a fiber optic illuminator (Cole-Parmer Series 41720) with a full VIS and NIR spectral emission. For each step of humic acid addition, aliquots of the tank water were removed and the powdered humic acid was thoroughly mixed with the aliquot. The aliquot, in turn, was thoroughly mixed into the tank before reflectance measures were taken. The lake water reflectance, prior to humic acid additions, was comparable to field measurements. The lake water prior to treatment had strong reflectance peaks centered at 566 and 708 nm and chlorophyll and phycocynanin induced reflectance minima (Figure 21). With each stepwise addition of humic acids, reflectance decreased within the VIS and lower NIR portions of the spectrum. There was no direct evidence that CDOM fluorescence caused an increased signal return at any wavelength, thus any CDOM fluorescence effect was masked by the increased absorption. Although all wavelengths were affected, reflectance decreases were most severe at the green and NIR peaks with the effects greatest, overall, at lower wavelengths (Figure 21). For example, the percent reflectance measures at 566 and 708 nm decreased by 4.79 and 4.63 respectively (or reductions in magnitude of 88.1% and 72.1%). In comparison, the percent reflectance measures at the 438 (Soret band) and 675 (Q band) minima decreased by 0.82 and 1.77 (or reductions of 59.3% and 61.7%). Declines in green peak heights were especially sensitive to the initial eight additions (up to $ABS_{440} \sim 8\ m^{-1}$). These data agree with an analytical model of water reflectance by Dekker (1993), in which he found that the minimum of phytoplankton reflectance at 676 nm was less affected by varying CDOM levels than was the NIR peak region at 706. Additionally, in the Netherlands shallow drainages that cut through peatlands exhibited high CDOM ($ABS_{440} = 5.4$ to $16.5\ m^{-1}$), and their reflectance spectra resembled those with the higher CDOM levels shown in Figure 21 (Rijkeboer et al., 1998). These workers also measured cases of extreme CDOM levels in fens (ABS_{440} up to $66\ m^{-1}$) , and found all reflectance in the visible range to be less than 1% with a strong shift of peak reflectance to NIR wavelengths.

Humic acid additions had a greater impact on the 443 to 555 nm algorithm ratio than on the 670 to 700 nm ratio (see inset, Figure 21). The 443 to 555 ratio was more sensitive at lower CDOM levels, corresponding to the large decreases in reflectance observed in green peak reflectance together with the less sensitive blue region, resulting in relatively large increases in the ratio. At ABS_{440} values of 4 and 8 m^{-1}, the 443 to 555 nm ratio increased by 63.6% and 110.2% over a starting value of 0.274. In comparison, the 670 to 700 nm ratio increased by only 0.9% and 1.9% at these intervals, from a starting value of 0.521. At the final ABS_{440} value of 31.6 m^{-1}, the 443 to 555 nm and 670 to 700 nm ratios increased by 223.6% and 26.0% respectively (inset, Figure 21). Based on these data, the chl *a* algorithms using green and blue wavelength ratios are far more sensitive to CDOM than the red and NIR algorithms. However, as shown above, this sensitivity would be dependent on the phytoplankton induced magnitude of the difference between green and blue band reflectance. Smaller differences between these bands should result in reduced sensitivity for chl *a* prediction.

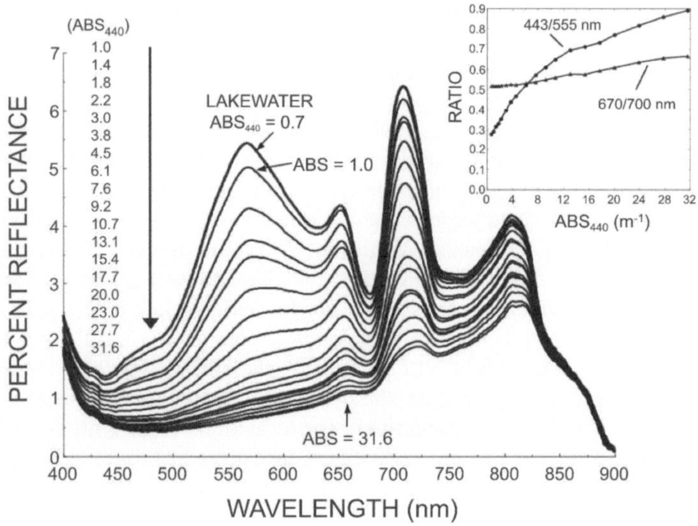

Figure 21. Effect of stepwise additions of humic acid on the spectral reflectance of a 16 liter lake water sample with an *Anabaena* sp. bloom held in a laboratory tank with lamp illumination (see text). Chl *a* concentration was 116.3 μg/l. Absorption (m^{-1}) values at 440 nm are shown for each step. Inset shows the effect of the humic acid additions on band ratios for two chl *a* algorithms.

Attenuation by CDOM is through absorption, although these materials intergrade with colloidal scale, detrital particles. The organic tripton (particulate detritus) fraction commonly has absorption spectra resembling the shape of CDOM absorption spectra (Dekker, 1993; Mobley, 1994; Rijkeboer, 1998; Brando and Dekker, 2003). The magnitude of organic tripton absorption may exceed that of CDOM in some coastal waters. In a transect of the St. Marys River in Georgia (Figure 22), CDOM dominated lower salinity, upstream stations (about 20-30 mg/l as dissolved organic matter) and organic tripton averaged about 2-3 mg/l in the same reach. However, organic tripton averaged about 10 - 15 mg/l in the lower estuary and CDOM levels were diluted to about 4-6 mg/l (Alberts et al., 2004). Typically, terrigenous organic matter is conservatively diluted in the estuarine mixing zone. At reduced flows and greatly increased residence times in the St. Marys estuary, CDOM dilution was non-conservative in the low to mid-salinity reach (Figure 22). Non-conservative mixing patterns were attributed to the formation of fine particulates as freshwater mixes with seawater (Alberts and Griffin, 1996). These particulates are commonly clay-detritus aggregates, with bacterial colonization and enriched nitrogen to carbon ratios (Lind and Davalos, 1990; Carlough, 1994). Many drainages have seasonal variations in discharge and CDOM concentrations. In Georgia coastal rivers, higher flow conditions are coincident with higher CDOM concentrations (Figure 22; also see Alberts and Filip, 1994), and the slope of the CDOM loading rate versus discharge relationship can exceed 1.0. Thus, higher discharge rates can create larger offshore plumes with higher CDOM concentrations. Estuarine and near shore waters may thus experience marked spatial and temporal variations in this optical constituent, resulting in strong effects on the water reflectance characteristics used to discriminate chlorophyll signals. Offshore waters generally have much lower CDOM levels than estuaries, with ABS_{440} values of about 0.01 to 0.1 m^{-1} in open oceans, and about an order of magnitude greater in coastal waters (Kirk, 1994). At these levels, the effect of CDOM on chl *a* estimation should be greatly reduced.

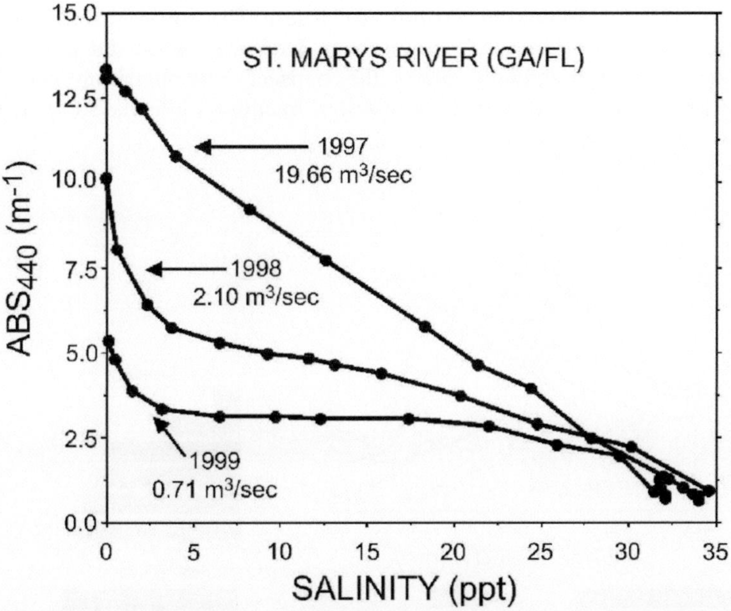

Figure 22. CDOM absorption values (m^{-1}) at 440 nm versus salinity for stations along the mixing zone and offshore plume of the St. Marys River on the Georgia/Florida border. Note the nearly conservative mixing during dilution in 1997, and non-conservative mixing in 1998 and 1999 at highly reduced flow conditions. Measurements of surface waters were taken each year in early summer (late June or early July). River flows (m^3/sec) are for a U.S. Geological Survey gauging station just above the estuarine mixing zone and the values represent averages of the daily values for the months of May and June in each year.

8. OACs and Reflectance Spectra Along Longitudinal Transects

Several examples from my research in Southeastern U.S. estuaries and near-shore waters illustrate the interactions of OACs and the composite impact of longitudinal gradients on water reflectance (Figures 23, 24). Sapelo Sound, at the north end of Blackbeard and Sapelo Islands, Georgia, receives only localized drainage. Sapelo Sound is relatively saline, has low levels of CDOM, and has less physical-chemical variation than the neighboring estuarine complex to the south, which is impacted by variability in flows of the Altamaha River. A transect from the Sound to adjacent near shore waters was made in July, 2000 aboard the R.V. Spartina (University of Georgia Marine Institute, Sapelo Island, Georgia). Salinity on the transect's six stations, beginning in mid-Sound, increased from about 30.9 to 35.3 ppt (Figure 23). Both chlorophyll and seston decreased by an order of magnitude (from 19.4 to 1.2 µg/l and 21.1 to 1.0 mg/l respectively). However, the percent of organic matter in the seston increased about 3 fold, from 26.7 to 74.2%. Secchi disk transparency increased from 0.78 to 5.30 m, while CDOM ABS_{440} values declined about two fold, from 0.43 to 0.20 m^{-1}

(not shown). As seen in Figure 23 water reflectance decreased with decreased scattering matter. The reflectance decrease was much stronger at the chl *a* red absorption band compared to the blue, Soret band. Peak reflectance was 4.8% at 580 nm in the Sound and declined to 2.1% at 548 nm offshore. The depth of the red absorption trough also declined sharply. Both the 443 to 555 nm and 670 to 700 nm ratios declined in a non-linear manner with increasing chlorophyll (see inset). The covariation of chl *a* and seston might suggest a Case 1 water condition. However, since the percentage of organic matter increased strongly along the transect, phytoplankton and detrital particles most likely dominated the offshore seston fractions while clays dominated the sound fraction (Figure 23).

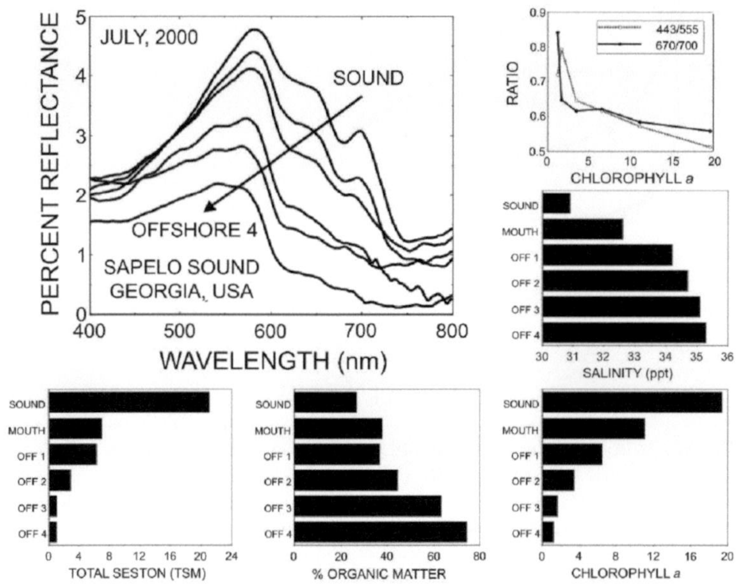

Figure 23. Reflectance curves, associated water measures, and calculated chl *a* band ratios at stations along a sampling transect from Sapelo Sound, Georgia to offshore waters of the Georgia Bight. Note: chl *a* values are μg/l, total seston values are mg/l dry weight.

A second longitudinal gradient example was measured in June, 2003 along a transect of 10 stations from the lower freshwater reach of the Edisto River to the mouth of St. Helena Sound in South Carolina (Hladik, 2004). In this transect, salinity increased from <0.1 to 27.2 ppt at the mouth of the Sound (Figure 24). Chlorophyll had a more complex pattern, with an initial decline from upriver values of 24.2 to 3.5 μg/l, a sharp increase to 26.5 in mid transect stations, and then a modest decline to about 18 μg/l in the Sound. Total seston generally increased and reached very high levels between mid transect and Sound stations. Seston averaged 5.7 mg/l in the first two upriver stations and 90.3 mg/l at the lower two Sound stations. CDOM (ABS_{440}) was very high upstream and decreased sharply along the mixing gradient, with the ABS_{440} values declining from 13.0 to 2.2 m^{-1}. Thus, the lowest CDOM absorption value in this transect was approximately five times the largest value for the previous Sapelo Sound transect. Reflectance generally increased along the transect, related to increasing scattering by seston and decreasing CDOM. Peak reflectance was 0.7% at about 700 nm at the upriver station and 4.4% at 648 nm at the mouth of the Sound. Depth of the Chl *a* red absorption trough was only partially correlated to chl *a* because the absorption of the high chl *a* concentration at the first, upriver station was largely masked by the high CDOM concentration (Figure 24) in a manner comparable to the masking effect at higher CDOM levels shown in Figure 21. Blue reflectance was low across the entire transect and very unresponsive to the large changes in both phytoplankton and total seston densities. Both ratio indices (443 to 555 nm and 670 to 700 nm) were relatively unresponsive and poorly linked to chl *a* concentrations (inset). In spite of nearly 5 times greater seston at the St. Helena Sound stations, peak reflectance values were roughly comparable with the Sapelo Sound stations, again illustrating the strong dampening effect of these elevated CDOM concentrations.

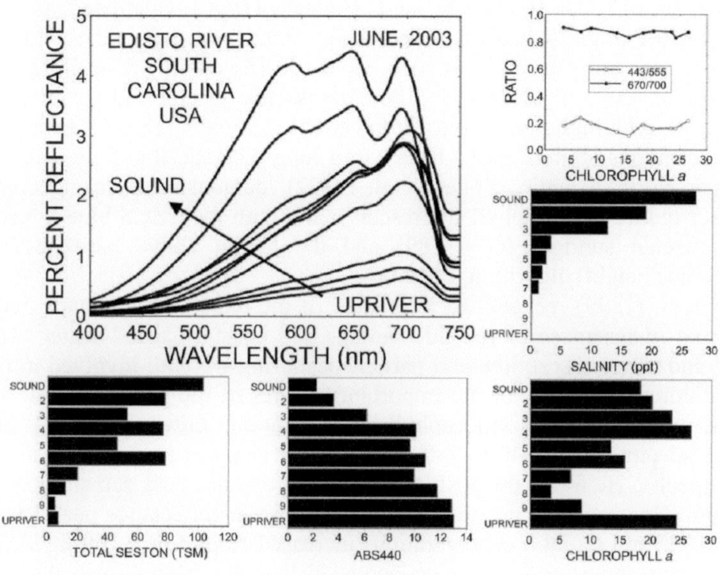

Figure 24. Reflectance spectra, associated water measures, and calculated chl *a* band ratios at stations along a sampling transect of the Edisto River in South Carolina. The transect extended from above the estuarine mixing zone to the mouth of St. Helena Sound, just south of Edisto Island. Note: chl *a* values are μg/l, CDOM absorption values at 440 nm are m^{-1}, and total seston values are mg/l dry weight.

9. The Reflectance Peak Near 700 nm: A Key Optical Feature in Case 2 Water Chlorophyll Algorithms

A confusing set of issues surrounds the interpretation of the reflectance maxima in the upper red and lower NIR spectral region (Figure 9). This feature is key to many of the emerging algorithms for Case 2 waters, and deserves additional consideration. Traditionally, ocean researchers equate this peak feature strictly to chlorophyll fluorescence (Carder and Steward, 1985; Gower and Borstad, 1990; Doeffer, 1993). Chlorophyll fluorescence is a well-studied phenomena that occurs universally in bacterial, protist, and higher plant chlorophylls (Butler, 1966; Rowan, 1989; Falkowski and Raven, 1997). The principal emission band is sharply focused at 685 nm, although a broader and less intense band is also present and centered near 730 nm (Butler, 1966). Chlorophyll fluorescence is inelastic, and the 685 nm emission center is independent of the excitation wavelength (Mobley, 1994). However, the strength of chl *a* fluorescence is dependent on the wavelength of an absorbed photon. The 685 nm bandwidth, at 50% of full maximum, is about 25 nm (i.e. 672.5 to 696.5 nm).

Fluorescence yield (a ratio of photons detected within the emission spectra to photons absorbed within the excitation spectra) varies from < 0.01 to 0.10, and typically falls between 0.01 and 0.05 (Mobley, 1994). The yield is affected by the light regime, its spectral composition, and the physiological and nutritional state of photosynthetic cells as well as algal taxomony (Doerffer, 1993; Babin et al., 1996). Algal fluorescence is also subject to reabsorption by chlorophyll and some accessory pigments, as well as by surrounding water and absorbing CDOM and tripton materials (Pozdnyakov et al., 2002). Gower and Borstad (1990) calculated that percent reflectance for the red/NIR peak increases about 0.02% per μg/l chl *a*, although this relationship, and hence the fluorescence yield, decreased in areas of less thermal stratification and lower algal density. The fluorescence to chlorophyll relationship is quite non-linear, with higher sensitivity at lower chl *a*, allowing remote detection below 1 μg/l concentrations (Babin et al., 1996). In a long distance flightline across Case 1 waters in the North Atlantic, Yoder et al. (1992), demonstrated strong correlations between laser induced chlorophyll fluorescence and both the CZCS blue to green ratio index of reflected sunlight ($r = 0.89$) and the height above baseline (FLH, or fluorescence line height) algorithm ($r = 0.96$).

Gower et al. (1999) reexamined the nature of the red/NIR reflectance peak. They concluded that in instances of red tide blooms and coastal Case 2 water conditions, chlorophyll and water absorption, and particle scattering were all involved in the peak, along with fluorescence. One of the important features of the peak is a shift to higher wavelengths with increased chlorophyll concentration. Gitelson (1992) measured fluorescence spectra (range 640 to 750 nm) of water samples immediately after taking reflectance spectra. He found the position of the fluorescence peak remained constant at about 680 nm. However, the position of the *in situ* water reflectance peak varied from about 682 to 715 nm as chl *a* concentrations increased from 3 to > 100 μg/l. Vos et al. (1986) noted this same behavior, and concluded that the minimum in the combined absorption of chlorophyll and water accounted for the peak and its shift with increasing pigment concentration. This behavior is clearly seen in the graphical model of the combined absorptions of different chl *a* concentrations and water (Figure 6). The peak position of fluorescence emission is constant as chl *a* concentrations increases and it is difficult to reconcile a fluorescence explanation and the shift in reflectance peak position. Indeed, as the fluorescence emission builds with increasing chl *a* concentration, fluorescence should overtake water absorption in magnitude and maintain the reflectance peak position near 685 nm.

The large increases in the NIR peak height with increased white and red clay concentrations (Figures 14 and 17) demonstrate the important role of particle scattering in the presence of pigment and water absorption. However, the minimal effect (Figures 13 and 16) of increased concentrations of mineral tripton on the NIR reflectance peak position (Schalles et al., 2001) indicates that scattering is relatively unimportant in this peak position. The interpretation of the peak as a region of minimum, aggregate absorption along with coincident, wavelength independent scattering can thus be applied to both the mid-500 nm green peak and the NIR peak.

The issue of the relative importance of fluorescence to the NIR peak was explored in a set of observations where blocking filters were used to control the excitation and emission phenomena associated with fluorescence (Schalles et al., In Prep). The rationale were: 1) to block incident light in the spectral range known to stimulate chlorophyll fluorescence and observe whether a decrease in the NIR peak reflectance occurs; and 2) to block incident light in the red and near-infrared range of the peak and observe whether an underlying fluorescence pattern was apparent. In these observations, light from a quartz halogen projector lamp obliquely illuminated an 8 liter sample container. Measurements were made with a model SE590 spectroradiometer (Spectron Engineering) with a 6° field of view foreoptic placed in nadir view 30 cm above the water surface. The instrument and foreoptics were calibrated for radiometric response using an integrating sphere at the NASA-Goddard SFC (Starks et al., 1995). The spectral composition of incident light was controlled with narrow pass filters placed over the projector lens. A "green" filter eliminated light below 620 nm and a red filter blocked light above 540 nm (Corion filters LS600-S-1288 and LL600-S-H942). The filters performed properly, but absorbed some light within their transmission ranges. Corrections for wavelength specific transmittance were applied to subsequent data measured with the filters. In the example presented here, an algal culture with a chl *a* concentration of 410 μg/l and a composition dominated by the green algae *Chlorella* and *Chlamydomas* sp. filled the sample container to a depth of 20 cm. When red and NIR light were excluded using the green narrow pass filter, a small emission feature, with a peak centered at the *in vivo* chlorophyll fluorescence wavelength of 685 nm, was evident (Figure 25). More importantly, the removal of red and NIR light by the green filter removed almost all of the NIR peak feature, indicating that most of the energy within the peak is reflectance and not fluorescence. When light energy below 600 nm, capable of exciting chlorophyll fluorescence, was excluded using the red filter, the NIR peak was strongly conserved, although the data indicated a small reduction of energy between about 670 and 720 nm (Figure 25). In other words, removal of most excitation energy for fluorescence had no significant effect on the shape or magnitude of the NIR peak.

In further analysis, the corrected radiance values obtained using the red narrow pass filter were subtracted from the unfiltered radiances, within the wavelength bounds of the NIR peak. The resultant curve had the same form as the fluorescence emission curve, but had almost twice the magnitude (see inset). Bandwidth (at 50% of full maximum) was 28 nm. Note that spectra isolated with both the red and green filters (inset) had maximum values at 685 nm, in agreement with the accepted wavelength for chlorophyll fluorescence, and that a shoulder (green filter spectra) or second, smaller peak (difference spectra) exists to the right of the NIR peak. The latter probably represents the secondary peak of chl *a* fluorescence near 730 nm (Butler, 1966). Although the precise magnitude of the contribution of chlorophyll fluorescence to the NIR peak remains uncertain in this case, it is clearly small in this example. Based on integration (data not shown) of the areas between 670 and 720 nm under the curves of the NIR peak (Figure 25 - unfiltered) and the chlorophyll emission feature (difference spectra - see upper curve of inset), fluorescence accounted for 7.9% of the energy of the entire NIR feature. However, at the NIR peak position, located at 709 nm, fluorescence accounted for only 3.7% of the total energy.

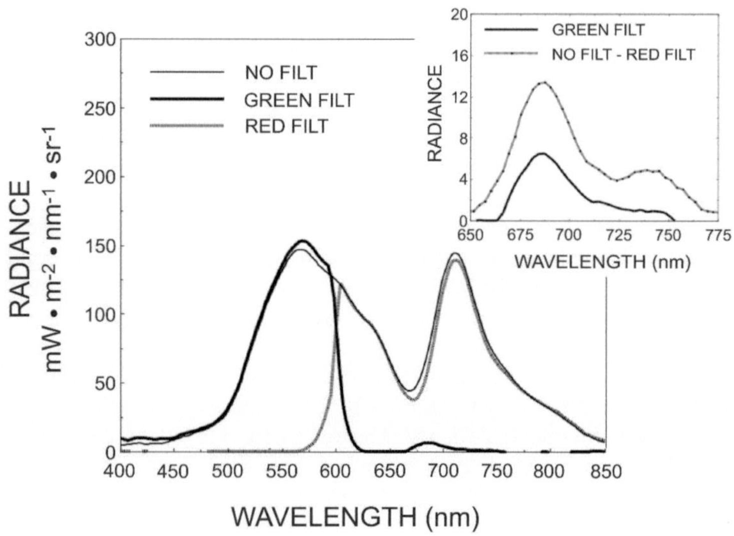

Figure 25. An experiment to isolate the contribution of algal chlorophyll fluorescence to the NIR reflectance peak of phytoplankton. The water leaving radiance, $L_w(O+)$, was measured at nadir above an algal culture in an 8 liter indoor tank with lamp illumination. Measurements were made with no filter place across the lamp source, with a narrow pass green (and blue) filter place across the lamp that prevented illumination of the tank at wavelengths above about 620 nm, and with a red filter that prevented lamp illumination below about 540 nm (indicated in figure). Data for the filter measurements was corrected for the transmission properties of each filter. The inset shows the curve of emissive, fluorescence energy measured with the green filter in place and for a second emission curve calculated from the difference between energy values without lamp filtration and with use of the red filter (see text for further information on the experiment).

Based on the evidence of field and laboratory measurements of reflectance and IOPs, the characteristics of pigment fluorescence, and modeling results which explicitly account for fluorescence yield, particle scattering, and water and pigment absorption, Gitelson (1992) concluded that algal scattering and absorption, coupled with the absorption properties of water, accounted for most of the NIR peak height and energy at chl *a* concentrations above 15 μg/l and that fluorescence is the dominant process below that threshold. The chlorophyll regime controls which processes are dominant. Thus, workers in oligotrophic Case 1 waters are correct in referring to the peak as a fluorescence phenomena but workers in eutrophic Case 2 waters are equally correct in referring to the peak as a scattering phenomenon governed by the interplay of chlorophyll and water absorption.

10. Comparison of Algorithms for Chlorophyll Estimation

As established above, the majority of passive (solar induced) remote sensing chlorophyll algorithms utilize: 1) a ratio of a band or bands within the chl *a* and carotenoid area of strong absorption between 440 and 510 nm versus the pigment minimum green region between at 550 to 555 nm; or 2) a ratio of a band in the NIR region of minimum combined water and pigment absorption and chlorophyll fluorescence between 685 and 710 nm and the red chlorophyll absorption band between about 670 and 675 nm.

Examples of chlorophyll algorithms for two extreme conditions of Case 2 waters are shown in Figures 26 and 27 (inset). Carter Lake (Figure 7; Schalles et al., 1998b) is hypereutrophic and dominated by a diatom bloom in spring and during warm seasons by a filamentous cyanobacterial bloom. Thus, major shifts in pigment composition and quantity occur at this site throughout the year (Schalles and Yacobi, 2000). This shallow, groundwater fed lake is prone to wind resuspension of bottom sediments (largely detritus and clay) but has no turbid surface inflows. The best relationship between chl *a* and reflectance information was with the height of the NIR peak above a reference baseline extended between the red absorption trough at 675 nm and a region beyond pigment activity at 750 nm (Figure 26, inset). The relationship, based on measurements in a chl *a* range of 36 to 244 μg/l, was a first order, linear fit with an r^2 value of 0.83 (Table 1). This same relationship was the strongest predictor, as well, for a set of northern glacial lakes in Iowa, and for Lake Kinneret and Haifa Bay in Israel (Table 1; Gitelson et al., 2000). The regression coefficient for slope varied almost 50% between these diverse sites.

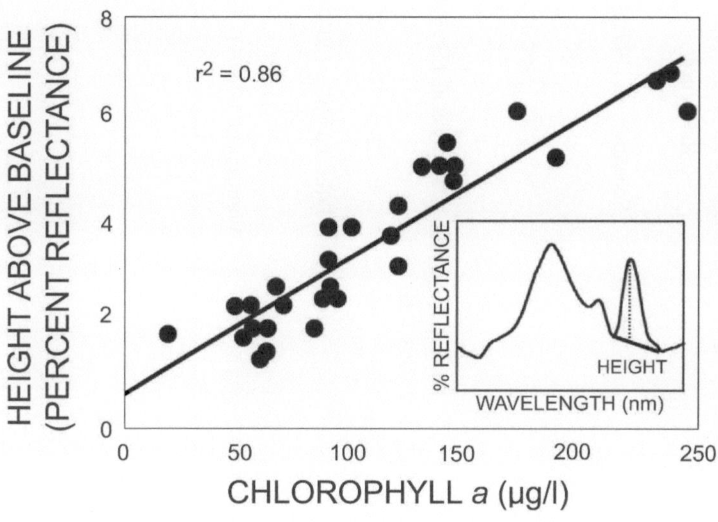

Figure 26. Chlorophyll *a* concentration versus NIR Height Above Baseline calculations from reflectance spectra in hypereutrophic Carter Lake, NE (Schalles et al., 1998). The inset graphically illustrates the basis for the height calculation.

Table 1. Comparison of retrieval algorithms for chlorophyll *a* in Case 1 and Case 2 waters. Rl is the reflectance value for a respective wavelength, and RLH is reflectance line height above baseline (see Figure 26), a_w=water absorption (m^{-1}), b_b = backscatter (m^{-1}), and int is the 10 nm bandwidth integral.

	Reference	r^2	Site(s)
Chl *a* = a + b*(RLH(670-750))	Schalles et al. (1998b)		
a = 1.77, b = 40.8		0.96	Lake Kinneret, Israel
a = 4.90, b = 47.2		0.93	Haifa Bay, Israel
a = 6.20, b = 31.8		0.83	Carter Lake, Nebraska, USA
Chl *a* $= 10^{(0.4708-3.8469R+4.5338R2-2.4434R3)} - 0.0414$	O'Reilly et al. (1998)	0.932	Global Oceans, some coastal waters
Where: $R = \frac{\text{(maximum value from R443, R490, or R510; see text)}}{\text{R555}}$			
Chl *a* $= 1.629\ (R490/R55)^{-2.551}$	D'Sa and Miller (2002)	0.81	Mississippi River plume - Gulf of Mexico
Chl *a* = log (-4.951*(R525/R554)) + 0.136	Vertucci and Likens (1989)	0.74	Adirondack Lakes, New York, USA

Table 1, cont'd.

	Reference	r^2	
Chl *a* = a + b*(R705/R662)	Kallio et al. (2003)		
a = -68.7, b = 108.5		0.98	Lake Lohjanjarvi, Finland
a = -77.1, b = 112.1		0.96	Lake Hiidenvesi, Finland
Chl *a* = -52.91 + 73.59*(R705/R678)	Thiemann and Kaufmann (2000)	0.89	Glacial Lakes, northern Germany
Chl *a* = 89* R705/R670 + 10*$(R705/R670)^2$ – 34	Mittenzwey et al. (1992)	0.98	Lakes and rivers, Germany
Chl *a* = R704/R672*(((a_w704+b_b)-a_w672)-b_b^p)/a*672	Gons (1999)	0.95	Lakes(Netherlands,China); Estuaries
Where: a_w704=0.630; a_w672=0.415; a*672=0.0176			
b_b776= 1.2*((R778*2.69)/ 0.187-R778); p =1.065			
Chl *a* = 56.7 + 161.0* -28.4*X^2	Dall 'Olmo et al. (2003)	0.94	Lakes and reservoirs, Nebraska, USA
Where: X = ((intR660-670^{-1}) - (int R720-$R730^{-1}$)) * (intR740-R750)			
Chl *a* = ((AVE R650+R700)-R675)/(AVE R440 + R550)	Hladik (2004)	0.80	Estuaries and near shore, Southeastern USA

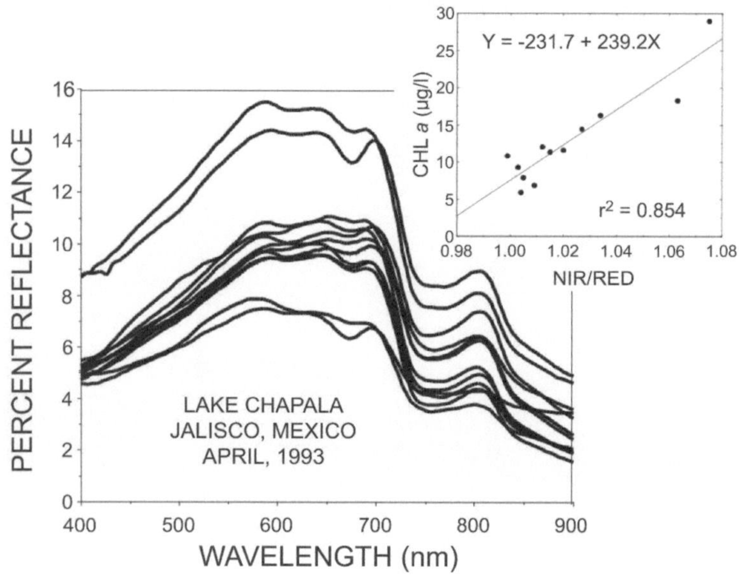

Figure 27. Reflectance spectra from twelve stations with high clay turbidities on Lake Chapala, Mexico in early April, 1993. The inset shows the relationship between measured chl *a* and a band ratio between the absolute height of the NIR peak and reflectance at the red, chlorophyll absorption trough near 675 nm.

The second example is Lake Chapala, a clay turbid tropical lake in the central highlands of Mexico (Lind et al., 1992). The largest natural lake in Mexico, Chapala has a surface area of about 1,100 km^2. The principal inflow is the Rio Lerma, which drains a watershed with a high degree of agricultural and industrial development. At low and at normal water levels, average depth varies from 4.5 to 7.2 m (Limón et al., 1989). Turbidity is primarily from fine clays (average 0.5 µ dia) and phytoplankton are typically light limited. In a detailed study carried out 1972 to 1984, Limón et al., 1989 reported an average Secchi depth transparency of 0.56 m and an average seston value of 29.7 mg/l (range 19-109). During one year of intensive study, chl *a* averaged 5.4 µg/l and the average annual primary production rate was only 80 g C/m^2 (Lind et al., 1992). In March and April, 1993, an optical study was conducted on this lake. Spectral reflectance was measured with the Spectron SE590 spectroradiometer at twelve stations, along with water sample collection for OAC estimates. The average and ranges for these stations were: chl *a* 12.8 µg/l (range = 5.9 - 29.0); seston = 84.0 mg/l (13.5 - 155.5); nephleometric turbidity = 47.7 NTU (33.4 - 67.0); and Secchi transparency = 0.24 m (0.09 - 0.52). The correlation between seston and chlorophyll was very weak (r = 0.105). Reflectance, measured at 50 cm above the surface, was high as a consequence of the high seston and turbidity (Figure 27). Maximum reflectance occurred near 585 nm, and ranged from 7.6 to 15.5%. The depth of the red chlorophyll absorption band varied in a general manner with chl *a* concentration. A second NIR peak, centered near 800 nm, represented a localized region of reduced water absorption in the presence of the high particle scattering. This second NIR peak is beyond the upper limits of algal pigment absorption activity. The best chl *a* regression model (r^2 = 0.854) was a ratio of the value of the NIR peak (positioned near 695 nm) to red trough (Figure 27, see inset). No relationship existed between blue and green region band ratios and chlorophyll in the Lake Chapala stations.

Morel and Prieur (1977) proposed a reflectance ratio of 440 to 560 nm for discrimination of ocean chlorophyll, based on observed absorption maximum and minimum values found at these two wavelengths. This relationship was found to be inverse and nonlinear. Gordon and Morel (1983) attributed this nonlinear relationship to an increase in the ratio of phytoplankton (and, hence, chl *a*) to detrital byproducts with increasing chl *a* concentrations. They reasoned that higher chl *a* concentrations are typically found in nutrient stimulated conditions with vigorous algal population growth. Grazing pressures presumably reduce algal levels and increase both particulate and dissolved detrital products . Gordon and Morel (1983) also noted that absorption at 440 nm included a large accessory pigment component, and that variations in the chl *a* to accessory pigment ratio would add variation to the blue to green ratio algorithm and decrease the accuracy of chl *a* estimation. Aiken et al. (1995) found significant inter-province variation in the chlorophyll to carotenoid ratio but concluded that the two band, blue to green ratio approach was still robust and that regional calibrations were possible.

Chlorophyll algorithms have evolved over time. The Coastal Zone Color Scanner (CZCS), onboard the Nimbus 7 satellite, was launched in late 1978 and provided eight years of large scale, synoptic estimates of algal pigment. In 1981, the sensor performance became somewhat degraded (Evans and Gordon, 1994), but useful data were still generated. The instrument had bands for bio-optical monitoring at 443, 520, 550, and 670 nm, with 20 nm bandwidths. A two band ratio of 443 nm to 550 nm was calibrated and routinely used for chl *a* estimation, generating 66,000 CZCS chl *a* images. Two newer, operational sensors (SeaWiFS and MODIS) were also designed to use bands in the blue and green regions for ocean chlorophyll surveillance. These instruments have increased numbers of bands, improved calibration and atmospheric correction capabilities, and higher bit rates (for radiometric resolution). These new satellite programs have also greatly benefited from extensive, close range analyses of ocean pigments and water column optics since the era of CZCS. A recent multi-band, optimization procedure termed OC4 (for Ocean Color 4) utilizes 4 wavelengths for chl *a* estimation (O'Reilly et al., 1998; Table 1). The approach is termed the maximum band ratio (MBR) approach, based on comparing the ratios of 443 to 555 nm, 490 to 555 nm, and 510 to 555 nm. The largest value of these three ratios is used in a third order polynomial regression equation as the exponential term in a power function equation to best represent the sigmoidal relationship between chl *a* and band ratio calculations. In general, the 443 to 555 ratio is maximal below 0.3 μg/l, the 490 to 555 ratio is maximal between 0.3 and 1.5 μg/l, and the 510 to 555 ratio is maximal above 1.5 μg/l (O'Reilley et al., 1998). Their model was parameterized with data from the SeaWiFS Bio-Optical Algorithm Mini-Workshop (hereafter, SeaBAM). The data were collected from 919 ocean and coastal observation stations and had a chl *a* range of 0.019 to 32.79 μg/l. The multiband, ocean color algorithm approach continues to be refined for differing bio-optical ocean provinces and for attempts to utilize data from coastal waters tending toward Case 2 characteristics (Wernand et al., 1998; Stumpf et al., 2000; Sathyendranath et al., 2001; Yoo et al., 2002; D'Sa and Miller, 2003; Blondeau-Patissier et al., 2004). Carder et al. (1999) found a significant reduction in prediction errors when MODIS ocean color algorithms were adapted to three sets of oceanic bio-optical conditions.

Vertucci and Likens (1989) studied a group of 44 low to moderate productivity Adirondack mountain lakes in which the non-acid corrected chl *a* averages for five lake classes ranged from 0.3 - 4.8 μg/l. These lakes ranged from Case 1, low productivity, blue water lakes to Case 2 conditions with moderate chl *a*, variable CDOM, and colors ranging from gray-brown to yellow-green to brown. Optimization procedures were used to determine the best wavelength combinations within visible and lower NIR regions. In these lakes, a two band ratio of 525 nm to 554 nm (Table 1) performed significantly better than the band ratio of 443 to 550 nm. A bio-optical model using pigment, seston, and CDOM data produced model spectra very similar to their field spectra.

A majority of the semi-empirical algorithms for chlorophyll estimation in Case 2 waters utilize some variant of NIR peak height to chl *a* red absorption (Table 1). In a number of these studies, a simple two band ratio of the NIR peak, at either a fixed wavelength or at its maximum value is normalized to the minimum of reflectance at or near 675 nm. Kallio (2003) used a first order, linear equation with a band ratio of 705 to 662 nm to predict chlorophyll from AISA imagery data in two Finnish lakes with a chl *a* range of 6 to 70 μg/l. Thiemann and Kaufmann (2000) used a similar band ratio of 705 to 678 nm ratio to calibrate aerial hyperspectral imagery of lakes in northern Germany (chl *a* range of 5 to 350 μg/l). Mittenzwey et al. (1992) fit a second order polynomial to a band ratio of 705 to 670 nm in a set of lakes and rivers in Germany with a much wider chl *a* range of 5 - 350 μg/l (Table 1).

As discussed previously, Schalles et al. (1998b) and Gitelson et al. (2000) determined that a model using the height of the NIR peak above a normalizing baseline from 675 to 750 nm (Figure 26) was the best algorithm for eutrophic midwestern lakes in the United States and for Lake Kinneret and Haifa Bay in Israel. Gons (1999) modified the simple two band ratio (704 to 672 nm) approach using a bio-optical model. His approach required an inversion model for reflectance at 776 nm to estimate the backscatter coefficient b_b. The backscatter term became an adjustment for extra scattering by tripton particles, when combined with the respective water absorption coefficients for the 672 and 704 nm bands. This approach provided a robust chlorophyll prediction model for a diverse set of inland and coastal Case 2 waters with a chl *a* range of 2 to 994 μg/l. Gons et al. (2002) and Ruddick et al. (2004) modified this bio-optical model to utilize available bands on the MERIS satellite sensor. Dall 'Olmo et al. (2003) proposed a novel technique, with an algorithm derived from pigment estimation in terrestrial vegetation (Table 1). Their model was parameterized with Case 2 data from turbid and productive lakes and reservoirs in Nebraska, USA with a chl *a* range of 7 to 194 and a seston range of < 0.1 to 214 mg/l. Their model, as with Gons et al. (2002), uses a NIR band (740 to 750 nm) to compensate for backscatter by non-algal particles. However, their model uses a difference in the reciprocals of two bands (660 to 670 nm and 720 to 730 nm) to quantify pigment specific absorption activity and doesn't utilize the contrast in the NIR peak and chl *a* red trough features.

My laboratory recently participated in remote sensing studies at five National Estuarine Research Reserves. We examined water reflectance patterns within estuarine longitudinal mixing gradients and across watersheds and coastal provinces (Hladik, 2004; Schalles and Hladik, 2004). Optical measurements and analysis of OACs were made at Apalachicola Bay, Florida (October, 2002), Sapelo Island, Georgia and vicinity (January and August, 2003 and June, 2004), ACE Basin, South Carolina (June, 2003), Grand Bay, Mississippi (October, 2003), and Delaware Bay near Dover, Delaware (July, 2004). Reflectance was estimated from upwelling and downwelling signals collected simultaneously with a pair of fiber optic cables connected to separate

USB2000 spectroradiometers (Ocean Optics Corp., Dunedin, FL). Water samples were filtered for pigment and seston analysis and filtrate was used for CDOM absorption (Table 2). Algal pigments were extracted using 90% buffered acetone with a tissue grinder. Spectral data were collected and processed with the CDAP software package (CALMIT, University of Nebraska). We sampled 155 stations. Eleven stations, with ratios of total depth to Secchi transparency less than 1.5, were eliminated from analysis to avoid bottom reflectance biases. We measured a 600 fold range for chl *a*, a 200 fold range for ABS_{400}, and a 90 fold range for seston (Table 2). Correlations (r values) between OACs were quite low: seston and chl *a* = 0.274, seston and ABS_{440} = 0.031, and chl *a* and ABS_{440} = 0.045. In Figure 28, a conventional ocean color ratio (490 to 555 nm) was calculated for our data and compared to the SeaBAM data set used to parameterize the Ocean Color 4 algorithm (O'Reilly et al., 1998; see above). In the SeaBAM data set, the 490 to 550 nm ratio, fitted to a power function, yielded the highest r^2 value (0.915) of any band ratio pair and regression model in the SeaBAM data. The SeaBAM data had a strong negative correlation between the 490:555 ratio and chl *a*; however, our coastal data clearly did not (Figure 28). However, when a Case 2 algorithm was applied the r^2 value was much higher (discussed below). Some overlap occurred between individual values of the two data sets, particularly with the SeaBAM Chesapeake Bay stations (area A in the Figure) and a higher chlorophyll site in the Canadian arctic (area B; also see Figure 3 in O'Reilly et al., 1998). Cunningham et al. (2001) found similar divergence, using a reflectance band ratio of 443 to 555 nm, for Scottish fjord stations with high CDOM.

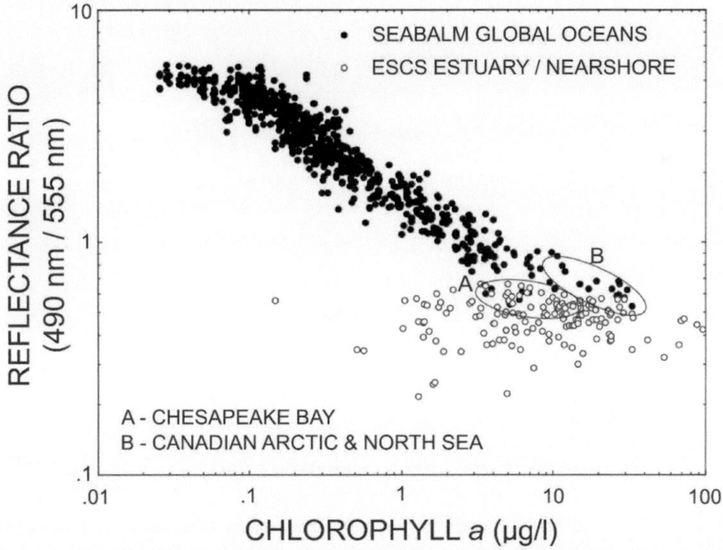

Figure 28. Chl *a* concentration versus water column reflectance ratio (490/553 nm) for the SEABAM data set (O'Reilly et al., 1998) used to calibrate the Ocean Color 4 (OC4) chlorophyll algorithm for the SEAWIFS satellite (closed circles) compared with ratio calculations from 144 sites in five National Estuarine Research Reserves in the eastern United States (open circles).

The series of Case 1 and 2 water algorithms listed in Table 1 were applied to our southeastern estuary data set. No relationships existed with any of the three Ocean

Color (OC4), two band ratios (Table 3; Figure 29a). There was also no relationship with the maximum band ratio values, where the numerator is the largest of the reflectance values at 443, 490, and 510 nm (Table 3). The simple 670 to 700 nm ratio had a relatively weak relationship to chl *a* ($r^2 = 0.482$; Table 3, Figure 29b). The bio-optical model of Gons et al. (2002) was revised for our data set, using reflectance and water absorption coefficient data at 674 nm and 698 nm, at the wavelengths best corresponding to our red trough and NIR values (Table 3). Although this model was robust in Northern Europe inland and coastal waters, we found a weaker relationship in our Atlantic Ocean and Gulf of Mexico coastal settings ($r^2 = 0.322$). The height of the NIR peak above a normalizing baseline between 675 and 750 nm was also unsatisfactory ($r^2 = 0.355$; Table 3).

Table 2. Summary of water analyses for 144 stations at four National Estuarine Research Reserves in the Southeastern United States (Hladik, 2004). Seston values are dry weight.

	Unit	Average	Median	Minimum	Maximum
chl *a*	μg/l	14.8	10.8	0.2	118.9
seston	mg/l	30.2	22.1	1.3	118.3
ABS_{440}	m^{-1}	4.9	3.7	0.1	21.1

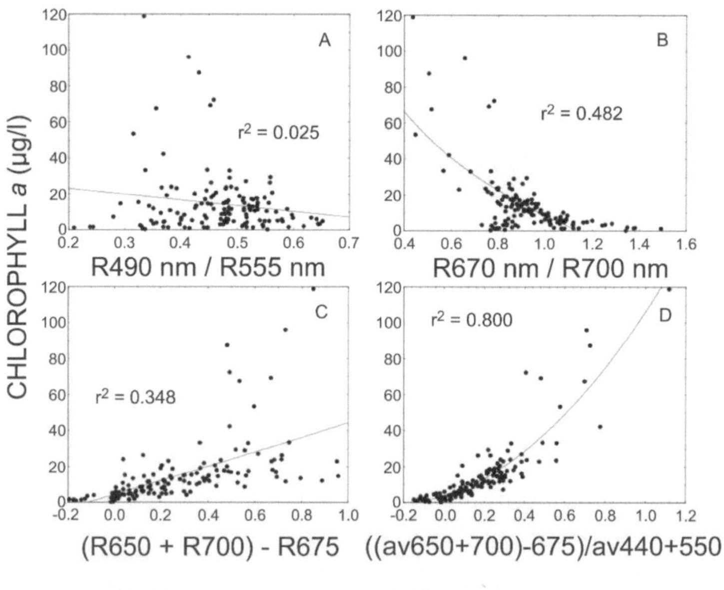

Figure 29. Comparison of four algorithms for chl *a* prediction. The data set is 144stations from five National Estuarine Research Reserves (Hladik, 2004; see Tables 1,2, and 3). (A) Ocean color ratio of 490 to 555 nm - see Figure 28, (B) Case 2 water ratio of 670to 700 nm, (C) new algorithm for depth of red trough below a normaling baseline from 650 to 700 nm, and (D) modification of this new algorithm using an average of reflectance at 440 and 550 nm as a denominator for correction of affects of tripton amplification and CDOM dampening.

The best fit between chl *a* and our data set for 144 stations was found for variants of a model which calculated the depth of the red trough feature, at 675 nm, below a reference line between 650 nm and 700 nm, normalized as a ratio (average of reflectance values at 650 and 700 nm minus reflectance at 675 nm - see Table 3). The reference line position extends between the right edge of the shoulder near 650 nm and the vicinity of the NIR peak. The r^2 value for this algorithm was 0.348 (Table 3, Figure 29c). A second feature of the model was inclusion of reflectance data from lower wavelengths (440 and 550 nm) to normalize between the amplifying effects of tripton scattering and CDOM suppression of reflectance. The lower wavelength areas are more sensitive to these factors (Schalles et al., 1998a; Bowers et al., 2004), as described previously, although the factors also clearly influence the red and lower NIR wavelengths used in models for Case 2 waters. Second order, nonlinear equations best described the results of the model variants. When the reflectance at 550 nm was used as a denominator term, the regression r^2 improved from 0.775 (Table 3). A modest further improvement (r^2 = 0.800) was obtained when an average of reflectances at 440 and 550 nm was used in the denominator (Table 3, Fig. 29d). This later approach greatly reduced the correlation of residuals with elevated CDOM and seston conditions. The largest remaining residuals came from stations with higher chl *a* in several Georgia and Delaware estuaries (> 25 μg/l). We consider this model provisional and are collecting additional data in Chesapeake Bay, -Maryland in 2005 to increase the number of higher chlorophyll observations and to further test for robustness.

Table 3. A comparison of the relationships between various semi-empirical pigment algorithms and measured chl *a* (μg/l) in the ECSC study at four National Estuarine Research Reserves in the Southeastern United States (Hladik, 2004). See Table 1 for explanation of model terms.

MODEL	EQUATION	r^2
R443 / R555	Y = 14.6 - 0.0079X	0.000
R490 / R555	Y = 29.41 - 31.77X	0.025
R510 / R555	Y = 14.58 - 0.0073X	0.000
(max R443,R490,R510) / R555	Y = 15.04 - 0.701X	0.000
R670 / R700	Y = 8.56 - 145.72*LOG(X)	0.482
$(R698/R675*a_w698-b_b)-a_w674-b_b{}^{1.063}$	Y = -1.73 + 44.42X	0.322
NIR Peak Height (RLH R675 to R750)[1]	Y = 1.61 + 11.97X	0.355
(ave R650+R700) - R675	$Y = 4.34 + 38.84X + 1.139X^2$	0.348
(ave R650 + R700) - R675) / 555	$Y = 3.93 + 59.64X + 131.04X^2$	0.775
(aveR650+R700)-R675)/(aveR555+R550)	$Y = 3.72 + 34.92X + 67.63X^2$	0.800

[1] see Figure 26

11. Inverse Models to Relate Inherent Optical Properties and Chlorophyll Using Reflectance Spectra

The problems with site and temporal specificity in the above approaches using empirical and semi-analytical algorithms have motivated numerous researchers to develop analytical model approaches to link IOPs with measured AOPs, including reflectance spectra (Doerffer and Fischer, 1991; Dekker et al., 1997; Lahet et al., 2001; Sathyendranath et al., 2001; Pozdnyakov et al., 2002; Brando and Dekker, 2003; Bowers et al., 2004; Gege and Albert, this volume). If IOPs are well characterized and field reflectance data are carefully collected, inverse models can accurately retrieve water constituents (Bukata et al., 1995). However, the inverse approach is highly sensitive to errors in the measured radiometric variables, and even small errors may invalidate the inversions (Mobley, 1994).

An obvious problem is finding unique solutions for inversion problems in which optical water constituents and boundary conditions vary. Potentially, different sets of boundary and IOP conditions may lead to the same solution. Iterative, Monte Carlo techniques are typically used to find an optimal solution. Boundary conditions (atmospheric aerosols, water surface conditions, sun angle) may be very dynamic, particularly in coastal areas. High accuracy in atmospheric correction is crucial to success in discriminating water constituents, especially with satellite data (Brando and Dekker, 2003). The ability to differentiate optically shallow versus optically deep waters and the contribution of bottom reflectance is also important. Lee et al. (1999) developed a model which can derive bottom depths and water column properties from reflectance spectra. Heterogeneity in the IOPs is a further challenge. For example, differences in the slope of the absorption spectra of CDOM and detrital tripton (Bricaud et al., 1991; Carder et al., 1989; Dekker, 1993; Bowers et al., 2003), variability in pigment composition and cell packaging effects, and differences in volume scattering functions of the different classes of sestonic particles (Dekker et al., 1997) can all reduce the robustness of the assumed IOPs in coastal settings. Thus, it becomes necessary to measure and define specific IOPs (SIOPs) for a given site, and perhaps for a specific time (Brando and Dekker, 2003). Many of the inverse procedures reduce the optically active constituents to the three primary classes reviewed here (algal pigment, seston, and CDOM). However, these classes are often, themselves, complex mixes of constituents. Tidal fluxes, river plumes, and other physical mechanisms, which create spatial and temporal variability, can complicate the application of inversion models to large areas measured with satellite sensors (Doerffer et al., 1994). In many cases, it may not be practical for ship based researchers to synoptically measure water properties for IOP estimation. In fairness, this set of considerations often applies to the simpler, band ratio algorithms discussed above.

In spite of the serious constraints of applying inverse model approaches to coastal waters, a recent study of Moreton and Deception Bays in southeast Queensland, Australia demonstrated impressive results (Brando and Dekker, 2003). The authors used hyperspectral Hyperion Imaging Spectrometer data from the NASA Earth Observing One (EO-1) satellite. A MODTRAN-4 based atmospheric correction for coastal waters was developed. Validation of the atmospheric correction was confirmed by establishing good agreement between Hyperion hyperspectral reflectance and subsurface reflectance measured by a field instrument. The SIOPs of three classes of optical constituents (chl *a*, tripton, and CDOM) were measured and validated using

Hydrolight simulations. An efficient, direct solution of the inversion model was obtained using a matrix inversion method (MIM). The MIM was performed on each pixel of a Hyperion scene for Deception Bay. Maps of chl *a*, tripton, and CDOM were produced which had impressive accuracy and precision. Sensitivity analyses demonstrated that Hyperion could resolve constituents at the following intervals: chl *a* = 2.32 μg/l, tripton = 12.5 mg/l, and CDOM = 0.21 m^{-1}. The authors concluded that the procedure will require a much more thorough measurement of the spatial and temporal variation in SIOPs before accuracy can be further improved and the technique can be applied more broadly to optically diverse coastal waters.

12. Summary and Recommendations

Remote sensing procedures for chl *a* are becoming operational in coastal waters. However, these waters may quite possibly represent the most diverse optical conditions, in space and time, of any aquatic habitat. The ability to detect chl *a* in coastal waters is complicated by: 1) the large dynamic range of pigment in coastal waters; 2) a dominance of water optics by CDOM and/or tripton in many situations; 3) the spatial and temporal dynamics of coastal hydrology; 4) the diverse nature of the optical constituents and their SIOPs; 5) atmospheric aerosols and the requirement for accurate atmospheric corrections; 6) water density gradients and light and nutrient induced, non-uniform vertical distributions; 7) bottom reflectance and the diversity of benthic substrates; and 8) the problem of mixed pixels related to spatial resolution and contributions from shore zones. These challenges have required a redirection of optics research and remote sensing operational schemes for the Case 2 water conditions prevalent in most coastal settings.

The ability to detect chlorophyll in Case 2 waters requires, especially, hyperspectral data with the resolution to detect sometimes subtle pigment absorption bands, different accessory pigments, shoulders and peaks in reflectance spectra related to scattering and fluorescence activities, and good calibration and accuracy of instruments. The red and lower NIR spectral regions are often the most favorable for chl *a* detection in Case 2 waters because: 1) accessory pigments have minimal contribution to total absorption at these wavelengths; 2) CDOM and organic tripton have reduced absorption; and 3) the NIR peak feature in reflectance is more sensitive to the higher chl *a* ranges found in coastal waters. On the other hand, this wavelength range is less practical for lower chl *a* ranges because of dominance by water absorption.

Ocean color algorithms utilizing blue and green spectral bands are appropriate when Case 1 conditions occur in coastal waters. Pixel classification procedures may be required to make an initial examination of reflectance pattern and the selection of the most appropriate algorithm and parameterizations. However, the techniques for Case 2 water measurements reviewed in this chapter may also hold promise for the open oceans and other Case 1 waters such as large lakes (International Ocean-Color Coordinating Group, 2000). The challenges of Case 2 water optics have stimulated significant advancements in bio-optical modeling and insights into the interactions of optical constituents. Case 1 algorithms rely on an assumed covariance of constituents, whereas the techniques for relating Case 2 constituents to reflectance signals often require multivariate, non-linear models (International Ocean-Color Coordinating Group, 2000).

The best procedures to apply in a given situation often are decided by constraints of time, money, and resolution requirements of the user. In some cases, detection of several general classes of water conditions and chl *a* levels may be sufficient. In other situations, change detection and the ability to resolve relatively small differences in chl *a* may be critical to bloom detection and reduction of damage to aquaculture operations, the ability to answer key ecological questions and improve modeling of primary production, and to meet regulatory criteria.

The problems remaining in order to achieve more routine, operational abilities to remotely measure chl *a* in coastal waters are relatively well identified. A number of research groups are actively engaged, as readily confirmed by the growing number of publications and technical presentations in Case 2 water optics research. Progress is evident, and further, rapid advancements are probable. Mobley (1994) was prescient in stating:

"The development of remote-sensing inversion algorithms for use in Case 2 waters scarcely has begun. As with other facets of the hydrologic optics of such waters, the remote sensing of case 2 waters will provide challenging research problems for a generation of scientists."

13. Acknowledgements

I have greatly benefited from several collaborations in the past 15 years in both tank mesocosm experiments and field observations in inland and coastal waters. Donald Rundquist and Anatoly Gitelson at the University of Nebraska, Yosef Yacobi at the Kinneret Limnology Laboratory, Frank Schiebe at the USDA/ARS National Agricultural Water Quality Laboratory, and Jim Alberts and Monika Takacs at the University of Georgia Marine Institute at Sapelo Island were particularly generous with their time, intellectual engagement, and companionship. The Lake Chapala observations were conducted with Laura Davalos-Lind, Owen Lind, and Frank Schiebe. My work and perspectives have also been greatly enhanced from interactions with Pat Brezonik, Arnold Dekker, Giorgio Dall 'Olmo, Frank (Jerry) DeNoyelles, Sharon Dewey, Rolland Fraser, Luoheng Han, Giorgio Dall 'Olmo, Jeffrey Peake, Laurie Richardson, Robbi Stark, Patrick Starks, Richard Stumpf, and Bill Troeger. I've received support for the studies reported here from the NASA-Nebraska Space Grant Office, NOAA-NCCOS Environmental Cooperative Science Center, USEPA, visiting scientist appointments with the USDA/ARS National Agricultural Water Quality Laboratory, University of Georgia Marine Institute, and CALMIT-University of Nebraska-Lincoln. Scientists and technicians at the following NOAA National Estuarine Research Reserves provided outstanding support: ACE Basin, South Carolina, Apalachicola Bay, Florida, Delaware Bay, Delaware, and Grand Bay, Mississippi. I've worked closely with two Creighton University graduate students, Christine Hladik and Kim Whitman and benefited greatly from their enthusiasm, dedication, assistance, and questioning. The following individuals were particularly helpful in field and laboratory work and logistical support: Patrick A'Hearn, Ambrose Anoruo, Jennifer Arnhold, Wallace Bell, Jennifer Cherrier, Kevin Dillon, Chunlei Fan, Barbara Hayes, Chris Hiemstra, Rhonda Horner, Mark Huggins, Lisa Hughes, Charles Jagoe, David Kennedy, Amy Kroenke, Bryan Leavitt, Lee Edmiston, Mark Harwell, Stuart McNally, Christine Michaud, Katherine Milla, Albert Miller, Denise Lani Pascual, Heather Peterson, Kelli Peterson, Mary Price, Mike Reiter, Robert Scarborough, Jim Schalles, Amy Sheil, James Tycast, George Walker, Latrincy Whitehurst, Betty Winner, and Mark Woodrey. Finally, I'd like to thank my wife Nancy

Edwards Schalles for her generous support, encouragement, and patience. This paper is Contribution Number 949 from the University of Georgia Marine Institute.

14. References

Ackleson, S.G., and V. Klemas.1986. Two-flow simulation of the natural light field within a canopy of submerged aquatic vegetation. Applied Optics, 25:1129-1136.

Aiken, J., G.F. Moore, C.C. Trees, S.B. Hooker, and D.K. Clark. 1995. The SeaWiFS CZCS-Type Pigment Algorithm. SeaWiFS Technical Report Series , NASA Technical Memorandum 104566, Vol. 29. Goddard Space Flight Center, Greenbelt, Maryland.

Albert, A., and C.D. Mobley. 2003. An analytical model for subsurface irradiance and remote sensing reflectance in deep and shallow case-2 waters. Ocean Optics Express, 11: 2873-2890.

Alberts, J.J., and Z. Filip. 1994. Humic substances in rivers and estuaries of Georgia, USA. Trends in Chemical Geology, 1:143-162.

Alberts, J.J., and C. Griffin. 1996. Formation of particulate organic carbon (POC) from dissolved organic carbon (DOC) in salt marsh estuaries of the southeastern United States. Archives fur Hydrobiologie Special Issues, Advances in Limnology, 47:401-409.

Alberts, J.J., M. Takacs, and J.F. Schalles. 2004. Ultraviolet-visible and fluorescence spectral evidence of natural organic matter (NOM) changes along an estuarine salinity gradient. Estuaries, 27:297-311.

American Public Health Association. 1998. Standard Methods for the Examination of Water and Wastewater (20th edition). Section 1200 - Chlorophyll. American Public Health Association.

Andre, J.-M. 1992. Ocean color remote-sensing and the subsurface vertical structure of phytoplankton pigments. Deep Sea Research, 39:763-779.

Annenberg, P. 2000. Analysis of CASI data - a case study from the Archipelago of Stockholm, Sweden. Proceedings of the Sixth International Conference on Remote Sensing for Marine and Coastal Environments. University of Michigan Press., 2:149-156.

Antoine, D., J.-M. Andre, and A. Morel. 1996. Oceanic primary production 2. Estimation of global scale from satellite (coastal zone color scanner) chlorophyll. Global Biogeochemical Cycles, 10:57-69.

Babin, M., A. Morel, and B. Gentili. 1996. Remote sensing of sea surface Sun-induced chlorophyll fluorescence: consequences of natural variations in the optical characteristics of phytoplankton and the quantum yield of chlorophyll *a* fluorescence. International Journal of Remote Sensing, 17:2417-2448.

Barnard, A.H., W.S. Pegau, and J.R.V. Zaneveld. 1998. Global relationships of the inherent optic properties of the oceans. Journal of Geophysical Research, 103:24,955-24,968.

Bidigare, R.R., M.E. Ondrusek, J.H. Morrow, and D.A. Kiefer. 1990. *In vivo* absorption properties of algal pigments. SPIE Ocean Optics X: 1302:290-302.

Blondeau-Patissier, D., G.H. Tilstone, V. Martinez-Vicente, and G.F. Moore. 2004. Comparison of bio-physical marine products from SeaWiFS, MODIS, and a bio-optical model with *in situ* measurements from Northern European waters. Journal of Optics A: Pure and Applied Optics, 6:875-889.

Boss, E., and J.R.V. Zaneveld. 2003. The effect of bottom substrate on inherent optical properties: Evidence of biogeochemical processes. Limnology and Oceanography, 48:346-354.

Bowers, D.G., D. Evans, D.N. Thomas, K. Ellis, and P.J. le B. 2004. Williams. Interpreting the colour of an estuary. Estuarine, Coastal, and Shelf Science, 59:13-20.

Brando, V.E., and A.G. Dekker. 2003. Satellite hyperspectral remote sensing for estimating estuarine and coastal water quality. IEEE Transactions on Geosciences and Remote Sensing, 41:1378-1387.

Bricaud, A., A. Morel, and L. Prieur. 1981. Absorption by dissolved organic matter in the sea (yellow substances) in the UV and visible domains. Limnology and Oceanography, 26:43-53.

Brown, C.W., and J.A. Yoder. 1994. Coccolithophorid blooms in the global ocean. Journal of Geophysical Research, 99:7467-7482.

Bukata, R.P., J.H. Jerome, K.Y. Kondatyev, and D.V. Pozdnyakov. 1995. Optical Properties and Remote Sensing of Inland and Coastal Waters. CRC Press, Boca Raton, 362 pp.

Butler, W.L. 1966. Spectral characteristics of chlorophyll in green plants. In: (L.P. Vernon and G.R. Seely, editors) The Chlorophylls. Academic Press, pp. 343-380.

Carder, K.L., and R.G. Steward. 1985. A remote-sensing reflectance model of a red-tide dinoflagellate off west Florida. Limnology and Oceanography, 30:286-298.

Carder, K.L., R.G. Steward, G.R. Harvey, and P.B. Ortner. 1989. Marine humic and fulvic acids: Their effects on remote sensing of ocean chlorophyll. Limnology and Oceanography, 34:68-81.

Carder, K.L., F.R. Chen, Z.P. Lee, S.K. Hawes, and D. Kamykowski. 1999. Semianalytic Moderate-Resolution Imaging Spectrometer algorithms for chlorophyll *a* and absorption with bio-optical domains based on nitrate-depletion temperatures. Journal of Geophysical Research, 104:5403-5421.

Carlough, L.A. 1994. Origins, structure, and trophic significance of amorphous seston in a blackwater river. Freshwater Biology, 31:227-237.

Casazza, G, C. Silvestri, and E. Spada. 2003. Classification of coastal waters according to the new Italian water legislation and comparison to the European Water Directive. Journal of Coastal Classification, 9:65-72.
Cole, B.E., and J.E. Cloern. 1987. An empirical model for estimating phytoplankton productivity in estuaries. Marine Ecology Progress Series, 36:299-305.
Cracknell, A.P., S.K. Newcombe, A.F. Black, and N.E. Kirbu. 2001. The ABDMAP (Algal Bloom Detection, Monitoring and Prediction) Concerted Action. International Journal of Remote Sensing, 22:205-247.
Cunningham, A., P. Wood, and K. Jones. 2001. Reflectance properties of hydrographically and optically stratified fjords (Scottish sea lochs) during the Spring diatom bloom. International Journal of Remote Sensing, 22:2885-2897.
Curran, P.J., and E.M.M. Novo. 1988. The relationship between suspended sediment concentration and remotely sensed spectral radiance: A review. Journal of Coastal Research, 4:351-368.
D'Sa, E.J., and R.L. Miller. 2003. Bio-optical properties in waters influenced by the Mississippi River during low flow conditions. Remote Sensing of the Environment, 84:538-549.
Dall' Olmo, G., A.A. Gitelson, and D.C. Rundquist. 2003. Towards a unified approach for remote estimation of chlorophyll-a in both terrestrial vegetation and turbid productive waters. Geophysical Research Letters, 30:1938-1941.
Dekker, A.G. Detection of the Optical Water Quality Parameters for Eutrophic Waters by High Resolution Remote Sensing. 1993. Ph.D. Thesis. Free University, Amsterdam, The Netherlands, 212 pp.
Dekker, A.G., Z. Zamurovic-Nenad, H.J. Hoogenboom, and S.W.M. Peters. 1996. Remote sensing, ecological water quality modeling, and in situ measurements: a case study in shallow lakes. Hydrological Sciences, 41:531-547.
Dekker, A.G., H.J. Hoogenboom, L.M. Goddijn, and T.J.M. Malthus. 1997. The relation between inherent optical properties and reflectance spectra in turbid inland waters. Remote Sensing Reviews, 15:59-74
Doerffer, R. 1993. Estimation of primary production by observation of solar-stimulated fluorescence. ICES Marine Science Symposium, 197:104-113.
Doerffer, R., and J. Fischer. 1994. Concentrations of chlorophyll, suspended matter, and gelbstoff in case II waters derived from satellite coastal zone color scanner data with inverse modeling methods. Journal of Geophysical Research, 99:7457-7466.
Evans, R.H., and H.R. Gordon. 1994.Coastal zone color scanner "system calibration": a retrospective examination. Journal of Geophysical Research, 99:7293-7307.
Falkowski, P.G., and J.A. Raven. 1997. Aquatic Photosynthesis. Blackwell Science, 375 pp.
Gallegos, C.L., D.L. Correll, and J.W. Pierce. 1990. Modeling spectral diffuse attenuation, absorpion, and scattering coefficients in a turbid estuary. Limnology and Oceanography, 35:1486-1502.
Gallegos, C.L., and T.E. Jordan. 2002. Impact of the spring 2000 phytoplankton bloom in Chesapeake Bay on optical properties and light penetration in the Rhode River, Maryland. Estuaries, 25:508-518.
Gitelson, A. 1992. The peak near 700 nm on radiance spectra of algae and water: relationships of its magnitude and position with chlorophyll concentration. International Journal of Remote Sensing, 13:3367-3373.
Gitelson, A.A., J.F. Schalles, D.C. Rundquist, F.R. Schiebe, and Y.Z. Yacobi. 1999. Comparative reflectance properties of algal cultures with manipulated densities. Journal of Applied Phycology, 11:345-354.
Gitelson, A.A. Y.Z. Yacobi, J.F. Schalles, D.C. Rundquist, L. Han, R. Stark, and D. Etzion. 2000. Remote estimation of phytoplankton density in productive waters. Limnology and Lake Management - Archives fur Hydrobiolia, Special Issues, Advances in Limnology, 55:121-136.
Gons, H.J. 1999. Optical teledetection of chlorophyll a in turbid inland waters. Environmental Science and Technology, 33:1127-1132.
Gons, H.J., M. Rijkeboer, and K.G. Ruddick. 2002. A chlorophyll-retrieval algorithm for satellite imagery (Medium Resolution Imaging Spectrometer) of inland and coastal waters. Journal of Plankton Research, 24:947-951.
Gordon, H.R. 1997. Atmospheric correction of ocean color imagery in the Earth Observing System era. Journal of Geophyical Research, 102:17,081-17,106.
Gordon, H.R., and A.Y. Morel. 1983. Remote Assessment of Ocean Color for Interpretation of Satellite Visible Imagery. A Review. Springer-Verlag, New York. 114 pp.
Gower, J.F.R, and G.A. Borstad. 1990. Mapping of phytoplankton by solar-stimulated fluorescence using an imaging spectrometer. International Journal of Remote Sensing, 11:313-320.
Gower, J.F.R., R. Doerffer, and G.A. Borstad. 1999. Interpretation of the 685 nm peak in water-leaving radiance spectra in terms of fluorescence, absorption and scattering, and its observation by MERIS. International Journal of Remote Sensing, 20:1771-1786.
Hladik, C.M. 2004. Close range, hyperspectral remote sensing of Southeastern estuaries and an evaluation of phytoplankton chlorophyll *a* predictive algorithms. M.S. Thesis, Creighton University, Omaha, Nebraska, USA.126 pp.

Han, L., and D.C. Rundquist. 1994. The response of both surface reflectance and underwater light field to various levels of suspended sediments: Preliminary results. Photogrammetric Engineering and Remote Sensing, 60:1463-1471.

International Ocean-Color Coordinating Group. (S. Sathyendranath, Ed.) 2000. Remote Sensing of Ocean Color in Coastal, and Other Optically Complex, Waters. IOCCG Report Number 3, 140 pp.

Jahnke, R.A. 1996. The global ocean flux of particulate organic carbon: areal distribution and magnitude. Global Biogeochemical Cycles, 10:71-88.

Jassby, A.D., B.E. Cole, and J.E. Cloern. 1997. The design of sampling transects for characterizing water quality in estuaries. Estuarine, Coastal and Shelf Science, 45:285-302.

Jerlov, N.G. 1976. Marine Optics, 2nd edition. Elsevier Scientific Publishing Company.

Jordan, T.E., D.L. Cornell, J. Miklas, and D.E. Weller. 1991. Long-term trends in estuarine nutrients and chlorophyll, and short-term effects of variation in watershed discharge. Marine Ecology Progress Series, 75:121-132.

Jumars, P.A. 1993. Concepts in Biological Oceanography. Oxford University Press.

Kahru, M., and B.G. Mitchell. 1998. Spectral reflectance and absorption of a massive red tide off southern California. Journal of Geophysical Research, 103:21,601-21,609.

Kallio, K., S. Koponen, and J. Pulliainen. 2003. Feasibility of airborne imaging spectrometry for lake monitoring - a case study of spatial chlorophyll *a* distribution in two meso-eutrophic lakes. International Journal of Remote Sensing, 24:3771-3790.

Kirk, J.T.O. 1994. Light and Photosynthesis in Aquatic Ecosystems. 2nd Edition. Cambridge University Press.

Kitchen, J.C., and J.R.V. Zaneveld. 1992. A three-layered sphere model of the optical properties of phytoplankton. Limnology and Oceanography, 37:1680-1690.

Klemas, V., and D.F. Polis. 1977. Remote sensing of estuarine fronts and their effects on pollutants. Photogrammetric Engineering and Remote Sensing, 43:599-612.

Krijgsman, J. 1994. Optical remote sensing of water quality parameters - Interpretation of reflectance spectra. Ph.D. Thesis. Delft University of Technology, Delft University Press.

Lahet, F., S. Ouillon, and P. Forget. 2001. Colour classification of coastal waters of the Ebro river plume from spectral reflectances. International Journal of Remote Sensing, 22:1639-1664.

Lee, Z., K.L. Carder, C.D. Mobley, R.G. Steward, and J.S. Patch. 1998. Hyperspectral remote sensing for shallow waters. 1. A semianlaytical model. Applied Optics, 37:6329-6338.

Lee, Z., K.L. Carder, C.D. Mobley, R.G. Steward, and J.S. Patch. 1999. Hyperspectral remote sensing for shallow waters. 2. Deriving bottom depths and water properties by optimization. Applied Optics, 38:3831-3843.

Lefevre, N., A.H. Taylor, F.J. Gilbert, and R.J. Geider. 2003. Modeling carbon to nitrogen and carbon to chlorophyll *a* ratios in the ocean at low latitudes: Evaluation of the role of physiological plasticity. Limnology and Oceanography, 48:1796-1807.

Limón, J.G., O.T. Lind, D.S. Vodopich, R. Doyle, and B.G. Trotter. 1989. Long- and short-term variation in the physical and chemical limnology of a large, shallow, turbid tropical lake (Lake Chapala, Mexico). Archives fur Hydrobiologie, Supplement 83 (Monographische Beiträge), 57-81.

Lind, O.T., and L.O. Dávalos. 1990. Clay, dissolved organic matter, and bacterial interactions in two reservoirs. Archives fur Hydrobiologie, Ergebn. Limnologie, 34:119-125.

Lind, O.T., R. Doyle, D.S. Vodopich, B.G. Trotter, J. Gualberto Limón, and L. Dávalos-Lind. 1992. Clay turbidity: regulation of phytoplankton in a large, nutrient-rich lake. Limnology and Oceanography, 37:549-565.

Maritorena, S., A. Morel, and B. Gentili. 1994. Diffuse reflectance of oceanic shallow waters: influence of water depth and bottom albedo. Limnology and Oceanography, 39:1689-1703.

Mittenzwey, K.-H., S. Ullrich, A.A. Gitelson, and K.Y. Kondratiev. 1992. Determination of chlorophyll a of inland waters on the basis of spectral reflectance. Limnology and Oceanography, 37:147-149.

Mobley, C.D. 1994. Light and Water: Radiative Transfer in Natural Waters. Academic Press.

Morel, A. 1974. Optical properties of pure water and sea water. In: (N. Jerlov and E. Steemann Nielsen, editors) Optical Aspects of Oceanography. Academic Press, p. 1-24.

Morel, A., and L. Prieur. 1977. Analysis of variations in ocean color. Limnology and Oceanography, 22: 709-722.

Morrow, J.H., B.N. White, M. Chimiente, and S. Hubler. 2000. A bio-optical approach to reservoir monitoring in Los Angeles. Limnology and Lake Management - Archives fur Hydrobiolia Special Issues, Advances in Limnology, 55:171-191.

Muller-Karger, F.E., C.R. McClain, T.R. Fisher, W.E. Esaias, and R. Varela. 1989. Pigment distribution in the Caribbean Sea: observations from space. Progress in Oceanography, 23:23-64.

Munday, J.C., and P.L. Zubkoff. 1981. Remote sensing of blooms in a turbid estuary. Photogrammetric Engineering and Remote Sensing, 47:523-531.

Myers, MR, J.T. Hardy, C.H. Mazel, and P. Dustan. 1999. Optical spectra and pigmentation of Caribbean reef corals and macroalgae. Coral Reefs, 18:179-186

Odhe, T., and H. Siegel. 2001. Correction of bottom influence in ocean colour satellite images of shallow water areas of the Baltic Sea. International Journal of Remote Sensing, 22:297-313.

O'Reilly, J.E., S. Maritorena, B.G. Mitchell, D.A. Siegel, K.L. Carder, S.A. Garver, M. Kahru, and C. McClain. 1998. Ocean color chlorophyll algorithms for SeaWiFS. Journal of Geophysical Research, 103:24,937-24,953.

Paerl, H.W. 1988. Nuisance phytoplankton blooms in coastal, estuarine, and inland waters. Limnology and Oceanography, 33:823-847.

Pettersson, L.H., D. Durand, O.M. Johannessen, E. Svendsen, and H. Soiland. 2000. Satellite observations and model predictions of toxic algae blooms in coastal waters. Proceedings of the Sixth International Conference on Remote Sensing for Marine and Coastal Environments, 1:48-55.

Pozdnyakov, D., A. Lyaskovsky, H. Grassl, and L. Pettersson. 2002. Numerical modeling of transpectral processes in natural waters: implications for remote sensing. International Journal of Remote Sensing, 23:1581-1607.

Quinby-Hunt, M.S., A.J. Hunt, K. Lofftus, and D. Shapiro. 1989. Polarized-light scattering studies of marine *Chlorella*. Limnology and Oceanography, 34:1587-1600.

Richardson, L.L. 1996. Remote sensing of algal bloom dynamics. Bioscience, 46:492-501.

Rijkeboer, M., A.G. Dekker, and H. J. Gons. 1998. Subsurface irradiance reflectance spectra of inland waters differing in morphometry and hydrology. Aquatic Ecology, 31:313-323.

Ritchie, J.C., F.R. Schiebe, C.M. Cooper, and J.A. Harrington, Jr. 1994. Chlorophyll measurements in the presence of suspended sediment using broad band spectral sensors aboard satellites. Journal of Freshwater Ecology, 9:197-206.

Rowan, K.S. 1989. Photosynthetic Pigments of Algae. Cambridge University Press.

Ruddick, K., Y. Park, and B. Nechad. 2003. MERIS imagery of Belgian coastal waters: mapping of suspended particulate matter and chlorophyll-*a*. Proceedings of MERIS user workshop, Frascati, Italy, ESA SP-249.

Rundquist, D.C., J.F. Schalles, and J.S. Peake. 1995. The response of volume reflectance to manipulated algal concentrations above bright and dark bottoms at various depths in an experimental pool. Geocarto International, 10:5-14.

Sathyendranath, S., L. Lazzara, and L. Prieur. 1987. Variations in the spectral values of specific absorption of phytoplankton. Limnology and Oceanography, 32:403-415.

Sathyendranath, S., G. Cota, V. Stuart, H. Maass, and T. Platt. 2001. Remote sensing of phytoplankton pigments: a comparison of empirical and theoretical approaches. International Journal of Remote Sensing, 22:249-273.

Schalles, J.F., F.R. Schiebe, P.J. Starks, and W.W. Troeger. 1997. Estimation of algal and suspended sediment loads (singly and combined) using hyperspectral sensors and experiments. Proceedings of the Fourth International Conf. on Remote Sensing of Marine and Coastal Environments, 1:247-258.

Schalles. J.F., A.T. Sheil, J.F. Tycast, J.J. Alberts, and Y.Z. Yacobi. 1998a. Detection of chlorophyll, seston, and dissolved organic matter in the estuarine mixing zone of Georgia coastal plain rivers. Proceedings of the Fifth International Conference on Remote Sensing for Marine and Coastal Environments, 2:315-324.

Schalles, J.F., A.A. Gitelson, Y.Z. Yacobi, and A.E. Kroenke. 1998b. Estimation of chlorophyll a from time series measurements of high spectral resolution reflectance in an eutrophic lake. Journal of Phycology, 34:383-390b.

Schalles, J.F. and Y.Z. Yacobi. 2000. Remote detection and seasonal patterns of phycocyanin, carotenoid, and chlorophyll pigments in eutrophic waters. Limnology and Lake Management - Archives fur Hydrobiolia Special Issues, Advances in Limnology, 55:153-168.

Schalles, J.F., D.C. Rundquist, and F.R. Schiebe. 2001. The influence of suspended clays on phytoplankton reflectance signatures and the remote estimation of chlorophyll. Vehr. Intern. Verein. Limnol. 27: 3619-3625.

Schalles, J.F., F.R. Schiebe, C.M. Hladik, F.R. Schiebe, J.F. DeNoyelles, and D.R. Rundquist. An analysis of the red/NIR fluorescence and reflectance peak (690-720 nm) of phytoplankton blooms. (in prep.)

Schiebe, F.R., J.A. Harrington, Jr., and J.C. Ritchie. 1992. Remote sensing of suspended sediments: the Lake Chicot, Arkansas project. International Journal of Remote Sensing, 13:1487-1509.

Schwarz, J.N., P. Kowalczuk, S. Kaczmarek, G.F. Cota, B.G. Mitchell, M. Kahru, F.P. Chavez, A. Cunningham, D. McKee, P. Gege, M. Kishino, D.A. Phinney, and R. Raine. 2002. Two models for absorption by coloured dissolved organic matter (CDOM). Oceaonologia, 44:209-241.

Siegel, D.A., M. Wang, S. Maritorena, and W. Robinson. 2000. Atmospheric correction of satellite ocean color imagery: the black pixel assumption. Applied Optics, 39:3582-3591.

Smith, R.C., and K. Baker. 1981. Optical properties of the clearest natural waters. Applied Optics, 20: 177-184.

Starks, P.J., F.R. Schiebe, and J.F. Schalles. 1995. Characterization of the accuracy and precision of spectral measurements by a portable, silicon diode array spectrometer. Photogrammetry and Remote Sensing, 61:1239-1246.

Stumpf, R.P, R.A. Arnone, R.W. Gould, P. Martinolich, V. Ransibrahmanakul, P.A. Tester, R.G. Steward, A. Subramaniam, M. Culver, and J.R. Pennock. 2000. SeaWiFS ocean color data for U.S. Southeast coastal waters. Proceedings of the Sixth International Conference on Remote Sensing for Marine and Coastal Environments, 1:25-27.

Talling, J.F. 1993. Comparative seasonal changes, and inter-annual variability and stability, in a 26-year record of total phytoplankton biomass in four English Lake District basins. Hydrobiologia, 268:65-98.

Thiemann, S., and H. Kaufmann. 2000. Lake water quality monitoring using hyperspectral airborne data – a multitemporal approach. Proceedings of the Sixth International Conference on Remote Sensing for Marine and Coastal Environments, 2:157-164.

Walsh, J.J. 1991. Importance of the continental margins in the marine biogeochemical cycling of carbon and nitrogen. Nature, 350:53-55.

Wang, M., D.R. Lyzenga, and V.V. Klemas. 1996. Measurement of the optical properties in the Delaware Estuary. Journal of Coastal Research, 12:211-228.

Wernard, M.R., S.J. Shimwell, S. Boxall, and H.M. van Aken. 1998. Evaluation of specific semi-empirical coastal colour algorithms using historic data sets. Aquatic Ecology, 32:73-91.

Vertucci, F.A., and G.E. Likens. 1989. Spectral reflectance and water quality of Adirondack mountain region lakes. Limnology and Oceanography, 34:1656-1672.

Vos, W.L., M. Donze, and H. Buiteveld. 1986. On the reflectance spectrum of algae in water: the nature of the peak and 700 nm and its shift with varying concentration. Communications on Sanitary Engineering and Water Management, Number 7, TU Delft, The Netherlands, 29 pp.

Yoder, J.A., J. Aiken, R.N. Swift, F.E. Hoge, and P.M. Stegmann. 1992. Spatial variability in near-surface chlorophyll a fluorescence measured by the Airborne Oceanographic LIDAR (AOL). Deep Sea Research II, 40:37-53.

Yoo, S.-J., H.-C. Kim, J.-a. Lee, and M.-O. Park. 2002. Validation of chlorophyll algorithm in Ulleung Basin, East/Japan Sea. Korean Journal of Remote Sensing, 18:35-42.

Zaneveld, J.R.V., D.M. Roach, and H. Pak. 1974. The determination of the refractive distribution of oceanic particulates. Journal of Geophysical Research, 79:4091-4095.

Chapter 4

A TOOL FOR INVERSE MODELING OF SPECTRAL MEASUREMENTS IN DEEP AND SHALLOW WATERS

PETER GEGE[1] AND ANDREAS ALBERT[2]
[1]*DLR, Remote Sensing Technology Institute, P.O.Box 1116, 82230 Wessling, Germany*
[2]*GSF, Institute of Soil Ecology, Ingolstaedter Landstr. 1, 85764 Neuherberg, Germany*
Email: peter.gege@dlr.de, andreas.albert@gsf.de

1. Introduction

Light is the primary energy source of life on Earth. Apart from some deep-sea ecosystems, the existence of all organisms depends directly or indirectly on the process of photosynthesis, which is driven by electromagnetic radiation in the spectral range from 400 to 700 nm (photosynthetically active radiation, or PAR), or at longer wavelengths for some photosynthetic bacteria. Solar irradiance and the transparency of water are maximal, and atmospheric extinction is low, at PAR wavelengths. Thus, optical measurements in this spectral range can provide valuable information to characterize biological processes in aquatic ecosystems.

Many different types of optical instruments have been developed which can be used to quantitatively determine certain parameters of aquatic ecosystems. These instruments are operated on buoy, off shore platforms, ships, aircraft, and satellites, and measure radiation in different spectral bands. Remote sensing instruments usually provide images, thus measuring *radiance spectra* of individual pixels. *In situ* optical instruments can measure radiation with angle-integration (irradiance), with normalization to incident illumination (reflectance), with alterations versus depth (attenuation), or the fractions which are absorbed, scattered, or emitted in the water or from the dissolved or suspended constituents.

Usually data from each instrument, or sensor type, are analyzed with software that is specifically tailored to that instrument or spectrum. However, operating a group of programs is a potential source of errors, because the data analysis programs must be consistent with each other with respect to the model formulations and input data. In addition, maintenance and data handling are time consuming, as is training new staff. For these reasons it is desirable to have a single integrative program. Such a tool, the "Water colour Simulator" WASI, was developed for optical *in situ* measurements. The program, together with a detailed user manual, is available at http://extras.springer.com. It can also be downloaded from an ftp server (Gege, 2002a).

WASI is designed as a sensor-independent spectra generator and spectra analyzer. The program has well documented calculation steps and automated, graphical visualization of results. It can also generate and analyze large series of spectra. In the forward mode, up to three parameters can be iterated simultaneously to produce a great variety of spectra, while in the inverse mode an unlimited number of spectra can be read from files and automatically analyzed. The supported spectrum types are listed in Table 1. Calculations are based on analytical models. The data provided with WASI were determined at Lake Constance (Gege, 1994, 1995; Heege, 2000), and are suited for calculating all spectral types at a range of at least 390 to 800 nm and with 1 nm spectral resolution.

L.L. Richardson and E.F. LeDrew (eds.), Remote Sensing of Aquatic Coastal Ecosystem Processes: Science and Management Applications, 81-109.

Table 1. Types of spectral measurements for which inverse modeling is implemented.

Spectrum type	Model options	Symbol	Equation
Absorption	Exclude pure water	$a_{WC}(\lambda)$	(1)
	Include pure water	$a(\lambda)$	(3)
Attenuation	For downwelling irradiance	$K_d(\lambda)$	(5)
Specular reflectance	Wavelength dependent	$R_{rs}^{surf}(\lambda)$	(13a)
	Constant	R_{rs}^{surf}	(13b)
Irradiance reflectance	For deep water	$R(\lambda)$	(14)
	For shallow water	$R^{sh}(\lambda)$	(16)
Remote sensing reflectance	Below surface for deep water	$R_{rs}^{-}(\lambda)$	(17)
	Below surface for shallow water	$R_{rs}^{sh-}(\lambda)$	(19)
	Above surface	$R_{rs}(\lambda)$	(20)
Bottom reflectance	For irradiance sensors	$R^{b}(\lambda)$	(21)
	For radiance sensors	$R_{rs}^{b}(\lambda)$	(22)
Downwelling irradiance	Above surface	$E_d(\lambda)$	(23)
	Below surface	$E_d^{-}(\lambda)$	(24)
Upwelling radiance	Below surface	$L_u^{-}(\lambda)$	(26)
	Above surface	$L_u(\lambda)$	(27)

All input and output files in WASI are in text format (ASCII), making it easy to adapt calculations to regional circumstances by replacing some default input spectra and changing material-specific constants. A well-designed graphical user interface allows intuitive operation. An example of the interface is shown in Figure 1. Alternatively, WASI can be operated in a background mode where all actions are controlled by an input file. In this mode other programs can utilize WASI as a slave to generate or analyze data according to their demands. This input file, WASI.INI, is also used to initialize and document all program settings. It is automatically read during program start up, and a copy with the actual settings is automatically stored in the relevant directory whenever outputs from calculations are saved.

An overall description of WASI was given in Gege (2004). This chapter focuses on data analysis for inverse modeling of spectral measurements. Implemented algorithms, including newly developed models for shallow water, are summarized. Problems associated with inverse modeling, and solutions offered for WASI, are discussed. Finally, some examples of how to apply the program effectively are presented.

2. Models

2.1 ABSORPTION

2.1.1 *Water constituents*

Absorption of a mixture of water constituents is the sum of the components' absorption coefficients:

$$a_{WC}(\lambda) = \sum_{i=0}^{5} C_i \cdot a_i^*(\lambda) + X \cdot a_X^*(\lambda) + Y \cdot a_Y^*(\lambda), \qquad (1)$$

where λ denotes wavelength. Three groups of water absorbing constituents are considered: phytoplankton, non-chlorophyllous particles, and Gelbstoff.

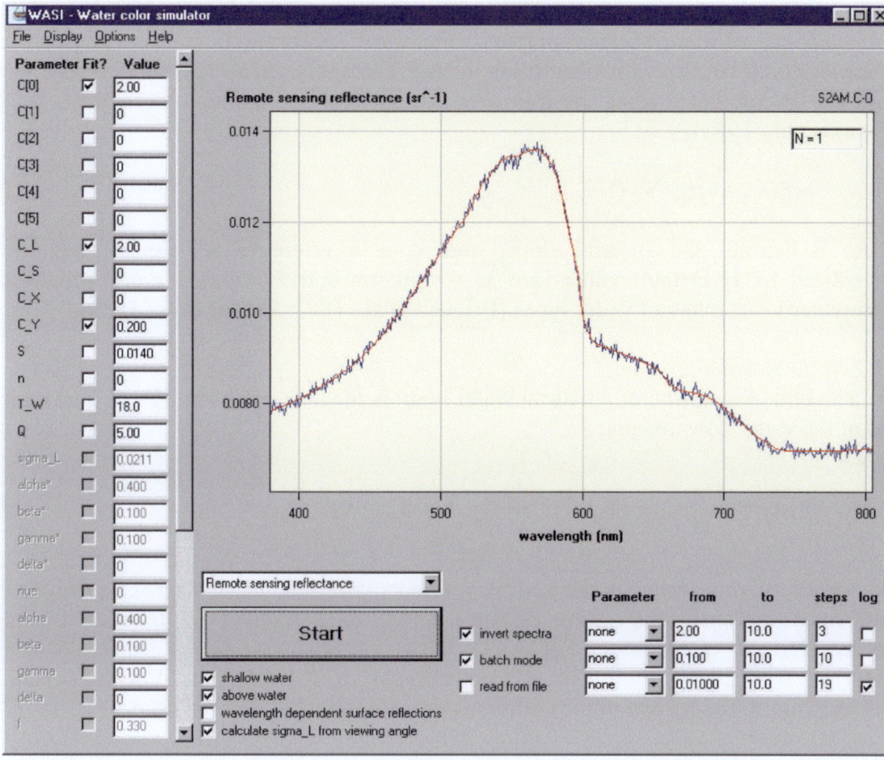

Figure 1. Graphical user interface of WASI in the inverse mode.

Phytoplankton. The high number of species that occur in natural waters causes variability in phytoplankton absorption properties. This is accounted for by the inclusion of 6 specific absorption spectra, $a_i^*(\lambda)$. If no phytoplankton classification is performed, the spectrum $a_0^*(\lambda)$ is selected to represent the specific absorption of phytoplankton. C_i indicates pigment concentration, where "pigment" is the sum of chlorophyll *a* and phaeophytin *a*.

The default spectra provided with WASI are based on measurements at Lake Constance. The five spectra $a_1^*(\lambda)\ldots a_5^*(\lambda)$ represent the lake's major optical classes "Cryptophyta type L", "Cryptophyta type H", "diatoms", "dinoflagellates", and "green algae" (Gege, 1994, 1995, 1998b). The spectrum $a_0^*(\lambda)$ is a weighted sum of these five spectra and represents a mixture which can be considered as typical for that lake. This was calculated by Heege (2000) using in-situ spectra of phytoplankton absorption which were derived from reflectance measurements (Gege, 1994, 1995) and pigment data from 32 days in 1990 and 1991. The spectrum $a_0^*(\lambda)$ was validated (by Heege) using 139 irradiance reflectance and 278 attenuation measurements from 1990 to 1996.

Non-chlorophyllous particles. Absorption is calculated as the product of concentration X and specific absorption $a_X^*(\lambda)$. The spectrum $a_X^*(\lambda)$ provided with WASI is taken from Prieur and Sathyendranath (1981). It is normalized to 1 at a reference wavelength λ_o of 440 nm.

Gelbstoff (dissolved organic matter). Gelbstoff absorption is the product of concentration Y and specific absorption $a_Y^*(\lambda)$. The spectrum $a_Y^*(\lambda)$ can either be read from file or calculated using the following exponential approximation (Nyquist, 1979; Bricaud et al., 1981):

$$a_Y^*(\lambda) = \exp[-S \cdot (\lambda - \lambda_o)], \qquad (2)$$

where S denotes the spectral slope, and λ_o is a reference wavelength with a_Y^* normalized to 1. Default values are λ_o = 440 nm and S = 0.014 nm^{-1}, which are representative of many of water types (Bricaud et al., 1981; Carder et al., 1989).

2.1.2 Natural water

The bulk absorption of a natural water body is the sum of absorption of pure water and of the water constituents:

$$a(\lambda) = a_W(\lambda) + (T - T_0) \cdot \frac{da_W(\lambda)}{dT} + a_{WC}(\lambda). \qquad (3)$$

Absorption of pure water is defined by a temperature-independent term a_W, which is valid for a reference temperature T_0, and a temperature gradient da_W/dT with T being the actual water temperature. For $a_W(\lambda)$, the spectrum measured by Buiteveld et al. (1994) at T_0 = 20°C is used for a spectral range of 391–787 nm. For $da_W(\lambda)/dT$ a spectrum measured by one of the authors (Gege, unpublished data) is used.

2.2 BACKSCATTERING

Backscattering (b_b) of a water body is the sum of backscattering by pure water (W) and suspended matter. For the latter, a distinction between large (≥5 µm, L) and small (<5 µm, S) particles is made. Thus, the following parameterization is chosen:

$$b_b(\lambda) = b_{b,W}(\lambda) + C_L \cdot b_{b,L}^* \cdot b_L(\lambda) + C_S \cdot b_{b,S}^* \cdot (\lambda/\lambda_S)^n. \qquad (4)$$

2.2.1 Pure water

For pure water, the empirical relation of Morel (1974) is used: $b_{b,W}(\lambda) = b_1 \cdot (\lambda/\lambda_1)^{-4.32}$. The specific backscattering coefficient, b_1, depends on salinity. It is b_1 = 0.00111 m^{-1} for fresh water and b_1 = 0.00144 m^{-1} for oceanic water with a salinity of 35–38 ‰, with λ_1 = 500 nm as the reference wavelength.

2.2.2 Large particles

Backscattering by large particles is calculated as the product of concentration C_L, specific backscattering coefficient $b_{b,L}*$, and normalized scattering function $b_L(\lambda)$. The user has several options for calculation:

- C_L can be treated either as an independent parameter, or $C_L = C_0$ can be assigned, where C_0 is the concentration of phytoplankton class no. 0 (see eq. 1). The latter is useful for Case 1 water types where the concentrations of particles and phytoplankton are highly correlated.

- $b_{b,L}*$ can be treated either as a constant with a default value of 0.0086 m^2 g^{-1} (Heege 2000), or as $b_{b,L}* = A \cdot C_L^B$. Such a non-linear dependency of scattering on concentration was observed for phytoplankton (Morel, 1980). It may be used for Case 1 water types, while $b_{b,L}*$ = constant is appropriate for Case 2 waters with significant sources of non-phytoplankton suspended matter. Typical values of the empirical constants are A = 0.0006 m^2 g^{-1} and B = –0.37 (Sathyendranath et al., 1989).

- $b_L(\lambda)$ can either be read from file, or it can be calculated as $b_L(\lambda) = a_0*(\lambda_L) / a_0*(\lambda)$, where $a_0*(\lambda)$ is the specific absorption spectrum of phytoplankton class no. 0 (see eq. 1), and λ_L denotes a reference wavelength. This method assumes that backscattering by large particles originates mainly from phytoplankton cells, and couples absorption and scattering according to the Case 1 waters model of Sathyendranath et al. (1989). However, such coupling may be used in exceptional cases only, since living algae have a negligible influence on the backscattering process by oceanic waters (Ahn et al., 1992), and in Case 2 waters particle scattering is weakly related to phytoplankton absorption in general. In WASI, $b_L(\lambda) = 1$ is set as default.

2.2.3 Small particles

Backscattering by small particles is calculated as the product of concentration C_S, specific backscattering coefficient $b_{b,S}*$, and a normalized scattering function $(\lambda/\lambda_S)^n$. The exponent n, which determines the spectral shape, depends on particle size distribution. The variable "n" is typically about –1 (Sathyendranath et al., 1989) and $b_{b,S}*$ is about 0.005 m^2 g^{-1} for λ_S = 500 nm.

2.3 ATTENUATION

The diffuse attenuation coefficient of irradiance E is defined as K = – (1/E) dE/dz, where z is the depth. Similarly, the attenuation coefficient of radiance L is defined as k = – (1/L) dL/dz. Attenuation is an apparent optical property (AOP) and depends not only on the properties of the medium, but additionally on the geometric distribution of the illuminating light field.

2.3.1 Diffuse attenuation for downwelling irradiance

The most important attenuation coefficient is K_d, which describes the extinction of downwelling irradiance E_d^-. The following parameterization is adapted from Gordon (1989), which largely eliminates the light field effect near the surface:

$$K_d(\lambda) = \kappa_0 \cdot \frac{a(\lambda)+b_b(\lambda)}{\cos\theta'_{sun}}. \quad (5)$$

$a(\lambda)$ is calculated according to eq. (3), $b_b(\lambda)$ using eq. (4). θ'_{sun} is the sun zenith angle in water. The coefficient κ_0 depends on the scattering phase function. Gordon (1989) determined a value of $\kappa_0 = 1.0395$ from Monte Carlo simulations in Case 1 waters, Albert and Mobley (2003) found a value of $\kappa_0 = 1.0546$ from simulations in Case 2 waters using the radiative transfer program Hydrolight (Mobley et al., 1993). Some authors use eq. (5) with $\kappa_0 = 1$ (Sathyendranath and Platt, 1988, 1997; Gordon et al., 1975). In WASI, κ_0 is read from the WASI.INI file; the default value is 1.0546.

2.3.2 Diffuse attenuation for upwelling irradiance

For upwelling irradiance two attenuation coefficients are used: K_{uW} for the radiation backscattered in the water, and K_{uB} for the radiation reflected from the bottom. The following parameterization is adopted from Albert and Mobley (2003):

$$K_{uW}(\lambda) = [a(\lambda)+b_b(\lambda)] \cdot [1+\omega_b(\lambda)]^{1.9991} \cdot \left[1+\frac{0.2995}{\cos\theta'_{sun}}\right]. \quad (6)$$

$$K_{uB}(\lambda) = [a(\lambda)+b_b(\lambda)] \cdot [1+\omega_b(\lambda)]^{1.2441} \cdot \left[1+\frac{0.5182}{\cos\theta'_{sun}}\right]. \quad (7)$$

The function $\omega_b(\lambda)$ depends on absorption $a(\lambda)$ and backscattering $b_b(\lambda)$ of the water body:

$$\omega_b(\lambda) = \frac{b_b(\lambda)}{a(\lambda)+b_b(\lambda)}. \quad (8)$$

Eqs. (6) and (7) are used in the model of irradiance reflectance in shallow waters.

2.3.3 Attenuation for upwelling radiance

For upwelling radiance two attenuation coefficients are used: k_{uW} for the radiation backscattered in the water, and k_{uB} for the radiation reflected from the bottom. The following parameterization is adopted from Albert and Mobley (2003):

$$k_{uW}(\lambda) = \frac{a(\lambda)+b_b(\lambda)}{\cos\theta'_v} \cdot [1+\omega_b(\lambda)]^{3.5421} \cdot \left[1-\frac{0.2786}{\cos\theta'_{sun}}\right], \quad (9)$$

$$k_{uB}(\lambda) = \frac{a(\lambda)+b_b(\lambda)}{\cos\theta'_v} \cdot [1+\omega_b(\lambda)]^{2.2658} \cdot \left[1+\frac{0.0577}{\cos\theta'_{sun}}\right], \quad (10)$$

where θ'_v is the viewing angle in water measured from the nadir direction. These equations are used in the model of remote sensing reflectance in shallow waters.

2.4 SPECULAR REFLECTANCE

An above-water radiance sensor looking down to the water surface measures the sum of two radiance components: one from the water body and one from the surface. The first comprises the desired information about the water constituents, the second is an unwanted add-on which requires correction. However, correction is difficult. For example, the method from the SeaWiFS protocols (Mueller and Austin, 1995), which is widely used in optical oceanography, leads to rms errors of the corrected water leaving radiance as large as 90% under typical field conditions (Toole et al., 2000). Thus, WASI offers different methods.

The radiance reflected from the surface, $L_r(\lambda)$, is a fraction σ_L of sky radiance $L_s(\lambda)$:

$$L_r(\lambda) = \sigma_L \cdot L_s(\lambda). \tag{11}$$

$L_s(\lambda)$ is the average radiance of that area of the sky that is specularly reflected into the sensor. It can be imported from file or calculated using eq. (25). σ_L is the Fresnel reflectance and depends on the angle of reflection. The value can either be specified by the user or it can be calculated from the viewing angle θ_v using the Fresnel equation for unpolarized light (Jerlov 1976):

$$\sigma_L = \frac{1}{2}\left|\frac{\sin^2(\theta_v - \theta'_v)}{\sin^2(\theta_v + \theta'_v)} + \frac{\tan^2(\theta_v - \theta'_v)}{\tan^2(\theta_v + \theta'_v)}\right|. \tag{12}$$

θ'_v is the angle of refraction, which is related to θ_v by Snell's law $n_W\ sin\theta'_v = sin\theta_v$, where $n_W \approx 1.33$ is the refractive index of water. For viewing angles near nadir, $\sigma_L \approx 0.02$.

The ratio of the radiance reflected from the water surface to the downwelling irradiance,

$$R_{rs}^{surf}(\lambda) = \frac{L_r(\lambda)}{E_d(\lambda)} = \sigma_L \cdot \frac{L_s(\lambda)}{E_d(\lambda)}, \tag{13a}$$

is called specular reflectance. $E_d(\lambda)$ and $L_s(\lambda)$ can either be imported from file, or one or both can be calculated using eq. (23) or (25). If the wavelength-independent model of surface reflection is chosen, it is

$$R_{rs}^{surf} = \frac{\sigma_L}{\pi}. \tag{13b}$$

Toole et al. (2000) showed that $R_{rs}^{surf}(\lambda)$ is nearly spectrally flat for an overcast day, but is not flat for clear-sky conditions. Thus, eq. (13a) should be used in general, and eq. (13b) only for overcast days.

2.5 IRRADIANCE REFLECTANCE

2.5.1 Deep water

The ratio of upwelling irradiance to downwelling irradiance in water, $R(\lambda) = E_u(\lambda)$ / $E_d(\lambda)$, is called irradiance reflectance (Mobley 1994). It is an AOP and depends not only on the properties of the medium, but also on the geometric distribution of the incoming light. A suitable parameterization which separates to a large extent the parameters of water and the illumination conditions was found by Gordon et al. (1975):

$$R(\lambda) = f \cdot \omega_b(\lambda). \tag{14}$$

The function $\omega_b(\lambda)$, which is given by eq. (8), depends only on inherent optical properties of the water body, absorption and backscattering. The factor f comprises the illumination dependencies. It can be treated either as an independent parameter with a default value of 0.33 according to Gordon et al. (1975), or the relationship of Albert and Mobley (2003) can be used:

$$f = 0.1034 \cdot \left(1 + 3.3586 \cdot \omega_b - 6.5358 \cdot \omega_b^2 + 4.6638 \cdot \omega_b^3\right) \cdot \left(1 + \frac{2.4121}{\cos \theta'_{sun}}\right). \tag{15}$$

θ'_{sun} is the sun zenith angle in water. Eq. (15) takes into consideration the fact that f depends not only on the geometric structure of the light field, expressed by the parameter θ'_{sun}, but also on the absorption and scattering properties of the water, which are included in ω_b. The weak dependence of f on the wind speed is neglected. Some alternate models of f are also included in WASI and can be used if desired, namely those of Kirk (1984), Morel and Gentili (1991), and Sathyendranath and Platt (1997).

Independently from Gordon, Prieur (1976) found the relation $R(\lambda) = f' \cdot b_b(\lambda)$ / $a(\lambda)$. It is also included in WASI. However, the Gordon algorithm (14) is favoured and set as the default, because it restricts the ω_b values to the physically reasonable range from 0 to 1, which is not the case for the Prieur equation.

2.5.2 Shallow water

For shallow water, the parameterization found by Albert and Mobley (2003) is used:

$$\begin{aligned} R^{sh}(\lambda) = {} & R(\lambda) \cdot \left[1 - 1.0546 \cdot \exp\{-(K_d(\lambda) + K_{uW}(\lambda)) \cdot z_B\}\right] \\ & + 0.9755 \cdot R^b(\lambda) \cdot \exp\{-(K_d(\lambda) + K_{uB}(\lambda)) \cdot z_B\} \end{aligned} \tag{16}$$

The first term on the right-hand side is the reflectance of a water layer of thickness z_B, and the second term is the contribution of the bottom. Bottom reflectance $R^b(\lambda)$ is calculated using eq. (21). The K's account for attenuation within the water layer and are calculated using eqs. (5), (6), and (7).

2.6 REMOTE SENSING REFLECTANCE

2.6.1 Deep water

The ratio of upwelling radiance to downwelling irradiance, $R_{rs}(\lambda) = L_u(\lambda) / E_d(\lambda)$, is called remote sensing reflectance (Mobley, 1994). It is an AOP and can be parameterized the same as $R(\lambda)$ (Albert and Mobley, 2003):

$$R_{rs}^-(\lambda) = f_{rs} \cdot \omega_b(\lambda). \tag{17}$$

Alternately, R_{rs}^- can be calculated as $R_{rs}^-(\lambda) = R(\lambda) / Q$, where $R(\lambda)$ is either calculated using eq. (14) or imported from file, and $Q \equiv E_u / L_u$ is treated as a parameter with a default value of 5 sr. A parameterization of the factor f_{rs}, which can be applied to both deep and shallow waters, was found by Albert and Mobley (2003):

$$f_{rs} = 0.0512 \cdot \left(1 + 4.6659 \cdot \omega_b - 7.8387 \cdot \omega_b^2 + 5.4571 \cdot \omega_b^3\right) \cdot \left(1 + \frac{0.4021}{\cos\theta'_v}\right) \cdot \left(1 + \frac{0.1098}{\cos\theta'_{sun}}\right). \tag{18}$$

Parameters of f_{rs} are ω_b of eq. (8), the sun zenith angle in water, θ'_{sun}, and the viewing angle in water, θ'_v. Alternately, f_{rs} can be calculated as $f_{rs} = f / Q$.

2.6.2 Shallow water

For shallow water, the following parameterization is chosen (Albert and Mobley, 2003):

$$\begin{aligned} R_{rs}^{sh-}(\lambda) = {} & R_{rs}^-(\lambda) \cdot \left[1 - 1.1576 \cdot \exp\{-(K_d(\lambda) + k_{uW}(\lambda)) \cdot z_B\}\right] \\ & + 1.0389 \cdot R_{rs}^b(\lambda) \cdot \exp\{-(K_d(\lambda) + k_{uB}(\lambda)) \cdot z_B\} \end{aligned} \tag{19}$$

The first term on the right-hand side is the reflectance of a water layer of thickness z_B, the second term the contribution of the bottom. Bottom reflectance $R^b_{rs}(\lambda)$ is calculated using eq. (22). K_d, k_{uW} and k_{uB} account for attenuation within the water layer and are calculated using eqs. (5), (9), and (10), respectively.

2.6.3 Above the surface

Above the surface, the user can select one of the following parameterizations:

$$R_{rs}(\lambda) = \frac{(1-\sigma)(1-\sigma_L^-)}{n_w^2} \cdot \frac{R_{rs}^-(\lambda)}{1 - \sigma^- \cdot Q \cdot R_{rs}^-(\lambda)} + R_{rs}^{surf}(\lambda), \tag{20a}$$

$$R_{rs}(\lambda) = \frac{(1-\sigma)(1-\sigma_L^-)}{n_w^2 \cdot Q} \cdot \frac{R(\lambda)}{1 - \sigma^- \cdot R(\lambda)} + R_{rs}^{surf}(\lambda), \tag{20b}$$

$$R_{rs}(\lambda) = \frac{(1-\sigma)(1-\sigma_L^-)}{n_w^2} \cdot \frac{R_{rs}^-(\lambda)}{1 - \sigma^- \cdot R(\lambda)} + R_{rs}^{surf}(\lambda). \tag{20c}$$

The three equations are formally identical. A derivation is given in Mobley (1994). The first term on the right-hand side of each equation describes reflection in the water, the second at the surface. Frequently, the first term alone is called remote sensing reflectance (e.g. Mobley, 1994). In WASI, the reflection at the surface is also included in the R_{rs} definition. It is calculated using eq. (13a) or (13b) and can easily be excluded by setting the reflection factor σ_L equal to zero.

$R_{rs}^-(\lambda)$ is calculated using eq. (17) or (19), $R(\lambda)$ using eq. (14) or (16). The factors σ, σ_L^-, and σ^- are the reflection factors for E_d, L_u^-, and E_u^-, respectively. σ depends on the radiance distribution and on surface waves. Typical values are 0.02 to 0.03 for clear sky conditions and solar zenith angles below 45°, and 0.05 to 0.07 for overcast skies (Jerlov, 1976; Preisendorfer and Mobley, 1985, 1986). The default value is $\sigma = 0.03$. σ_L^- can either be calculated as a function of θ_v using eq. (12), or a constant value can be inserted. σ^- is in the range of 0.50 to 0.57 with a value of 0.54 considered typical (Jerome et al., 1990; Mobley, 1999). The defaults of the other constants are set to Q = 5 sr and $n_W = 1.33$.

Selection of the equation to use depends on the application:

- Eq. (20a) links remote sensing reflectance in water to that in air. Since the same spectrum type is used above and below the water surface, it is the most convenient parameterisation. This equation is used by default.
- Eq. (20b) is useful when $R_{rs}(\lambda)$ is linked to $R(\lambda)$, for example if *in situ* measurements of $R(\lambda)$ were performed as "ground truth" for a remote sensing instrument.
- Eq. (20c) avoids the use of the factor Q, which is difficult to assess. The equation is useful, for example, for optical closure experiments which investigate the consistency of measurements above and below the water surface by measuring simultaneously the spectra $R_{rs}(\lambda)$, $R(\lambda)$, and $R_{rs}^-(\lambda)$.

2.7 BOTTOM REFLECTANCE

The irradiance reflectance of a surface is called albedo. When N different surfaces of albedo $a_n(\lambda)$ are viewed simultaneously, the measured albedo is the following sum:

$$R^b(\lambda) = \sum_{n=0}^{N-1} f_n \cdot a_n(\lambda), \tag{21}$$

where f_n is the areal fraction of surface number n within the sensor's field of view; it is $\Sigma f_n = 1$. This equation is implemented in WASI for N = 6 bottom types. Three of the spectra $a_n(\lambda)$ provided with WASI represent bare bottom, the other green makrophytes: 0 = a constant reflectance of 10%, 1 = sand, 2 = silt, 3 = Chara *aspera*, 4 = Potamogeton *perfoliatus*, 5 = Potamogeton *pectinatus*. All spectra were measured by Pinnel (2005). The sand spectrum is from a coastal shallow area in South Australia (Bolivar), the other spectra were measured at German lakes (Lake Constance and Starnberger See).

When the upwelling radiation is measured by a radiance sensor, the corresponding remote sensing reflectance can be expressed as follows:

$$R_{rs}^{b}(\lambda) = \sum_{n=0}^{N-1} f_n \cdot B_n \cdot a_n(\lambda)\,, \tag{22}$$

where B_n is the proportion of radiation which is reflected towards the sensor. In WASI, the B_n's of all surfaces are assumed to be angle-independent. The default values are set to $B_n = 1/\pi = 0.318\ sr^{-1}$, which represents isotropic reflection (Lambertian surfaces).

2.8 DOWNWELLING IRRADIANCE

2.8.1 Above the water surface

An analytic model of the downwelling irradiance spectrum $E_d(\lambda)$, using only a few parameters, was developed by Gege (1994, 1995). It fits to measured spectra with a high degree of accuracy (average rms error of 0.1%). The radiation illuminating the water surface is parameterized as the sum of four spectrally different components: (1) the direct solar radiation; (2) the blue sky (Rayleigh) scattering; (3) radiation scattered by aerosols (Mie scattering); and (4) clouds. Each component is expressed in terms of a wavelength-dependent fraction of the extraterrestrial solar irradiance $E_0(\lambda)$:

$$E_d(\lambda) = [\ \alpha \cdot t_A(\lambda) + \beta \cdot (\lambda/\lambda_R)^{-4.09} + \gamma \cdot (\lambda/\lambda_M)^{v} + \delta \cdot t_C(\lambda)\ ] \cdot E_0(\lambda). \tag{23}$$

The four functions $t_i(\lambda) = \{t_A(\lambda), (\lambda/\lambda_R)^{-4.09}, (\lambda/\lambda_M)^{v}, t_C(\lambda)\}$ are transmission spectra which spectrally characterize the four light sources. Their weights α, β, γ, and δ, may change from one measurement to the next, but the $t_i(\lambda)$ functions are assumed to be constant over time.

In order to make the weights α, β, γ, and δ relative intensities of the four light sources, each $t_i(\lambda)$ is normalized as $\int t_i(\lambda)\ E_0(\lambda)\ d\lambda = \int E_0(\lambda)\ d\lambda$ where the default integration interval is 400 to 800 nm. The functions $(\lambda/\lambda_R)^{-4.09}$ and $(\lambda/\lambda_M)^{v}$ are calculated during run-time. Normalization yields their scaling factors: λ_R = 533 nm, and λ_M is typically between 563 nm (v=-1) and 583 nm (v=1). The exponent v parameterizes the wavelength dependency of aerosol scattering. The remaining functions $t_A(\lambda)$ and $t_C(\lambda)$ are read from file. After import they are normalized. The two provided with WASI were determined from measurements at Lake Constance.

2.8.2 Below the water surface

The downwelling irradiance in water, E_d^-, is related to the downwelling irradiance in air, E_d, through $E_d^-(\lambda) = (1-\sigma) \cdot E_d(\lambda) + \sigma^- \cdot E_u^-(\lambda)$. σ is the reflection factor for downwelling irradiance in air, σ^- for upwelling irradiance in water, and E_u^- is the upwelling irradiance in water. Using the irradiance reflectance $R = E_u^- / E_d^-$ yields the following expression:

$$E_d^-(\lambda) = \frac{1-\sigma}{1-\sigma^- \cdot R(\lambda)} \cdot E_d(\lambda)\,. \tag{24}$$

This equation is used in WASI for calculating $E_d^-(\lambda)$. $R(\lambda)$ is calculated using eq. (14). $E_d(\lambda)$ can either be calculated according to eq. (23), or a measured spectrum can be taken. Default values of the reflection factors are $\sigma = 0.03$ and $\sigma^- = 0.54$.

2.9 SKY RADIANCE

The parameterization used for $E_d(\lambda)$ is also implemented for $L_s(\lambda)$:

$$L_s(\lambda) = [\ \alpha^* \cdot t_A(\lambda) + \beta^* \cdot (\lambda/\lambda_R)^{-4.09} + \gamma^* \cdot (\lambda/\lambda_M)^{\nu} + \delta^* \cdot t_C(\lambda)\ ] \cdot E_0(\lambda). \qquad (25)$$

The functions $E_0(\lambda)$, $t_A(\lambda)$, $(\lambda/\lambda_R)^{-4.09}$, $(\lambda/\lambda_M)^{\nu}$, and $t_C(\lambda)$ are those used with eq. (23). Parameters of $L_s(\lambda)$ are the weights α^*, β^*, γ^*, and δ^*, which represent the relative intensities of the four above-mentioned light sources for a radiance sensor, and the exponent ν.

This model of $L_s(\lambda)$ is included for modeling specular reflection at the water surface. Its usefulness has been demonstrated (Gege, 1998b). Capillary waves at the water surface, and moreover gravity waves, increase the sky area that is reflected into the sensor, and change the angle of reflection. Consequently, measurements of $L_s(\lambda)$ are frequently not reliable. For these cases, and if no $L_s(\lambda)$ measurement is available, eq. (25) can be applied. If the user selects the wavelength-independent model of surface reflections, $L_s(\lambda) = E_d(\lambda)/\pi$ is utilized.

2.10 UPWELLING RADIANCE

The upwelling radiance is that part of the downwelling irradiance which is reflected back from the water into a down-looking radiance sensor. Calculation is based on a model of R_{rs} and a model or a measurement of E_d.

In water, eq. (24) is used for calculating $E_d^-(\lambda)$, and eq. (17) or (19) for $R_{rs}^-(\lambda)$. The upwelling radiance is then calculated as follows:

$$L_u^-(\lambda) = R_{rs}^-(\lambda) \cdot E_d^-(\lambda)\ . \qquad (26)$$

In air, the upwelling radiance after crossing the water-air boundary is related to L_u^- as follows:

$$L_u(\lambda) = \frac{1-\sigma_L^-}{n_W^2} \cdot L_u^-(\lambda) + L_r(\lambda)\ . \qquad (27)$$

The first term on the right-hand side is the radiance upwelling in the water, weakened at the interface by Fresnel reflection (factor $1-\sigma_L^-$) and refraction (flux dilution by widening of the solid angle, factor $1/n_w^2$). $L_u^-(\lambda)$ is obtained from eq. (26), $L_r(\lambda)$ from eq. (11). σ_L^- can either be calculated as a function of θ_v using eq. (12), or a constant value can be used. Default values of the constants are $\sigma_L^- = 0.02$ and $n_W = 1.33$.

3. Inverse Modeling

Inverse modeling is the determination of model parameters for a given spectrum. The complete list of model parameters for all spectrum types is given in Table 2. These can be iterated in the forward mode to generate series of spectra, and their values can be determined in the inverse mode. The user defines which parameters are determined during inversion and which are kept constant. The former are called *fit parameters*. The actual number of fit parameters depends on the spectrum type, on model options, and on the user's choice of which parameters to fit and which to fix during inversion.

Table 2. List of WASI parameters.

Symbol	WASI	Units	Description
C_i	C[i]	µg/l	Concentration of phytoplankton class no. i, i = 0..5
C_L	C_L	mg/l	Concentration of large suspended particles
C_S	C_S	mg/l	Concentration of small suspended particles
X	C_X	m^{-1}	Concentration of non-chlorophyllous particles
Y	C_Y	m^{-1}	Concentration of Gelbstoff
S	S	nm^{-1}	Exponent of Gelbstoff absorption
n	n	-	Exponent of backscattering by small particles
T	T_W	°C	Water temperature
f	f	-	Proportionality factor of reflectance ("f-factor")
Q	Q	sr	Anisotropy factor ("Q-factor")
θ_{sun}	sun	°	Sun zenith angle
θ_v	view	°	Viewing angle (0 = nadir)
σ_L	sigma_L	-	Reflection factor of sky radiance
z_B	zB	m	Bottom depth
ν	nue	-	Exponent of aerosol scattering
α	alpha	-	Fraction of irradiance due to direct solar radiation
β	beta	-	Fraction of irradiance due to molecule scattering
γ	gamma	-	Fraction of irradiance due to aerosol scattering
δ	delta	-	Fraction of irradiance due to cloud scattering
α^*	alpha_s	sr^{-1}	Fraction of radiance due to direct solar radiation
β^*	beta_s	sr^{-1}	Fraction of radiance due to molecule scattering
γ^*	gamma_s	sr^{-1}	Fraction of radiance due to aerosol scattering
δ^*	delta_s	sr^{-1}	Fraction of radiance due to cloud scattering
f_n	fA[n]	-	Areal fraction of bottom surface type no. n, n = 0..5

3.1 IMPLEMENTED METHOD

3.1.1 Curve fitting

The fit parameters are determined iteratively using the method of nonlinear curve fitting. In the first iteration, a model spectrum is calculated using initial values for the fit parameters. This model spectrum is compared with the measured spectrum by calculating the residuum as a measure of correspondence. Then, in the further iterations, the values of the fit parameters are altered, resulting in altered model curves and altered residuals. The procedure is stopped after the best fit between the calculated and measured spectrum is found. The best fit corresponds to the minimum residuum, and these values are the estimates of fit parameters.

3.1.2 Search algorithm

Since an infinite number of possible parameter combinations exists, an effective algorithm of the iteration process is needed to select a new set of parameter values from the previous sets. WASI uses the Simplex algorithm (Nelder and Mead, 1965; Caceci and Cacheris, 1984). It has two advantages compared to other customary algorithms like Newton-Ralphson and Levenberg-Marquardt: it always converges, and computations are rapid since no matrix operations are required.

In the Simplex algorithm, a virtual space of M+1 dimensions is constructed, where M dimensions represent the M fit parameters, and one dimension the residuum. Each model curve corresponds to one point in that space, and the set of all possible model curves (obtained by all combinations of parameter values) forms an M dimensional surface. That point on the surface where the residuum is minimal represents the solution of the fit problem. The Simplex can be compared to a spider which crawls on the surface searching for the minimum. It consists of M+1 legs, where each leg (vertex) represents a model curve that has already been calculated. The decision regarding which set of parameter values is chosen in the next step (i.e. where the Simplex moves to) is made according to a strategy explained using Figure 2.

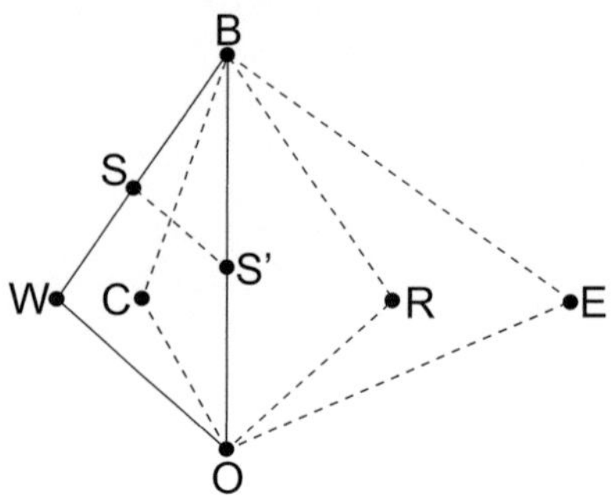

Figure 2. The Simplex and its potential contours in the next step. After Caceci and Cacheris (1984).

The triangle WBO represents the Simplex. W corresponds to the worst residuum, B to the best, and O to all others. Four new positions in the next step are considered: (1) reflection of W at the line OB so that RBO is the new Simplex; (2) contraction towards this line so that CBO is the new Simplex; (3) expansion beyond this line to the point E; (4) shrinkage parallel to the line WO so that SBS' is the new Simplex. Not all of these positions are always calculated: they are tested in this order, and the first position is taken where the new vertex is better than W. Usually the Simplex is trapped in a minimum after less than $20 \cdot M^2$ iterations (Caceci and Cacheris, 1984). However, if the surface contains local minima, the Simplex may be captured in one of these. In such cases it is important to start the search at a point not too far away from the global minimum.

3.1.3 Modes of operation

Three modes of operation are implemented in WASI:

- *Single spectrum mode.* Fitting is performed for a single spectrum. After inversion, an overlay of the spectrum and fitted curve is automatically shown on screen and the resulting fit values, number of iterations, and residuum are displayed. This mode allows the user to inspect the fit results of individual measurements. This mode is useful for optimizing the choice of initial values and the fit strategy before starting a batch job.

- *Batch mode.* A series of spectra from file is fitted. After each inversion, an overlay of the imported spectrum and fitted curve is automatically shown on screen. This mode is useful for processing large data sets.

- *Reconstruction mode.* Combines forward and inverse modes. Inversion is performed for a series of forward calculated spectra which are not necessarily read from file. The model parameters can be chosen differently for forward and inverse calculations. This mode is useful for performing sensitivity studies.

3.2 INVERSION PROBLEMS

3.2.1 Ambiguity

When different sets of model parameters yield similar spectra, the inversion problem is ambiguous. In such a case, no algorithm can reliably find the correct values of the fit parameters. The problem is model specific and increases drastically with the number of fit parameters.

An example is given in Figure 3. Three absorption spectra $a_{WC}(\lambda)$ were calculated using eq. (1) by summing the absorptions of phytoplankton chlorophyll (concentration C_0) and Gelbstoff (concentration Y, spectral slope S). The concentrations C_1, ..., C_5 and X were set to zero. The curves are almost identical from 400 to 600 nm, but have very different parameter values: the parameter set (C_0, Y, S) is (2, 0.2, 0.014) for curve A, (1, 0.232, 0.0124) for curve B, and (4, 0.132, 0.020) for curve C. Thus, although phytoplankton concentration differs by a factor of 4, the three curves can hardly be distinguished between 400 to 600 nm. It is consequently not possible to determine all three parameters C_0, Y, S from measurements in this spectral range, since any inversion method compensates for error with one parameter by using erroneous values for the two other parameters. There are two solutions to this type of problem: 1) at least one of the parameters must be known and kept constant during inversion, 2) the spectral range must be extended to wavelengths above 600 nm.

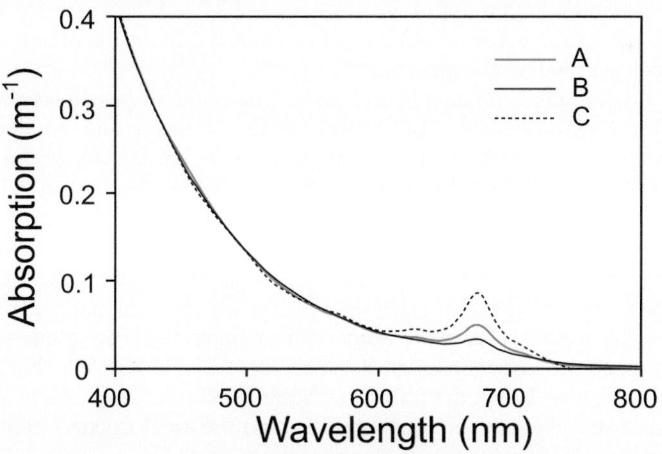

Figure 3. Illustration of the ambiguity problem. Although phytoplankton concentration C_0 differs by a factor of 4, the three spectra are very similar from 400 to 600 nm. The changes caused by C_0 are compensated by changes of the Gelbstoff parameters Y and S.

3.2.2 Failure to converge

An inversion algorithm doesn't always find a minimum. Several conditions can cause such a failure to converge:

- *Initial increments are too small.* If the initial steps of the search algorithm are too small, the differences in the residuals are too small to indicate an improvement for a particular parameter combination compared to others. Depending on the criterion for termination, this may cause premature conclusion, or travelling in the wrong direction in the multidimensional parameter space.

- *Acceptable errors are too large.* If the criterion for terminating inversion is chosen as too weak, the inversion algorithm may stop too early, before the minimum is found.

- *Unsuited initial values.* If the initial values of the fit parameters are too different from the correct values, the search for the minimum may start in a wrong direction. The greater the number of fit parameters, the more difficult it is to find the correct region in the multidimensional parameter space.

3.3 PROBLEM SOLUTIONS OF WASI

3.3.1 Use of pre-knowledge

In WASI, one can make use of expected parameter values. Typical values of all parameters and constants are stored in the file WASI.INI, which is read during start up of the program. The user can change them all. The parameters which are fitted during inversion can be initialised either with these expected values, or with estimates calculated by using analytic approximations (see 3.3.3). When a series of spectra is analyzed, the parameter values may be similar. Thus, the fit results of one measurement can be taken as start values for the next.

The parameter range can also be modified. The range of possible values is known in general for each parameter, and within WASI a lower and an upper value is attributed to each fit parameter. The defaults of these border values are read during program start from the WASI.INI file, and the user can change them. They are used whenever the search algorithm attempts to assign an out-of-range value to a parameter.

When a well-known correlation exists between model parameters, it may be useful to restrict a parameter search to values which depend on the actual values of one or more other parameters. No general scheme of this method (regularization) is implemented in WASI. However, some correlations between parameters can be utilized. For example, suspended matter can be correlated to phytoplankton chlorophyll by setting $C_L = C_0$ (see 2.2.1); the reflection factor for sky radiance can be related to the viewing angle using the Fresnel equation (12); and the areal fraction f_n of one bottom type can be related to the fractions of the other types using $\Sigma f_n = 1$ (see 2.7).

3.3.2 Adjust calculation of the residuum

The residuum, Δ, is the measure of the difference between a measured spectrum and its fitted curve. The inversion procedure's task is to find its minimum. The residuum can be envisaged as a surface in the M+1 dimensional space formed by the M fit parameters and by Δ. The shape of that surface depends on which construction law is used. Thus, the search for the minimum can be optimized by adjusting the construction law to the inversion problem.

Algorithms. The user has the choice between 6 algorithms to calculate Δ:

$$\Delta = \frac{1}{N}\sum_{i=1}^{N} g_i \cdot \left| m_i - f_i \right|^2 \tag{28a}$$

$$\Delta = \frac{1}{N}\sum_{i=1}^{N} g_i \cdot \left| m_i - f_i \right| \tag{28b}$$

$$\Delta = \frac{1}{N}\sum_{i=1}^{N} g_i \cdot \left| 1 - \frac{f_i}{m_i} \right|^2 \tag{28c}$$

$$\Delta = \frac{1}{N}\sum_{i=1}^{N} g_i \cdot \left| \ln(m_i) - \ln(f_i) \right|^2 \tag{28d}$$

$$\Delta = \frac{1}{N}\sum_{i=1}^{N} g_i \cdot \left| \ln(m_i) - \ln(f_i) \right| \tag{28e}$$

$$\Delta = \frac{1}{N}\sum_{i=1}^{N} g_i \cdot \left| 1 - \frac{\ln(f_i)}{\ln(m_i)} \right|^2 \tag{28f}$$

The residuum is a weighted sum over N spectral channels. The subscript i indicates the channel number, m_i denotes the measured value, f_i the fit value, and g_i the weight. Eq. (28a) in combination with $g_i = 1$ is the classical least-squares fit. During inversion the f_i values are changed, but not the m_i and g_i values.

The impact of residuum algorithm selection on the shape of the surface in the parameter space is illustrated in the example of Figure 4. Two contour plots of the residuum are shown for an inversion of absorption spectra, which were calculated using eq. (1) by summing the absorptions of phytoplankton and Gelbstoff. In this example the concentrations $C_1,...,C_5$ and X were set to zero. Phytoplankton chlorophyll concentration C_0 was set to 2 µg/l during forward and inverse calculation. The Gelbstoff parameters were set to $S = 0.014\ nm^{-1}$ and $Y = 0.3\ m^{-1}$ during forward calculation and then were iterated from $0.01\ nm^{-1} \leq S \leq 0.02\ nm^{-1}$ and $0 \leq Y \leq 0.6\ m^{-1}$ during inversion. No fit was performed during inversion; only the residuum was calculated for each parameter combination with equal weights $g_i = 1$.

The major difference between the two plots of Fig. 4 is the orientation of the valley which forms the minimum: the valley is almost parallel to the S axis for the classical least-squares fit (eq. 28a), i.e. the inversion cannot determine S reliably. When the logarithms of the m_i and f_i are taken (eq. 28d), the valley is oriented diagonally in the S-Y-plane, and thus fitting of both S and Y is feasible. For the concentrations chosen, eq. 28d is more appropriate because the absorption spectrum is dominated by the exponential function of Gelbstoff.

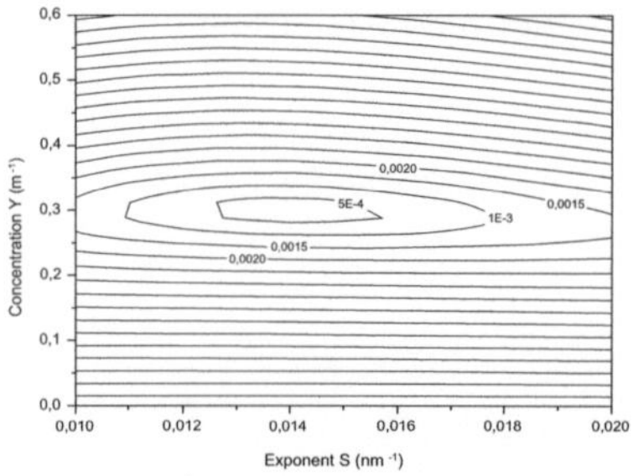

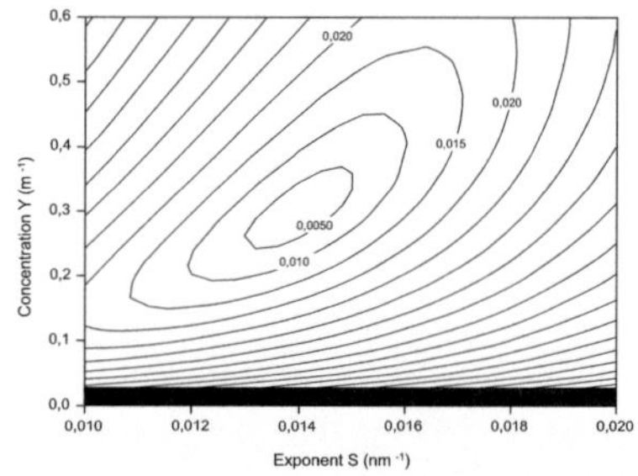

Figure 4. Contour plots of the residuum at inversion of absorption spectra. Both plots correspond to a least-squares fit. Left: Linear weighting of absorption values (eq. 28a). Right: Logarithmic weighting of absorption values (eq. 28d).

Spectral range and data interval. The user can select the channels i which will be taken for residuum calculation by specifying the upper and lower boundaries and channel interval. Modern instruments frequently provide hundreds of spectral channels. It is usually not necessary to use each channel for inversion. Reducing the number of channels reduces calculation time.

Spectral weighting. The channels are weighted individually such that their weights g_i are read from file. The default file is EINS.PRN, which sets all g_i to unity. The selection of different g_i's allows the user to exclude certain spectral regions or to weight the information spectrally. This feature is useful if the measurement or the model is not reliable in certain spectral intervals. For example, since the models do not include chlorophyll fluorescence at 685 nm, it may be useful to exclude channels around 685 nm or give them low weights.

3.3.3 Automatic determination of initial values

Making a good guess for the fit parameters' initial values is the best way to reduce all types of inversion problems. Thus, for operational data analysis it is desirable to have an automatic algorithm which estimates initial values for the most relevant parameters with acceptable errors. Such automatic methods are implemented for R and R_{rs} spectra. The case for R in deep water is described below, while those for R and R_{rs} in shallow waters are explained in Albert (2004) and Albert and Gege (2005).

Example: R in deep water. Irradiance reflectance R(λ) is one the most frequently measured spectrum types in optically deep waters. The most common parameters determined from these measurements are the concentrations of phytoplankton (C_0), Gelbstoff (Y) and large suspended particles (C_L). Gege (2002b) investigated the sensitivity of these fit parameters to errors. The study demonstrated a very small sensitivity for C_L, some sensitivity for Y, but very high sensitivity for C_0. Considering error propagation, the study suggested a two-steps procedure for initial values determination. The procedure has been further optimised, resulting in the four-step procedure summarized in Table 3 and presented below.

Table 3. Four-step procedure for initial values determination of irradiance reflectance spectra in deep water.

Step	Determine	Algorithm	Description
1	C_L, C_S	analytical	Estimate C_L and C_S from an analytic equation at a wavelength in the near infrared.
2	Y, C_0	analytical	Estimate Y and C_0 from analytic equations at two wavelengths; for C_L and C_S the values from step 1 are taken.
3	C_L, C_S, Y	fit	Determine initial values of C_L, C_s and Y by fit. C_0 is kept constant at the value from step 2; C_L, C_S and Y are initialized using the values from steps 1 and 2.
4	C_0, Y, S	fit	Determine initial values of C_0, Y and S by fit. C_L is kept constant at the value from step 3, Y is initialized using the value from step 3, S is initialized by the user-setting.

Step 1. Suspended matter backscattering, $B_0 = b_b - b_{b,W}$, can be calculated analytically from R at any wavelength λ_{IR} for which absorption $a(\lambda_{IR})$ is known. For λ_{IR} a wavelength in the near infrared is chosen since absorption of water constituents is generally very low compared to absorption of pure water at $\lambda > 700$ nm (Babin and Stramski 2002). Ideally, $\lambda_{IR} > 750$ nm should be used, since phytoplankton absorption $a_0^*(\lambda)$ is zero above 750 nm. The equation of determination is obtained from eqs. (8) and (14):

$$B_0 = \frac{a(\lambda_{IR}) \cdot R(\lambda_{IR})}{f - R(\lambda_{IR})} - b_{b,W}(\lambda_{IR}). \quad (29)$$

f is calculated using the selected f model, e.g. eq. (15), with the user-defined initial values as parameter values. If a B_0-dependent f model is selected, B_0 is calculated in two iterations, i.e. the B_0 value from the first iteration is taken to calculate f again, and using this f B_0 is calculated a final time.

Conversion from optical units B_0 to gravimetric concentrations C_L, C_S is based on eq. (4) assuming $b_L(\lambda) = 1$. Accordingly it is $B_0 = C_L \cdot b_{b,L}^* + C_S \cdot b_{b,S}^* \cdot (\lambda/\lambda_S)^n$. If $C_S = 0$, it is $C_L = B_0 / b_{b,L}^*$, otherwise the user-defined ratio $r_{SL} = C_S/C_L$ is retained, and C_L and C_S are calculated as follows:

$$C_L = \frac{B_0}{b_{b,L}^* + r_{SL} \cdot b_{b,S}^* \cdot \left(\frac{\lambda_{IR}}{\lambda_S}\right)^n}; \qquad C_S = r_{SL} \cdot C_L . \quad (30)$$

If $\lambda_{IR} > 750$ nm, the accuracy of the analytically estimated parameters C_L and C_S depends only on λ_{IR}, Gelbstoff absorption at λ_{IR}, and on C_L and C_S themselves. The dependence of the relative C_L error on Y and C_L is shown in Fig. 5 for two values of λ_{IR}, 750 and 870 nm. It was calculated as $100 \cdot (C_L/C_L^{fwd} - 1)$, where C_L^{fwd} is the value of forward calculation and C_L the value obtained from eq. (30). Values of $S = 0.014$ nm^{-1} and $C_S = 0$ were used. As Figure 5 shows, C_L can be determined using eqs. (29) and (30) with an accuracy of about 1%.

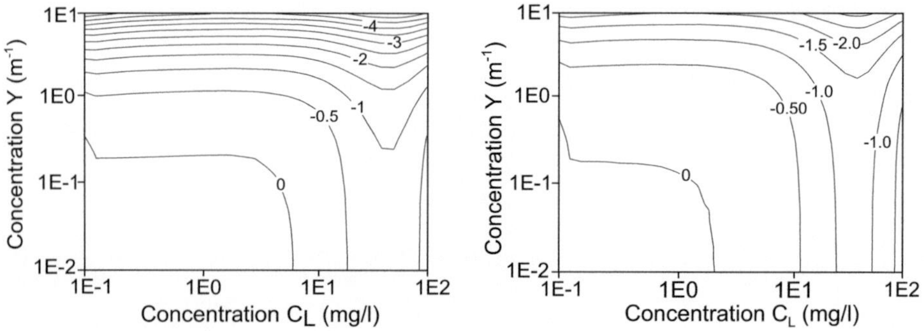

Figure 5. Relative errors in percent for analytic determination of C_L. Left: λ_{IR} = 750 nm, Right: λ_{IR} = 870 nm.

Step 2. If $b_b(\lambda)$ is known with little error, e.g. from step 1, C_0 and Y can be estimated analytically from two wavelengths λ_1, λ_2. The equations of determination are obtained by using the $R(\lambda)$ equation (14) and setting $C_1...C_5 = 0$, $T = T_0$, $X = 0$. This eq. (14') is solved for the sum $Y \cdot a_Y^*(\lambda) + C_0 \cdot a_0^*(\lambda)$, and the ratio R_A is taken for two wavelengths:

$$R_A := \frac{Y \cdot a_Y^*(\lambda_1) + C_0 \cdot a_0^*(\lambda_1)}{Y \cdot a_Y^*(\lambda_2) + C_0 \cdot a_0^*(\lambda_2)} = \frac{f \cdot \dfrac{b_b(\lambda_1)}{R(\lambda_1)} - a_W(\lambda_1) - b_b(\lambda_1)}{f \cdot \dfrac{b_b(\lambda_2)}{R(\lambda_2)} - a_W(\lambda_2) - b_b(\lambda_2)}. \tag{31}$$

Since all functions on the right-hand side are known, R_A can be calculated. Division of the nominator and denominator of the center expression by C_0 leads to an equation which has a single unknown parameter, the ratio Y/C_0. Rewriting this equation yields the following expression:

$$\frac{Y}{C_0} = \frac{R_A \cdot a_0^*(\lambda_2) - a_0^*(\lambda_1)}{a_Y^*(\lambda_1) - R_A \cdot a_Y^*(\lambda_2)}. \tag{32}$$

The ratio Y/C_0 is calculated using this equation. By inserting $Y = (Y/C_0) \cdot C_0$ into eq. (14') and solving it for C_0 at a wavelength λ_3, the following expression is obtained:

$$C_0 = \frac{f \cdot \dfrac{b_b(\lambda_3)}{R(\lambda_3)} - a_W(\lambda_3) - b_b(\lambda_3)}{a_0^*(\lambda_3) + \dfrac{Y}{C_0} \cdot a_Y^*(\lambda_3)}. \tag{33}$$

This equation is used to calculate C_0. Y is then calculated as $Y = (Y/C_0) \cdot C_0$.

The accuracy of the analytically estimated parameters C_0 and Y depends on λ_1, λ_2, λ_3, C_0, Y, C_L, C_S, and on the errors of C_L and C_S, as determined from step 1. Simulations were performed to optimize the choice of the wavelengths λ_1, λ_2, λ_3. These suggest: $\lambda_1 < 470$ nm, $\lambda_2 < 500$ nm, $\lambda_3 < 550$ nm. In each case, preference should be given to shorter wavelengths. A good choice is $\lambda_2 = \lambda_0$ since S errors don't affect Gelbstoff absorption at λ_0. For λ_3 no separate wavelength must be chosen, and it can be set as $\lambda_3 = \lambda_2$. Consequently, selection of only two wavelengths is implemented in WASI. Their default values are: $\lambda_1 = 413$ nm, $\lambda_2 = 440$ nm.

In order to illustrate the parameter dependencies and magnitudes of the C_0 and Y errors, Figure 6 presents two examples of these errors. R(λ) spectra were calculated using eq. (14) with $\omega_b(\lambda)$ from eq. (8) and f from eq. (15). All non-iterated parameters were set equal during forward and inverse calculation: $C_0 = 2$ µg/l, $C_1...C_5 = 0$, $X = 0$, $Y = 0.2$ m^{-1}, $C_S = 0$, $S = 0.014$ nm^{-1}, $\theta_{sun} = 30°$, $T = 18°C$. C_L was iterated from 0.1 to 100 mg/l, and for Figure 6A Y was iterated from 0.01 to 10 m^{-1}, for Figure 6B C_0 was iterated from 0.1 to 100 µg/l. All iterations were done in 51 steps such that the logarithmic values were equidistant. Thus, both plots in Figure 6 consist of $51 \cdot 51 = 2601$ data points, where each point represents the absolute error $100 \cdot |c/c^{fw} - 1|$ of the concentration c, Y for Figure 6A and C_0 for Figure 6B. c^{fw} is the concentration at forward calculation, and c is the retrieved value. Figure 6 indicates that Y can be determined analytically with a typical accuracy of < 30 %, and C_0 of about 30–100 %. This is sufficient for initial values.

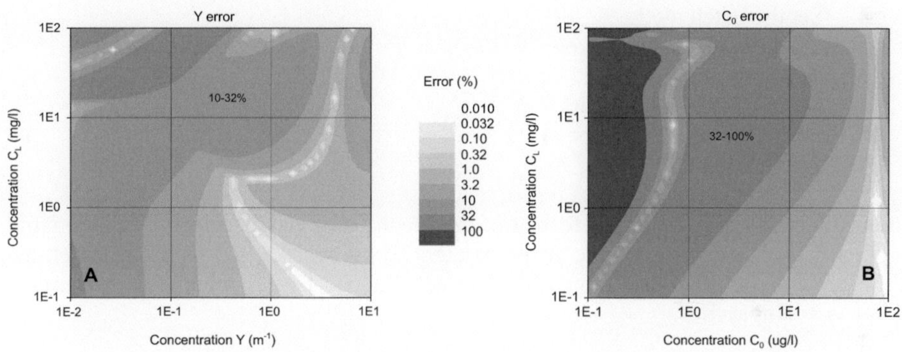

Figure 6. Relative errors of analytic determination of (A) Y error and (B) C_0 error.

Steps 3 and 4. These steps were suggested in the study mentioned above (Gege, 2002b). Newly developed Steps 1 and 2 make them unnecessary in most cases. However, they may be useful under certain conditions: if no suitable infrared channel is available for accurate determination of C_L or C_S, or if S is a fit parameter. Steps 3 and 4 improve the estimates for C_0, C_L, C_S and Y by including additional spectral information, and a start value of S can be determined.

3.3.4 Initialize Simplex

The search algorithm's dynamic memory, the Simplex, is a set of M+1 vectors. Each vector (or vertex) contains the actual values of the M fit parameters and the corresponding residuum. When the fit routine is started, the M+1 vertices are initialized. The fit parameters' initial values and the corresponding residuum form one vertex and the other M vertices are calculated using incremental changes of the initial values. These increments are set to 0.1 x the initial values. They cannot be changed by the user.

3.3.5 Terminate search

The fit is stopped when either the termination criterion is fulfilled or the maximum number of iterations is reached. The termination criterion is such that the differences between the actual parameter values must be less than a threshold for each parameter. Each parameter has its specific threshold, which is set to 10^{-5} times the initial value. It cannot be changed by the user. The user defines the maximum number of iterations, which should be set high enough that a forced stop is exceptional.

4. Applications

4.1 DATA ANALYSIS

Automatic data analysis of a series of spectra is performed in the batch mode within WASI. The spectra must be in ASCII format. The following user actions are required:

1. Copy all spectra into the same directory
2. Specify the directories of input and output data
3. Specify file names and file format of the input data
4. Set the batch mode
5. Specify the spectrum type
6. Tune the fit procedure
7. Set the model parameters
8. Select the fit parameters
9. Start calculation

The results are stored in ASCII format in the specified output directory. A single table, FITPARS.TXT, summarizes the fit parameters of all spectra. All calculation settings are documented in an actual copy of the file WASI.INI. The fit curves of all spectra can be saved as single files.

4.2 ERROR ANALYSIS

No measurement is perfect, no model is exact, and no input data set is complete. Thus the results of data analysis are unavoidably affected by errors. Sensitivity studies are designed to estimate the errors for a determined parameter caused by the different error sources. The best way to perform a sensitivity study is to simulate a large number of measurements using a reliable model, and to then analyze these subsequently as if they were real measurements. Well-defined discrepancies between the forward and inverse models can be introduced which cause errors in the retrieved parameters. In this way, parameter errors can be attributed quantitatively to different error sources. The advantages of using simulated spectra rather than measured spectra are that all studied effects are under control, and that the entire expected parameter interval can be covered without a gap.

4.2.1 Errors from the sensor

The number of parameters that can be derived from a measurement and the accuracy of the estimates depends very much on the sensor and on data quality. The following instrument characteristics are relevant: number, center wavelengths and spectral resolutions of spectral channels; radiometric resolution; accuracy of radiometric and spectral calibration; noise; and drift.

For estimating the limits of accuracy caused by the sensor, spectra of the given sensor are simulated for different combinations of model parameters, and spectra of a "perfect" sensor are calculated for the same combinations. Both sets of spectra are inverted, and the accuracies of the fit parameters are compared.

Sensor characteristics which are readily defined and changed easily using WASI are number and center wavelengths of the channels, radiometric resolution, and statistical noise. For studying the influence of spectral resolution and calibration accuracy, WASI can be used to calculate spectra which are sensor-adjusted with respect to the number and center wavelengths of the channels, radiometric resolution, and statistical noise. These spectra are then modified according to the instrument characteristics: the spectra are smoothed if the instrument has a lower spectral resolution than the input data, error spectra are added and/or multiplied to the spectra if radiometric calibration errors are investigated, and the wavelength values are changed if spectral calibration errors are analyzed. This is done using separate software, such as a spread sheet program. Finally, the sensor-adjusted spectra are inverted using WASI.

An example is given in Figure 7. Absorption spectra of water constituents were simulated for 4 sensors which differ in the number of spectral channels and in the noise level: spectra $a_{WC}(\lambda)$ were calculated using eq. (1) from 400 to 800 nm for wavelength intervals of 2 and 20 nm and statistical noise of 0.002 and 0.02 m^{-1} standard deviation. The Gelbstoff parameters chosen were $Y = 0.2\ m^{-1}$ and $S = 0.014\ nm^{-1}$. Concentrations $C_1, \ldots, C_5$ and X were set to zero. Phytoplankton chlorophyll concentration C_0 was changed from 0.1 to 100 µg/l in 51 steps. 20 spectra $a_{WC}(\lambda)$ were calculated for each C_0 value by applying a simple trick: a parameter not used in the actual model was iterated during forward calculation. Each of the $51 \cdot 20 = 1020$ spectra for each sensor was inverted with C_0 and Y as fit parameters. Figure 7 compares the C_0-dependency of the relative C_0 errors for the 4 sensor specifications. C_0 errors are more sensitive to noise than to the number of channels in this example.

4.2.2 Errors from the model

The radiative transfer equation for an absorbing and scattering medium like water cannot be solved analytically, hence observations which depend on the radiation field (AOPs) can only be approximated. WASI uses analytic approximations based on parameters which can be measured with relative ease. Advantages are that inversion is relatively simple, altered input data sets are included quickly, and computing is fast.

From a numerical point of view, any desired accuracy can be achieved by using converging methods such as Monte Carlo, invariant imbedding, matrix operator, successive order of scattering, finite elements, etc. However, high accuracy is at the expense of computing time, which, for these methods, is by far too long for inverting a set of spectra. In order to estimate errors introduced by the approximations of WASI, a set of spectra must be calculated using a numerically exact program, and for the same conditions a second set of spectra is calculated using WASI. Both sets of spectra are inverted, and the differences of the fit parameters reflect the errors introduced by the simplified model.

An example is given in Figure 8. Remote sensing reflectance spectra of shallow water were calculated using both HYDROLIGHT (Mobley et al., 1993, Mobley, 1994) and WASI (eq. 19). All input data and parameters for the two models were identical.

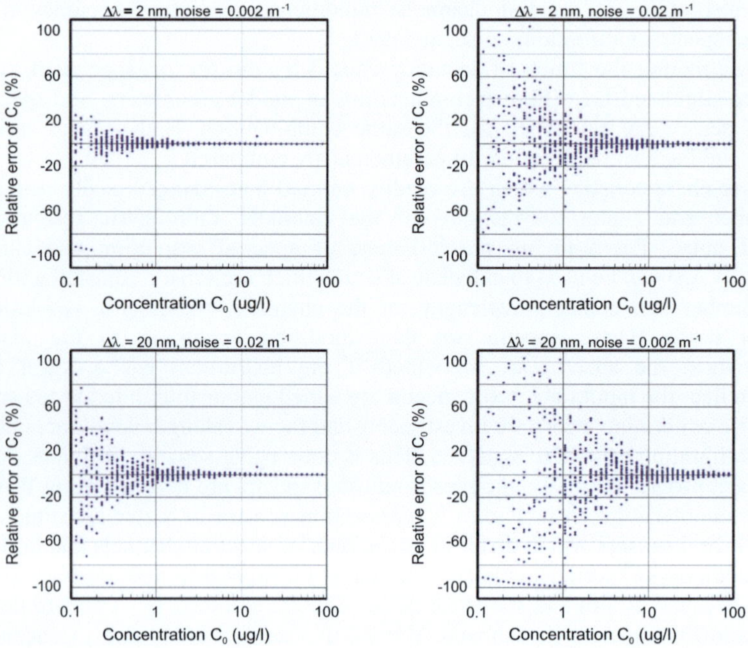

Figure 7. Illustration of errors from a sensor for an example of absorption measurements. The plots show errors for inverse modelling of phytoplankton concentration resulting from adjusting spectral width of channels and sensor noise.

Water constituent values were: C_0 = 2 μg/l; Y = 0.3 m^{-1}; S = 0.014 nm^{-1}. C_L was iterated from to 1 to 5 mg/l. The lake sediment spectrum provided with WASI was used for the bottom albedo. Computing times of HYDROLIGHT, which utilizes the invariant imbedding method, were typically 10^6 times longer than those for WASI.

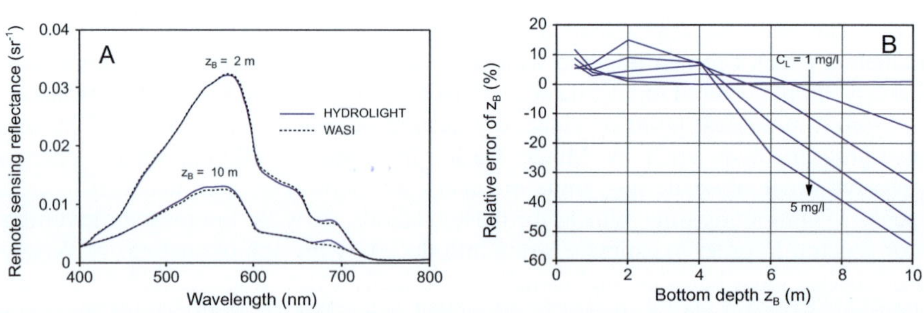

Figure 8. Illustration of errors from the model using an example of remote sensing reflectance spectra. A: Comparison of spectra from the numerically extensive program HYDROLIGHT with WASI. B: Errors of bottom depth when WASI is used for inverting HYDROLIGHT spectra.

Figure 8A compares the HYDROLIGHT and WASI spectra at 2 m and 10 m bottom depth for C_L = 2 mg/l. The differences can be attributed to Gelbstoff absorption and chlorophyll *a* fluorescence, which is accounted for in HYDROLIGHT but not in WASI. Both sets of spectra were inverted using WASI (eq. 19). During inversion only the bottom depth z_B was a fitted parameter; and all other parameters were fixed at their correct values. Figure 8B shows the relative errors $100 \cdot (z_B^{fit} / z_B - 1)$ as a function of z_B for C_L = 1, 2, 3, 4, 5 mg/l, where z_B^{fit} are the results from inverting the HYDROLIGHT spectra. The influence of the bottom albedo on remote sensing reflectance decreases with increasing depth and increasing concentration of suspended matter. Thus, the accuracy of z_B determination decreases accordingly. The corresponding errors from inverting the WASI spectra were close to zero and are not shown. Since only z_B was unknown, the errors in Figure 8B demonstrate the lower limit for model errors. More realistic error estimates are obtained by fitting additional variables such as C_0, Y and C_L along with z_B. However, the errors obtained are a mixture of errors from the model and error propagation (see 4.2.4). These two effects can be separated only by performing more detailed studies.

4.2.3 Errors from input data

Each spectrum type has a specific set of input data. Since all input data have uncertainties, it is useful to study the influence of their anticipated errors on the retrieved parameters. Three types of input data can be distinguished: (1) input spectra from a data base, (2) input spectra from actual field measurements, and (3) input parameters. Examples of input data with potentially significant errors are: $a_i^*(\lambda)$, $a_W(\lambda)$, $b_L(\lambda)$, $a_n(\lambda)$, $t_A(\lambda)$, and $t_C(\lambda)$ for type (1); $L_s(\lambda)$, $E_d(\lambda)$, and $R(\lambda)$ for type (2); and S, $b_{b,L}^*$, $b_{b,S}^*$, σ, σ_L, f, Q, B_n, and v for type (3).

The implications of input data errors on the accuracy of data analysis are studied by simulating measurements using a certain set of input data, and then inverting these simulations with altered input data. The method of altering input data depends on its type: for type (1) other files must be used, for type (2) the input measurements can be simulated, and for type (3) individual parameters have to be changed.

An example is given in Figure 9. Irradiance reflectance was calculated for shallow water using eq. (16). The following water constituent values were used: C_0 = 2 µg/l, C_L = 2 mg/l, Y = 0.3 m^{-1}, and S = 0.014 nm^{-1}. For Figure 9A the wavelength-independent spectrum of bottom albedo, $a_0(\lambda)$ = 0.1, was chosen. Its absolute value was changed from 0.1 to 0.3 during forward calculation by iterating the parameter f_0 of eq. (22) from 1 to 3, but kept constant at 0.2 during inversion. In this way relative errors of the bottom albedo from -50 % to 100 % were simulated. During inversion only the bottom depth z_B was a fit parameter. Its relative error is shown in Figure 9A as a function of z_B and of the relative albedo error. For Figure 9B, the silt spectrum provided with WASI was used as bottom albedo for forward calculation, and the five other bottom types were used during inversion. No fit of the incorrect parameters (areal fractions of the bottom types) was allowed. As expected, the errors depend on the bottom type and decrease with z_B.

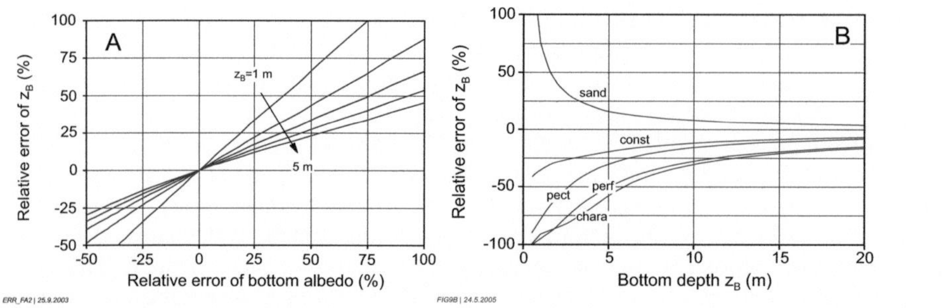

Figure 9. Illustration of errors from input data at the example of irradiance reflectance spectra. A: Errors caused by wrong scaling factor of the bottom albedo. B: Errors caused by wrong bottom type.

4.2.4 Error propagation

The influence of an incorrect model parameter value on the accuracy of the fit parameters can be analyzed effectively using the reconstruction mode of WASI, which combines forward and inverse modeling. The parameter of interest is iterated from a low to a high value during forward calculation and kept constant during fitting. With the exception of this parameter, the decision for which parameters to fit and which to fix is made in the same manner as during data analysis. All fixed parameters are kept equal in the forward and inverse mode. When the calculation is started, a series of spectra is calculated and subsequently inverted with well-defined errors for one parameter. A table is generated which lists the values of the iterated parameter, the residuum, the results of all fit parameters, and the relative errors of user-specified parameters.

An example is given in Figure 10. Absorption of water constituents was calculated using eq. (1). During forward calculation, phytoplankton concentration C_0 was changed from 0.5 to 8 µg/l. During inversion C_0 was fixed at 2, 1, and 4 µg/l for curves A, B, and C, respectively. Gelbstoff concentration, Y, and spectral slope S were estimated using inversion. Their relative errors are shown as a function of the relative C_0 error. The plots illustrate how C_0 errors induce Y (Figure 10A) and S (Figure 10B) errors.

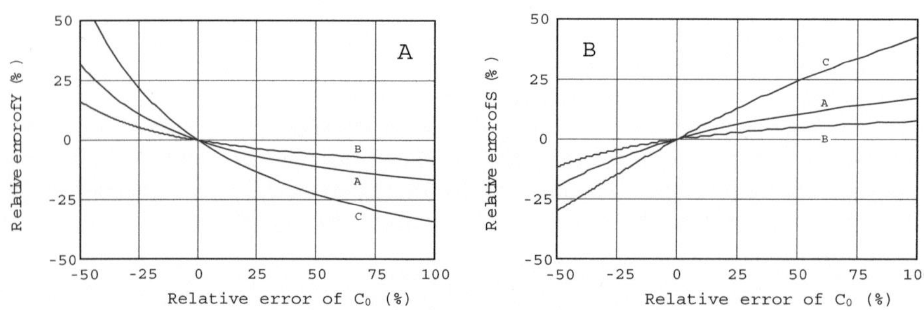

Figure 10. Illustration of error propagation at the example of absorption spectra. A: Errors of Gelbstoff concentration Y caused by C_0 errors. B: Errors of exponent S of Gelbstoff absorption caused by C_0 errors.

5. Conclusions

The Water Colour Simulator WASI is a user-friendly program for forward and inverse modeling of optical *in situ* measurements in aquatic environments. It supports all common types of spectral data obtained from shipborne instruments deployed above and below the water surface. Computationally, the program uses analytic models which are suited for all water types: deep and shallow, inland, coastal, and oceanic. Region-specific optical properties can be accounted for by exchanging the relevant input data.

The main application for WASI is data analysis. WASI is designed to automatically invert large series of spectra in a reasonable time, i.e. on the order of seconds per spectrum. In addition, it is well-suited to analyze errors from different sources by means of simulations. Since a consistent set of models is implemented and the same input data are used for the different spectrum types, data from different instruments can be compared easily (optical closure studies). Vice versa, optical properties of water constituents can be derived indirectly from non-specialized instruments, such as absorption of water constituents from reflectance measurements. Further applications of WASI are visualization of spectral changes upon parameter variation, data simulation, and student training.

WASI does have some restrictions. The implemented models are analytic approximations and do not account for all physical effects, such as fluorescence and Raman scattering or for certain water constituents such as detritus and bubbles. Vertical profiles cannot be calculated, nor can images be processed. Data from instruments on satellite and aircraft cannot be analyzed directly, since no atmospheric model is included. However, the latter can be done indirectly by coupling WASI with such a program.

6. Acknowledgements

We are grateful to Nicole Pinnel (Technical University Munich) for providing unpublished measurements of bottom albedo, and to Laurie Richardson (Florida International University) for improving the English of our manuscript.

This work was supported by the German Federal Ministry for Education and Research, Project ID 07 UFE 16/1. It is part of the Special Collaborative Program SFB 454 “Lake Constance” littoral funded by the German Research Foundation DFG.

7. References

Ahn, Y.H., A. Bricaud and A. Morel. 1992. Light backscattering efficiency and related properties of some phytoplankton. Deep-Sea Research, 39:1835-1855.

Albert, A., and C. D. Mobley. 2003. An analytical model for subsurface irradiance and remote sensing reflectance in deep and shallow case-2 waters. Optics Express 11, 2873-2890. http://www opticsexpress.org/abstract.cfm?URI=OPEX-11-22-2873.

Albert, A. 2004. Inversion technique for optical remote sensing in shallow water. Ph.D. thesis, University of Hamburg. http://www.sub.uni-hamburg.de/opus/volltexte/2005/2325/

Albert, A. and P. Gege. 2005. Inversion of irradiance and remote sensing reflectance in shallow water between 400 and 800 nm for calculations of water and bottom properties. Applied Optics (submitted).

Babin, M., and D. Stramski. 2002. Light absorption by aquatic particles in the near-infrared spectral region. Limnology and Oceanography, 47(3), 911-915.

Bricaud, A., A. Morel and L. Prieur. 1981. Absorption by dissolved organic matter of the sea (yellow substance) in the UV and visible domains. Limnology and Oceanography, 26:43-53.

Buiteveld, H., J.H.M. Hakvoort and M. Donze. 1994. The optical properties of pure water. Ocean Optics XII, SPIE, Vol. 2258:174-183.

Caceci, M.S. and W.P. Cacheris. 1984. Fitting Curves to Data. Byte May 1984: 340-362.

Carder, K.L., G.R. Harvey and P.B. Ortner. 1989. Marine humic and fulvic acids: their effects on remote sensing of ocean chlorophyll. Limnology and Oceanography, 34:68-81.

Gege, P. 1994. Gewässeranalyse mit passiver Fernerkundung: Ein Modell zur Interpretation optischer Spektralmessungen. Ph.D. thesis, University of Hamburg. DLR-Forschungsbericht 94-15, 171 pp.

Gege, P. 1995. Water analysis by remote sensing: A model for the interpretation of optical spectral measurements. Technical Translation ESA-TT-1324, 231 pp., July 1995.

Gege, P. 1998a. Characterization of the phytoplankton in Lake Constance for classification by remote sensing. Archives fur Hydrobiologia. Special Issues Advances in Limnology, 53, p. 179-193, Dezember 1998: Lake Constance, Characterization of an Ecosystem in Transition.

Gege, P. 1998b. Correction of specular reflections at the water surface. Ocean Optics XIV, 10-13 Nov. 1998, Kailua-Kona, Hawaii, USA. Conference Papers, Vol. 2.

Gege, P. 2002a. The Water Colour Simulator WASI. User manual for version 2. DLR Internal Report IB 564-01/02, 60 pp. – The actual version of the manual and of the software can be loaded from the ftp server *ftp.dfd.dlr.de*. Login: *anonymous*, password: *[email address]*, directory: */pub/WASI*.

Gege, P. 2002b. Error propagation at inversion of irradiance reflectance spectra in case-2 waters. Ocean Optics XVI Conference, November 18-22, 2002, Santa Fe, USA.

Gege, P. 2004. The water colour simulator WASI: An integrating software tool for analysis and simulation of optical in-situ spectra. Computers and Geosciences, 30:523-532.

Gordon, H.R., O.B. Brown and M. M. Jacobs. 1975. Computed Relationships between the Inherent and Apparent Optical Properties of a Flat Homogeneous Ocean. Applied Optics, 14:417-427.

Gordon, H. R. 1989. Can the Lambert-Beer law be applied to the diffuse attenuation coefficient of ocean water? Limnology and Oceanography, 34(8):1389-1409.

Heege, T. 2000. Flugzeuggestützte Fernerkundung von Wasserinhaltsstoffen am Bodensee. Ph.D. thesis, Free University of Berlin. DLR-Forschungsbericht 2000-40, 134 p.

Kirk, J.T.O. 1984. Dependence of relationship between inherent and apparent optical properties of water on solar altitude. Limnology and Oceanography, 29:350-356.

Jerlov, N.G. 1976. Marine Optics. Elsevier Scientific Publ. Company.

Jerome, J.H., R P. Bukata and J.E. Bruton. 1990. Determination of available subsurface light for photochemical and photobiological activity. Journal for Great Lakes Research, 16(3):436-443.

Mobley, C.D., B. Gentili, H.R. Gordon, Z. Jin, G. W. Kattawar, A. Morel, P. Reinersman, K. Stamnes and R. H. Stavn. 1993. Comparison of numerical models for computing underwater light fields. Applied Optics, 32: 7484-7504.

Mobley, C.D. 1994. Light and Water. Academic Press, 592 pp.

Mobley, C. D. 1999. Estimation of the remote-sensing reflectance from above-surface measurements. Applied Optics, 38:7442-7455.

Morel, A. 1974. Optical Properties of Pure Water and Pure Sea Water, p. 1-24. In N. G. Jerlov and E. Steemann Nielsen [eds.], Optical Aspects of Oceanography. Academic Press London.

Morel, A. 1980. In water and remote measurements of ocean colour. Boundary-Layer Meteorology, 18: 177-201.

Morel, A. and B. Gentili. 1991. Diffuse reflectance of oceanic waters: its dependence on Sun angle as influenced by the molecular scattering contribution. Applied Optics, 30:4427-4438.

Mueller, J.L. and R W. Austin 1995. Volume 25 of Ocean Optics Protocols for SeaWiFS Validation, Revision 1. S.B. Hooker, E.R. Firestone, and J. G. Acker, eds., NASA Tech. Memo. 104566. NASA Goddard Space Flight Center, Greenbelt, Md.

Nelder, J.A., and R. Mead. 1965. A simplex method for function minimization. Computer Journal 7:308-313.

Nyquist, G. 1979. Investigation of some optical properties of seawater with special reference to lignin sulfonates and humic substances. Ph.D. Thesis. Göteborgs Universitet, 200 pp.

Pinnel, N. 2005. Spectral discrimination of submerged macrophytes in lakes using hyperspectral remote sensing data. Ph.D. thesis. Limnological Institute of the Technical University Munich (in preparation).

Preisendorfer, R.W. and C.D. Mobley. 1985. Unpolarized irradiance reflectances and glitter patterns of random capillary waves on lakes and seas, by Monte Carlo simulation. NOAA Tech. Memo. ERL PMEL-63, Pacific Mar. Environ. Lab., Seattle, WA, 141 pp.

Preisendorfer, R. W. and C. D. Mobley. 1986. Albedos and glitter patterns of a wind-roughened sea surface. Journal of Physical Oceanography, 16:1293-1316.

Prieur, L. 1976. Transfers radiatifs dans les eaux de mer. Ph.D. thesis. Doctorat d'Etat, Univ. Pierre et Marie Curie, Paris, 243 pp.

Prieur, L. and S. Sathyendranath. 1981. An optical classification of coastal and oceanic waters based on the specific spectral absorption curves of phytoplankton pigments, dissolved organic matter, and other particulate materials. Limnology and Oceanography, 26:671-689.

Sathyendranath, S. and T. Platt. 1988. Oceanic Primary Production: Estimation by Remote Sensing at Local and Regional Scales. Science, 241:1613-1620.

Sathyendranath, S., L. Prieur and A. Morel. 1989. A three-component model of ocean colour and its application to remote sensing of phytoplankton pigments in coastal waters. International Journal of Remote Sensing, 10:1373-1394.

Sathyendranath, S. and T. Platt. 1997. Analytic model of ocean color. Applied Optics, 36:2620-2629.

Toole, D.A., D.A. Siegel, D.W. Menzies, M.J. Neumann and R.C. Smith. 2000. Remote-sensing reflectance determinations in the coastal ocean environment: impact of instrumental characteristics and environmental variability. Applied Optics, 39(3):456-469.

Chapter 5

INTEGRATION OF CORAL REEF ECOSYSTEM PROCESS STUDIES AND REMOTE SENSING

JOHN BROCK, KIMBERLY YATES, AND ROBERT HALLEY
U.S. Geological Survey, 600 4th Street South, St. Petersburg, FL 33701 USA
jbrock@usgs.gov

1. Introduction

Worldwide, local-scale anthropogenic stress combined with global climate change is driving shifts in the state of reef benthic communities from coral-rich to micro- or macroalgal-dominated (Knowlton, 1992; Done, 1999). Such phase shifts in reef benthic communities may be either abrupt or gradual, and case studies from diverse ocean basins demonstrate that recovery, while uncertain (Hughes, 1994), typically involves progression through successional stages (Done, 1992). These transitions in benthic community structure involve changes in community metabolism, and accordingly, the holistic evaluation of associated biogeochemical variables is of great intrinsic value (Done, 1992).

Effective reef management requires advance prediction of coral reef alteration in the face of anthropogenic stress and change in the global environment (Hatcher, 1997a). In practice, this goal requires techniques that can rapidly discern, at an early stage, sub-lethal effects that may cause long-term increases in mortality (Brown, 1988; Grigg and Dollar, 1990). Such methods would improve our understanding of the differences in population, community, and ecosystem structure, as well as function, between pristine and degraded reefs. This knowledge base could then support scientifically based management strategies (Done, 1992).

Brown (1988) noted the general lack of rigor in the assessment of stress on coral reefs and suggested that more quantitative approaches than currently exist are needed to allow objective understanding of coral reef dynamics. Sensitive techniques for the timely appraisal of pollution effects or generalized endemic stress in coral reefs are sorely lacking (Grigg and Dollar, 1990; Wilkinson, 1992). Moreover, monitoring methods based on population inventories, sclerochronology, or reproductive biology tend to be myopic and may give inconsistent results. Ideally, an improved means of evaluating reef stress would discriminate mortality due to natural causes from mortality due to anthropogenic causes (Brown, 1988).

Models of coral reef ecosystems, parameterized by process measurements and scaled in time-space using remote sensing, have the potential to address pressing research questions that are central to devising valid management strategies (Grigg et al., 1984; Hatcher, 1997b). To attain this goal, ecosystem-level models that integrate studies of physical and chemical forcing with observed biological and geological responses are required. This interdisciplinary approach to understanding reef biogeochemical dynamics can allow investigations that integrate the scales of time and space (Hatcher, 1997a), thereby enabling prediction of coral reef change (Andréfouët and Payri, 2001). In turn, prediction of holistic ecosystem function within various environmental forcing scenarios has substantial promise in mitigating future disturbance. Indeed, management of coral reefs at the ecosystem level has been

L.L. Richardson and E.F. LeDrew (eds.), Remote Sensing of Aquatic Coastal Ecosystem Processes: Science and Management Applications, 111-131.

suggested as the only meaningful approach to preserving coral reefs (Bohnsack and Ault, 1996; Christensen et al., 1996).

1.1 OBJECTIVES

The goal of this chapter is to present a framework for the use of remote sensing methods in investigations of the biogeochemical function of reef systems. The objectives are to: 1) present a rationale for estimation of the community metabolism of reef systems that is scaled in time and space by remote sensing; 2) provide a conceptual model of reef system excess organic carbon production and calcification based on the integration of multiple-source synoptic remote sensing with local metabolic functions determined through *in situ* process measurements; and 3) provide an example of the use of remote sensing in the spatial scaling of metabolic measurements.

2. Rationale

One of the aims of ecosystem studies is to understand apparently disjointed individual observations by expanding the analytical scale, resulting in a capacity for collective prediction (Levin, 1992). Typically, within ecosystem studies the physical environment is regarded as the primary driver of variation in biogeochemical processes and fluxes, and biological factors are regarded as secondary. A central aspect in this approach is the reduction of complexity, achieved through abstraction, which shifts the conceptualization of the ecosystem to higher levels of organization (Hatcher, 1997a). This implies up-shift in time-space scales to estimate processes at higher levels in the hierarchy, thereby eliminating detail and achieving an understanding of ecosystem-level function (Hatcher, 1997a). Overall, most ecosystem studies seek to integrate seemingly diverse observations, and reduce complexity, by quantifying key variables that emerge as stable holistic properties at higher levels of system organization (Holling, 1992; Hatcher, 1997a).

Model estimation of key rate variables that identify coral reef ecosystem state is a means to evaluate net biogeochemical status (Arias-Gonzalez et al., 1997), determine the limits of potential response to human exploitation (Grigg et al., 1984), and predict the impacts of global change (Smith and Buddemeier, 1992; Hatcher, 1997a). Brown (1988) noted that the assessment of community organic-carbon production and calcification has potential for the indication of sub-lethal perturbation. Smith and Buddemeier (1992) recognized that measures of community metabolism provide large-scale functional status and indicate incipient change. The rates of such processes on unperturbed reefs are remarkably consistent, and consequently, departures from the norm are readily recognized (Kinsey, 1983).

Net primary production (P_n) is defined as the remainder after subtraction of autotrophic respiration (R_a) from the gross primary production (P_g) of autotrophs (Hatcher, 1988). The balance of total ecosystem respiration (R), inclusive of both R_a and respiration by heterotrophs (R_h), and P_g, is called net community production (NCP). Net community production is always significantly lower than net primary production (Hatcher, 1997b). In this chapter, the organic-carbon-flux of interest is net community production (NCP), to avoid the inference that the metabolism of a component zone can be generalized to the entire ecosystem, whose domain may be a matter of perspective. Ultimately, NCP integrated over diurnal, seasonal, or annual time periods is termed excess production (E), a key variable because it represents the potential for biomass accumulation or export. Therefore, a significant absolute value of *E* implies non-steady state conditions at the time-space scale under consideration (Gattuso et al., 1998; Kinsey, 1985; Larkum, 1983). A community with positive *E* is defined as *net autotrophic*, and if the reverse condition (negative *E*) exists, the community is called *net heterotrophic* (Smith, 1988).

E is a measure of community metabolism that has been recognized by numerous researchers as a key rate variable that describes the holistic trophic state of coral reef ecosystems or their functional components (Hatcher, 1997a; Hatcher, 1997b). Accordingly, *E*, in addition to net calcification (*G*), has been selected herein as a currency for integrating process measurements and remote sensing within coral reef ecosystem models. Estimating the excess production of reefs requires the evaluation of basic ecosystem processes such as community photosynthesis, respiration, remineralization, and organic carbon transfers. Together, these data can support a holistic understanding of reef system ecological function (Grigg et al., 1984; Hatcher, 1997a).

Field investigations of reef metabolic processes using established methods are limited by the requirement to track change in water chemistry in large controlled volumes over defined time periods (Hatcher, 1997b). Consequently, biogeochemical fluxes are determined only during brief expeditions, and the resulting data sets generally do not provide spatial detail, diurnal variation, seasonality, or replicate measurements (Kinsey, 1985). Crossland et al. (1991) noted that, given uncertainty in flow respirometry measurements of reef metabolism, the *E* of whole reef systems must be uncertain by at least +/- 0.3 g C m^{-2} d^{-1}. This degree of uncertainty casts doubt on the conclusions of Smith and Buddemeier (1995) regarding *E*, and also those of geographically specific studies. For example, the investigation of Grigg et al. (1984) reported an *E* of 0.29 g C m^{-2} d^{-1} for French Frigate Shoals, Hawaii, which is about equal to the uncertainty proposed by Crossland et al. (1991). However, after one-half century of investigations (for the most part in the western Pacific and Atlantic Basins) it seems reasonable to conclude that the global *E* of reef systems is not high, and that reefs are generally about as net autotrophic as the oligotropic tropical oceans they inhabit.

Flow or chamber respirometry can provide estimates of the P_g and *R* of individual reef communities, but not the *E* of interlinked community mosaics in near trophic balance, the generic condition on coral reefs. The inability to determine if *E* differs <4% from zero using these methods (Smith and Buddemeier, 1995) is a major shortcoming, because it implies that even if practical restraints were eliminated, it would still be impossible to quantify the actual net metabolism of entire reef systems based solely on flow or chamber respirometry (Smith, 1983; Smith, 1985). Furthermore, the metabolism of reef-flats, where most measurements have been made, cannot be generalized to reef-slopes, lagoons, or entire reef systems (Hatcher, 1990). Yet, even very minor deviation of *E* from zero has great significance in determining trophic balance, nutrient import or export, and harvestable yield (Kinsey, 1985; Smith, 1983; Smith, 1985). Clearly, an alternate approach that admits variability in time-space scales is required for estimation of the net community metabolism of entire reef systems (Smith and Buddemeier, 1995).

Coral reef ecosystems have geographical extents in the tens of kilometers, and the coral reef provinces within which these ecosystems nest have even larger space scales. Aside from the measurement uncertainties cited above, the direct *in situ* measurement of whole ecosystem properties such as excess production is an intractable problem due to the breadth and geospatial variability of reef systems (Hatcher, 1997b). As an alternative, the integration of process measurements and remote sensing within a model whose geographical domain encompasses the entire reef system is a promising approach to the estimation of net metabolic properties at the ecosystem level (Andréfouët and Payri, 2001; Hatcher et al., 1987; Hatcher, 1997a; Hatcher, 1997b).

3. A Conceptual Model for the Use of Remote Sensing in Reef Metabolic Studies

Synoptic ecological models scaled by remote sensing are now in wide use in the investigation of both terrestrial and marine ecosystems (Gurney et al., 1993). We suggest that this approach has substantial potential to improve understanding of the biogeochemistry of reefs. The conceptual framework that we propose to evaluate the organic carbon metabolism of reef systems uses multiple-source remote sensing to: 1) assess ecosystem structure at the spatial scales of reef zones and intra-zone biotope distributions; and 2) model environmental forcing variables (limited to light level and nutrient concentration in this discussion) (Figure 1).

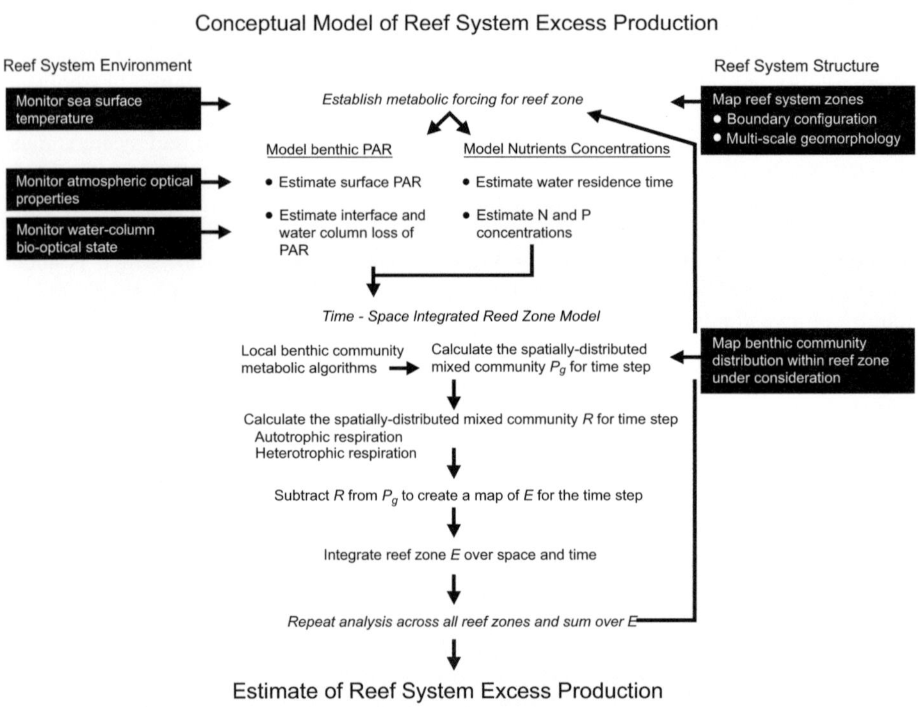

Figure 1. Conceptual model of reef system excess organic carbon production. Capabilities that rely upon remote sensing are surrounded by a black background.

In our proposed integration framework, the reef system under investigation is:

- segmented into biogeochemically distinct geomorphological structural zones based on moderate spatial resolution multispectral satellite imaging
- establishment of time-space patterns of light and nutrients for the reef zones (defined by the initial landscape segmentation) based on benthic landscape irradiance and hydrodynamic models driven by low spatial - high temporal resolution oceanographic and meteorological satellite sensors
- mapping of spatial distributions of intra-zone biotopes using high spatial resolution airborne or satellite-borne sensors
- measurement of *in situ* process to determine benthic community metabolic functions for the recognized biotopes

Guided by the regional distributions of environmental conditions modeled through indirect remote sensing and the reef zone biotope maps obtained by mapping with high resolution sensors, the spatially-distributed mixed community P_g, *R*, *E,* and *G* for the selected reef zone is calculated for a predetermined time increment. Iteration over time increments and then over reef zones, followed by time and space integration, results in seasonal or annual estimates of reef system *E* and *G*.

3.1 REMOTE SENSING OF REEF SYSTEM STRUCTURE

Coral reefs are optimized for high productivity and self-sufficiency within the nutrient-depleted surface layers of tropical oceans (Darwin, 1897; Hatcher, 1988; Larkum, 1983; Lewis, 1977). Their biological functioning is broadly mediated by physical processes that control transfers of organic material and inorganic nutrients. Therefore, the degree of hydrodynamic closure at internal and external boundaries fundamentally defines the ecosystem and establishes reservoirs for budgets of biotic and abiotic materials. Inherently, internal compartment processes dominate over the trans-boundary processes that control fluxes between reservoirs within the ecosystem (Hatcher, 1997a; Hatcher, 1997b). An essential concept is that boundaries within coral reef ecosystems separate compartments whose dominant processes differ and operate on different characteristic scales. Consequently, commonly recognized reef geomorphic zones, for example, lagoon, reef-flat, reef-crest, and forereef slope, normally represent a biological zonation that is paralleled by the spatial pattern of dominant physical processes. Geomorphologic break lines that can be recognized by moderate resolution satellite mapping sensors trace process-defined boundaries, sites of strong gradients in advection, water-mass mixing, and wave-energy dissipation (Done, 1983; Hatcher, 1997a).

3.2 METABOLISM AND CORAL REEF GEOMORPHOLOGY

Although investigation of entire reef systems, and in particular reef-slopes and lagoons, is incomplete, the existing data set of carbon-flux measurements reveals that reef structure is a major determinant of internal variation in *E* within coral reef ecosystems. The shallow perimeters of reefs are usually net autotrophic, and while a portion of the implied excess production is used for local growth, some is exported from the reef system, and some is transported inward (Hatcher, 1990). Typically, the outer-reef-flat exhibits the highest metabolic rates, and high excess production here and on the algal-dominated reef crest appears to be advected as detritus to backreef sinks. This gradient in metabolic activity inward from the reef periphery implies a general shift away from net autotrophy towards close mixotrophic balance as hydrodynamic closure increases, and parallels the obvious biological and morphological zonation seen on reefs. Moreover, a general correlation between calcification and gross photosynthesis is embedded in this spatial pattern (Kinsey, 1977).

These broad geomorphic patterns in organic metabolism led Crossland et al. (1991) to hypothesize that reef-perimeter zones are coupled to interior reef zones by hydrodynamics and the feeding behaviour of organisms. Kinsey (1983) observed that within reef systems that are in overall trophic balance, *P*/*R* is usually greater than one in high-energy zones that are hydrodynamically open and act as sources of organic matter. Quiescent backreef settings with higher water-residence times were recognized as sinks for organic matter (Kinsey, 1985). Essentially, Kinsey (1983, 1985) proposed that the positive excess production on reef slopes, crests, and flats is carried downstream, and subsequently deposited, on low-energy backreef platforms, or, perhaps sporadically due to severe storms, in lagoons. This functional association between neighboring producer and consumer communities acts to retain production within the ecosystem (Hatcher, 1990; Smith and Buddemeier, 1995). A consequence for the modeling of whole-reef organic-carbon metabolism is that it is not valid to generalize the plentiful measurements acquired on the easily-worked reef-flats to the entire reef system.

The use of moderate resolution satellite remote sensing to segment reef systems based on boundaries defined by hydrodynamic closure is a key element in integrating *in situ* process measurements with remote sensing. If so organized, the development of excess production models may proceed from a simple formulation for backreef areas, where the advection of nutrients and organic detritus is comparatively steady and predictable, toward more complex designs for fringing reefs that admit the effect of their more open boundaries. As model complexity increases to allow trans-boundary transfers, the required diversity of metabolism measurements in the field, the image mapping of benthic community structure, and the sensing of environmental forcing all scale up in unison. This is typically a valid paradigm for open ocean reef systems, clearly established by the observed differing metabolic performance of reef-slopes, flats, and lagoons (Hatcher, 1997a).

3.3 REMOTE SENSING OF REEF SYSTEM ZONES

Globally, coral reefs cover about 285,000 km^2. Improved definition of zonation within this total area would result in a first order improvement in reef carbon-flux estimates (Crossland et al., 1991). Fortunately, remote sensing affords the capability to non-intrusively map large coastal areas repeatedly, synoptically, and at various levels of detail dictated in part by the spatial, temporal, spectral, and radiometric resolutions of the chosen sensor (Green et al., 1996). Using the interpretation of reef-flat, knoll, and lagoon areas on aerial photographs, Atkinson and Grigg (1984) acquired representative measurements of metabolic performance for each zone in an investigation of benthic net community production at French Frigate Shoals, Hawaii. That study was an early attempt to use a form of aerial remote sensing to discriminate net heterotrophic and net autotrophic reef zones, and to model them as distinct internal entities in order to attain valid holistic estimates of *E* for the atoll under investigation.

Two decades ago, the remote sensing capabilities for landscape definition that are required to apply the approach of Atkinson and Grigg (1984) across entire reef systems were not available. In contrast, at present a new generation of aircraft and satellite-based sensors has excellent potential to discriminate both hydrodynamically bounded landscape components and the internal fine-scale spatial distribution of their metabolically distinct benthic communities (Green et al., 1996; Green et al., 2000; Maltus and Mumby; 2003; Mazel, 1999). Available mapping sensors may be categorized according to their appropriate application within the framework for the proposed reef system *E* model (Figure 1). Following Green et al. (2000), we divide the direct remote sensing of reefs into two categories: 1) moderate resolution imaging of landscape zones with corresponding functional modes (Andréfouët et al., 2001b); and 2) the finer-scale mapping of intra-zone biotopes (Mumby et al., 1997a; Mumby and Edwards, 2002).

Moderate spatial resolution satellite sensors (>10 m to <100 m) are useful in creating maps of reef system zones (see Chapter 12). Compared to other digital sensors, these instruments have the longest history of use in coastal studies. This sensor class includes the Landsat Multispectral Scanner (MSS), Thematic Mapper (TM), and Enhanced Thematic Mapper Plus (ETM+), the SPOT High Resolution Visible Scanner (HRV), and the Indian Remote Sensing Linear Imaging Self-Scanning Sensor (LISS) (Green et al., 2000). Most studies based on these moderate resolution satellite sensors have provided geographical information on geomorphological zones, such as forereef, reef crest, algal rim, lagoon, or emergent atoll (Green et al., 1996).

Khan et al. (1992) used a principal components approach to classify sub-tidal habitats in the Arabian Gulf on a Landsat TM image. At a Dominican Republic site, Michalek et al. (1993) examined the use of Landsat TM images for delineating geomorphology and three basic bottom classes - coral, seagrass, and sand. Ahmad and Neil (1994) compared the capabilities of Landsat TM and Landsat MSS for mapping coral reef zonation at Heron Reef, within the Great Barrier Reef of Australia, and found that TM provides greater geomorphological detail than MSS. More recently, Matsunaga and Kayanne (1997) used Landsat TM to investigate coarse-scale temporal habitat change on fringing reefs at Lshigaki Island in the Ryukyu chain. Peddle et al. (1995) carried out an analysis of radiative transfer in the use of SPOT HRV for mapping shallow habitats at the Fiji Islands. LeDrew et al. (1995) also studied the coral reef ecosystem at the Fiji Islands using SPOT HRV, and determined that for depths less than 10 m, SPOT bands 1 and 2 were useful in the spectral discrimination of coral reef features. Mumby et al. (1997b) determined that Landsat TM and SPOT HRV have limited capabilities to map seagrass standing crop at broad spatial scales, provided that extensive field verification efforts are undertaken.

Bainbridge and Reichelt (1988) concluded that moderate resolution satellite imagery is more appropriate for studying reef geomorphology than reef biology. The more recent studies cited above have substantiated this early judgment on the limitations of intermediate resolution satellite sensors, concluding that moderate spatial/ spectral resolution satellite sensors are useful only for the coarse descriptive mapping of reef systems (Malthus and Mumby, 2003). In a comparative study of the capabilities of various sensors for coral reef habitat mapping conducted for reefs of the Turks and Caicos Islands, Mumby et al. (1997a) found that for all moderate resolution satellite scanners, map accuracy typically dropped significantly from coarse- to fine-scale habitat discrimination. Moderate resolution satellite imagery is not sufficient for the detailed mapping of benthic habitats primarily due to intra-pixel mixing (Andréfouët et al., 2003a). Additional limitations are posed by the loss of radiometric contrast due to atmospheric effects (Lubin et al., 2001) and limited spectral resolution (Hochberg and Atkinson, 2003; Malthus and Mumby, 2003). Intermediate resolution satellite sensors do not provide reliable quantitative estimates of *in situ* coral cover, nor can they discriminate live coral from fleshy algae (Dustan et al., 2000; Green et al., 1996).

In contrast, regional landscape definition, i.e. the synoptic discrimination of geomorphic zones using imagery from Landsat, SPOT, and similar satellites, is of great value from a reef system *E* modeling perspective (Bouvet et al., 2003). This coarse-scale mapping is the first step in the implementation of our proposed synoptic *E* model strategy, because the landscape units thus identified form a mosaic of compartments within the model domain. The required model complexity for each compartment may then be matched to its openness to abiotic and biotic exchange processes.

A recent SPOT-based study of atoll rims in French Polynesia (Andréfouët et al., 2001b) illustrates the role proposed by the authors for moderate resolution satellite sensors within the suggested synoptic *E* model framework. Andréfouët et al. (2001b) used SPOT HRV multi-spectral images to characterize the structure and degree of hydrodynamic closure of atoll rims, distinguishing nine rim types based on the dominance of vegetated, submerged, intertidal, and emerged domains. Knowledge of rim structure and climatological ocean swell were combined to estimate the water residence time for the lagoons, yielding a quantitative measure of closure that has major consequences for internal biological processes (Andréfouët et al., 2001b). In closely related studies, SPOT images were used to assess atoll reef-flat spillways bordering lagoons in the Tuamotu Archipelago, and to thereby investigate the use of water renewal time as a criterion for categorizing the adjacent lagoons (Andréfouët and Payri, 2001; Pages and Andréfouët, 2001). Potential flow rates and water-renewal times under various swell conditions were calculated for the spillways observed by SPOT, and a positive correlation between estimated water-renewal times and lagoonal concentrations of dissolved organic matter (DOM) was identified (Pages and Andréfouët, 2001). Similarly, Andréfouët and Payri (2001) found a significant relationship between phytoplankton biomass and water-renewal time for lagoons greater than 25 km^2 in surface area. Pages and Andréfouët (2001) concluded that SPOT images are useful in the rapid regional assessment of lagoon hydrodynamics.

3.4 MAPPING INTRA-ZONE BENTHIC BIOTOPES

The use of habitat mapping based on remote sensing in the estimation of reef system *E* assumes that metabolic performance is correlated with the composition of the benthic communities within reef zones (Hatcher, 1997b). Aside from the traditional descriptive interpretation of aerial photography, remote sensing studies of coral reefs prior to the mid-1990s were mainly attempts to map geomorphic zones (Green et al., 1996). The discrimination of ecologically meaningful assemblages is more challenging, because high spatial detail is required (Maeder et al., 2002; Mumby et al., 1997a), and because coral, algae, and seagrass all contain photosynthetic pigments and exhibit similar spectral reflectance (Holden and LeDrew, 1998, 1999; Myers et al., 1999; Hochberg and Atkinson, 2000; Schalles et al., 2000; Hedley and Mumby, 2002; Hochberg et al., 2003; Kutser et al., 2003).

Aircraft-based sensing has significant advantages in the ecological mapping of reef system components. Tropical coastal ecosystems are highly complex geographically, and the high resolution images acquired from aircraft can capture intricate spatial detail. High resolution aircraft imaging, either aerial photography or hyperspectral scanning and imaging, can provide habitat information at the level of coral colonies and patch reefs (Catt and Hopley, 1988). Moreover, aircraft can fly below the cloud cover that typically obscures features on satellite images in the tropics (Chauvaud et al., 1998).

Measurements of the reflectance spectra of reef biota and inorganic substrates indicate that the dominant types can be distinguished *in situ* (Knight et al., 1997; Holden and LeDrew, 1999). In spite of the confounding effect of the overlying water column (Lee et al., 1994; Holden and LeDrew, 1999, 2001, 2002; Purkis and Pasterkamp, 2004), this inherent capability for spectral discrimination applies in some degree to hyperspectral scanners mounted on aircraft flying low enough to minimize atmospheric contamination (Clark et al., 1997; Gordon and Wang, 1994; Mumby et al., 2001; Hu and Carder, 2002).

3.5 REMOTE SENSING OF THE REEF SYSTEM ENVIRONMENT

Aside from the control exerted by varying substrate composition, the gross correlation between benthic metabolic performance and hydrodynamic closure is forced by associated environmental gradients (Hatcher, 1990; Crossland et al., 1991). Variation in the benthic light, temperature, and nutrient environments across a reef are key axes for the determination of E, and are related to hydrodynamic closure, and more broadly, latitude and land proximity (Crossland et al., 1991).

Low spatial resolution/high temporal resolution (>100 m, <10 days) oceanographic and meteorological satellite sensors provide capabilities for monitoring the ocean environment surrounding reef systems (Green et al., 2000; Mazel, 1999). Such observation of the broader reef environment has also been termed "indirect" remote sensing, distinct from the "direct" observation of reef substrates (Mazel, 1999). Low spatial/high temporal resolution satellite sensors (Strong et al., 2000) are a class of remote sensing instruments distinct from the *moderate* spatial and temporal resolution satellite sensors routinely applied to the mapping of reef zonation and intra-zone biotopes (Andréfouët et al. 2001a; Andréfouët et al., 2003b; Andréfouët et al., 2004). Designed to serve the oceanographic and meteorological communities, sensors such as the NASA SeaWiFS (McClain et al., 2004), NASA MODIS (Barnes et al., 2003), and the NOAA AVHRR (Green et al., 2000) have excellent capabilities to enable the modeling of environmental forcing variables (Gattuso and Jaubert, 1985; Gattuso et al., 1993) for models of coral reef benthic metabolism (Andréfouët and Payri, 2001). The discussion presented below briefly addresses the use of indirect remote sensing in the modeling of reef system carbon and carbonate metabolism. [The reader is referred to Chapter 11 (Newman et al.) in this volume for a comprehensive discussion of the remote sensing capabilities for coastal environmental analysis.]

Meteorological satellite imaging allows the estimation of global or regional surface irradiance fields, and oceanographic satellite imaging of ocean color provides the synoptic concentrations of optically active sea water constituents on the same temporal (daily) time scale. These multiple source satellite-derived products can be combined within submarine light models to estimate photosynthetically available radiation through the photic zone and at shallow benthic substrates. Various models for the estimation of incident solar radiation based on satellite observations and routine meteorological observations have been developed and evaluated (Bishop and Rossow, 1991; Chertock et al., 1991; Darnell et al., 1988; Dedieu et al., 1987; Frouin et al., 1988; Pinker and Ewing, 1985). Estimates of downwelling surface shortwave radiation can now be routinely made at accuracies within 20 W m^{-2} on monthly time scales, and at spatial scales that meet the requirements of reef system biogeochemical modeling (Pinker et al., 1995).

Algorithms for estimating fluxes of photosynthetically available radiation that rely on satellite visible wavelength imaging are now used routinely in models of terrestrial

(Gu and Smith, 1997) and marine mixed layer (Sathyendranath and Platt, 1988) biogeochemistry. Similarly, models of the diel light regime on reefs can be used to force benthic community photosynthesis versus irradiance functions developed by measuring metabolism of defined biotopes under conditions of diurnally or seasonally varying illumination (Larkum, 1983). Morel (1988) and numerous other researchers have developed optical models for Case 1 waters that are typical of the tropical oceans adjacent to reef systems, defined as those waters for which phytoplankton and their derivatives predominate in determining optical properties. These pigment-dependent optical models allow the propagation of sunlight in the upper ocean to be predicted as a function of the local algal content, a variable that in turn may be estimated from satellite observations of ocean spectral reflectance, informally called "ocean color" (McClain et al., 2004).

Sathyendranath and Platt (1993) presented a general approach to the estimation of integrated ocean primary production that could readily be adapted to the modeling of benthic gross photosynthesis across reef systems. This approach is based on combining satellite ocean color sensing of global mixed layer algal biomass with spectral models for the computation of ocean surface and interior irradiance, and local photosynthesis/ irradiance algorithms (Sathyendranath and Platt, 1993). Clearly, the benthic light regime driving gross photosynthesis on coral reefs can be obtained by combining models of atmospheric and submarine light transmission with synoptic multi-temporal information on atmospheric and sea water opacity from multiple source satellite observations. A complicating factor in this application is the need to remove the component of upwelling radiance that is derived from refection at the sea floor from the total spectral water-leaving radiance that is observed by ocean color sensors following atmospheric correction. Given knowledge of the spatial pattern of bottom albedo on reef systems from high resolution mapping sensors or *in situ* measurements, we regard this task as tractable.

The structure and function of coral reef communities reveal geographic differences that appear to be greatly impacted by nutrient inputs (Birkeland, 1988). Further, there is evidence that reef algae, animals with zooxanthellae, and coral reef communities show nutrient-limited responses for gross photosynthesis and net primary production (Atkinson, 1988). Apparently, aspects of the community productivity of coral reefs can be nutrient-limited (Atkinson, 1981; Atkinson, 1988), but the conception of nutrient limitation by a single nutrient is an oversimplification for reef systems (Lewis et al., 1985).

Hydrodynamics are important in the nutrient-regulation of coral reef communities because N versus P limitation of the net production of aquatic ecosystems is largely a function of the degree of confinement (Smith, 1984). Remote sensing can aid hydrodynamic studies that are aimed at determining nutrient-regulation across reef systems by setting the initial boundary conditions for simulations of circulation, in flow model validation, and in the calibration of particle transport models (Acker et al., 2004; Gibbs and Shaw, 2002; Ouillon et al., 2004; Young et al., 1994). Sea surface temperature (SST) fields derived from Advanced Very High Resolution Radiometer (AVHRR) observations acquired by NOAA polar orbiting satellites (Li et al., 2001) are commonly used to establish surface boundary conditions for regional hydrodynamic circulation models. Normally, these satellite thermal infrared images are used to determine initial SST conditions, and then additional satellite SST fields are assimilated during the simulation of circulation to calibrate or validate the model. For example, Gibbs and Shaw (2002) presented a methodology for the use of remotely sensed SST fields to set the initial surface conditions for a regional numerical model of circulation around the Kaikoura Canyon on the coast of New Zealand.

4. Scaling Up Coral Reef Metabolism Using Remote Sensing: A Case Study

We estimated net calcification (*G*) and excess production (*E*) for a representative study area on the carbonate platform of the northern Florida reef tract by integrating benthic chamber respirometry with habitat mapping based on aircraft hyperspectral imaging. The study area is located approximately 5 km east of Caesar's Creek and Elliot Key, on the carbonate platform seaward of Hawk Channel in the northern Florida reef tract within Biscayne National Park (Figure 2) (Ginsburg and Shinn, 1993). This study area consists of a lagoonal platform that lies landward of Ajax Reef, a shelf edge bank reef, and is covered by carbonate sand, mostly capped by a *Thallasia testudinum* seagrass meadow, and contains numerous scattered patch reefs surrounded by bare sandy halos (McPherson and Halley, 1996; Shinn et al., 1989).

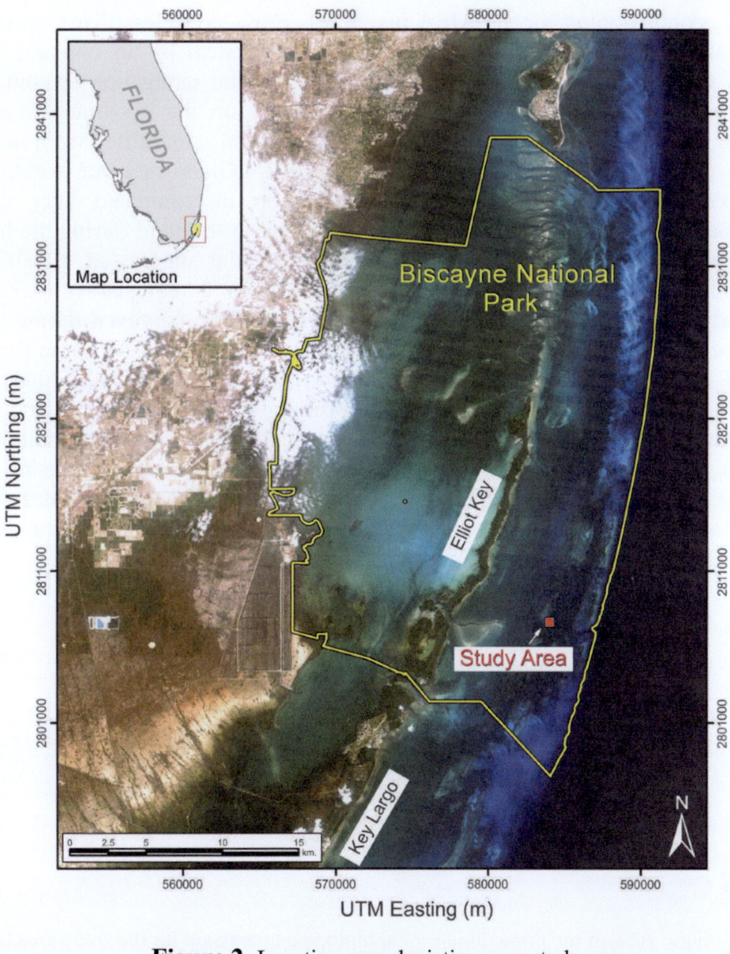

Figure 2. Location map depicting case study area.

4.1. BIOTOPE MAPPING

The Advanced Imaging Spectrometer for Applications (AISA) sensor was flown at an altitude of approximately 1500 m on January 7, 2001 to collect hyperspectral imagery over the study area at 1.5 m spatial resolution in 25 10 nm wide spectral channels (personal communication, Oliver Weatherbee). To avoid sun glint, overflights were conducted in the early morning and late afternoon when solar illumination angle was less than 30°. Concurrent with the collection of upwelling spectral radiance, a Fiber Optic Downwelling Irradiance System (FODIS) incorporated in the AISA sensor package measured downwelling irradiance in the same spectral channels. Subsequent to the overflights, the apparent at-platform remote sensing reflectance (R_{rs}) was calculated for each spectral channel. A combined rudimentary atmospheric-water column radiance correction of the AISA spectral R_{rs} images was performed by use of the "darkest pixel" method (Gordon, 1978; Gordon and Clark, 1981), using adjacent deep water pixels within the AISA coverage region. The AISA system was calibrated prior to the overflights with reference to NIST traceable integrating sphere (personal communication, Oliver Weatherbee).

Ground geopositioning for the AISA pixels was provided to less than 5 m error in the Universal Transverse Mercator (UTM) projection system by an onboard OMNI STAR GPS receiver integrated with a C-Migits II inertial navigation system. The AISA image underwent supervised classification based on the maximum-likelihood decision rule (Mumby et al., 1997a, 1998), using training sites identified from over 200 field survey points collected in March 2001 and 2002. Given that our investigation evaluated reef-zone carbon fluxes, the classification discriminated very distinct substrate types that could be unambiguously recognized in the field during the benthic chamber deployments. The benthic cover types that the supervised classification discriminated were limited to sand, seagrass, and two coral reef classes (dense live substrate and sparse live substrate). The study area was also surveyed during a two-week time period in early August 2002 by Experimental Advanced Airborne Research lidar (EAARL) overflights staged from Marathon, Florida (Brock et al., 2004). The EAARL is a temporal waveform-resolving, airborne **li**ght **d**etection and **r**anging (lidar) instrument that is designed to measure the fine scale topography of shallow reef substrates at a vertical resolution of about 15 cm at surface spots, and densities of at least one 20 cm diameter spot per m^2 (Wright and Brock, 2002). The lidar-derived submarine topography was merged with the AISA classification to create a three-dimensional biotope map for the study area (Figure 3).

Figure 3. Oblique view of the three dimensional biotope map created for the study area by draping the AISA image classification on the lidar-based digital elevation model.

4.2 BENTHIC PROCESS MEASUREMENTS

The Submersible Habitat for Assessing Reef Quality (SHARQ) was deployed over various habitats in the study area during June 27-July 1, 2000, May 30-June 2, 2001, and July 12-13, 2002, to measure diurnal spring-summer rates of calcification, photosynthesis, and respiration (Yates and Halley, 2003). The SHARQ is a large (4.9 m long by 2.4 m wide by 1.2 m high) portable incubation system that acquires *in situ* measurements of community-scale metabolic rates by isolating the water mass overlying the substrate, thereby allowing the monitoring of changes in water chemistry over time (Figure 4). Following the method of Kinsey (1978), total alkalinity (*TA*), dissolved oxygen (O_2), pH, salinity, and water temperature were measured and used to calculate community metabolism. Net calcification (*G*) was determined for each of six 4 h periods within each 24 h incubation period using the alkalinity anomaly technique of Smith and Key (1975). Net photosynthesis and respiration were calculated from dissolved O_2 at 15 min intervals. Net metabolic rates per unit area of substrate were obtained by multiplying the O_2 concentration change by the volume to basal surface area ratio of the chamber. Diurnal curves for net photosynthesis and respiration were developed based on the determinations for 15 min intervals. Daily gross production was calculated by integrating the daytime photosynthesis curves established for instantaneous gross production. A 24 h respiration estimate was based on the mean rates of respiration measured during each incubation period (Yates and Halley, 2003). Mean metabolic rates were calculated for representative sites selected for each benthic class defined in the supervised classification of the AISA image. This procedure established the rates of daily spring-summer photosynthesis, respiration, and calcification for the mapped substrates that were subsequently used in estimating spatially integrated carbon and carbonate fluxes.

Figure 4. Underwater photograph depicting a deployment of the Submersible Habitat for the Assessment of Reef Quality (SHARQ).

4.3 CALCULATION OF LANDSCAPE METABOLISM

A raster model implemented at a 1.5 m grid cell resolution was used to extrapolate the SHARQ metabolic measurements across the AISA-based benthic class map. This operation resulted in a set of maps for each mapping sensor that depicted the estimated spatial patterns of spring-summer daytime calcification (*G*), daily gross photosynthesis (*P*), and 24 hour respiration (*R*), all in units of gm C m^{-2}. An excess production map (*E*) (gm C m^{-2}) was created for each sensor type by differencing the corresponding estimated spatial distributions of *P* and *R* (Figure 5). The flux distributions shown on the AISA-based reefscape metabolism maps were integrated both spatially and across all benthic cover types to allow carbon and carbonate benthic flux estimates for the entire study area (Figure 6).

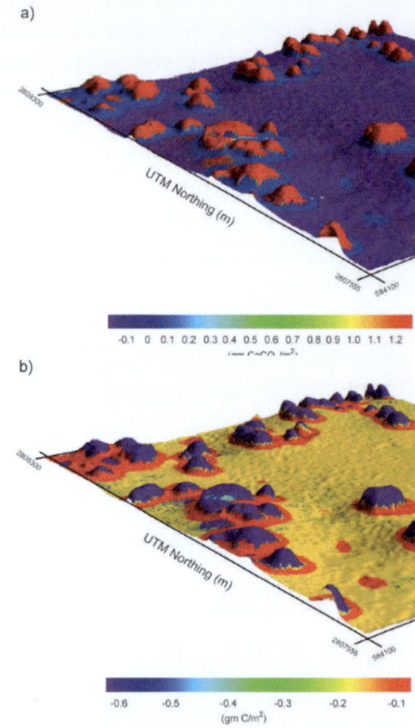

Figure 5. Oblique view of the three dimensional daily net calcification (a) and daily excess production (b) maps created for the study area.

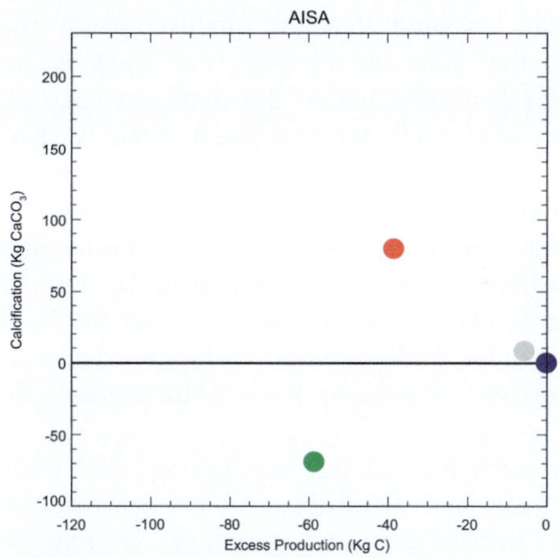

Figure 6. Plot of total daily excess production (Kg C) versus daily net calcification (Kg $CaCO_3$) spatially integrated by benthic class over the study area. Determinations for sand, seagrass, dense live substrate, and sparse live substrate are plotted as gray, green, red, and blue dots, respectively.

5. Discussion

Historically, community-scale measurements of reef metabolic processes have been performed using a flow respirometry approach based on the lagrangian monitoring of the chemistry of a water mass passing over a reef zone. Flow respirometry requires that water circulation at the study site be well characterized using current meters or other water-mass tracking techniques, assumes the conservation of water mass along transects, and requires unidirectional currents (Yates and Halley, 2003). Flow respirometry is also limited by the resolution of geochemical measurements, and is difficult or impossible to employ at night. Further, the complicated topography of coral reefs complicates the use of flow respirometry, resulting in the potential for large errors. Therefore, traditional flow respirometry is insufficient for the estimation of reef metabolic processes at the ecosystem process level.

In contrast, the integration of multiple-source synoptic remote sensing with local metabolic functions determined through *in situ* process measurements allows the modeling of reef system excess organic carbon production and calcification on several spatial scales. At present, remote sensing methods can not provide local metabolic algorithms for coral reef communities. This was demonstrated by Joyce and Phinn (2003) in a study that examined the relationship between spectral reflectance, chlorophyll *a* content, and photosynthetic capacity for common coral reef substrates. Joyce and Phinn (2003) found that photosynthetic capacity did not exhibit statistically significant correlations with spectral reflectance or absorption, the optical variables that influence remote sensing signals based on reflected sunlight. However, large portable incubation systems such as the SHARQ afford the capability to measure rates of photosynthesis, respiration, and calcification on substrates. Benthic chamber experiments under varying light and nutrient conditions enable the definition of biotope-specific metabolic algorithms. Further, deployment is not limited due to current patterns, and the large footprint enables measurement of community-scale fluxes (Yates and Halley, 2003).

Our case study has revealed that the shallow patch reef community on the carbonate platform of the northern Florida reef tract is net calcifying and net heterotrophic. Indeed, taken as a whole the platform segment studied, covered by seagrass meadows, bare sand, and scattered patch reefs, is also net heterotrophic, but exhibits net carbonate dissolution. Spatial extrapolation of community metabolism based on mapping with airborne hyperspectral and lidar scanning allows us to infer that within this reef zone, much of the carbonate sediment produced on the patch reef crests is deposited on the adjacent platform and consumed by dissolution. Therefore, although the live substrate classes exhibit positive rates for G, the likelihood of future reef accretion is limited. This finding, based on biogeochemical function, is consistent with the observed dominance of octocorals over stony corals in the present-day patch reef community.

Although near the perimeter of the reef system, the carbonate platform in the northern Florida reef tract was determined to be net heterotrophic, as are all it's component biotopes. In the classical paradigm (Hatcher, 1990), this platform immediately landward of the northern Florida bank-barrier reef zone might be expected to be a net autotrophic shallow reef system perimeter region. The expected high rates of P_g at this outer reef system location would result in a P/R ratio in excess of 1, with the advection of excess production towards more closed backreef sinks that are in close mixotrophic balance (Hatcher, 1990). Based on our determination of a P/R ratio less than 1, we infer that metabolism on the carbonate platform of the Florida reef tract is dominated by external inputs.

Although this location is near a large urban population, the net autotrophy that might be expected to result from associated eutrophication does not exist. Accordingly, we infer that at the study site the inward advection of inorganic nutrients is not the dominant forcing mechanism for benthic biogeochemical function. Indeed, the observed net heterotrophy suggests that influxes of organic detritus, most likely of land origin, followed by *in situ* remineralization, is driving overall benthic metabolic function. This result implies that nutrient loading is occurring at the substrate/water column interface. Therefore, the monitoring of water column nutrient concentrations should be focused at the benthic boundary layer to be of maximal value in the management of this reef system zone.

6. Conclusions

Coral reefs are extremely high in habitat complexity, with a diverse assemblage of plants and animals embedded across a substrate that is convoluted topographically (Hatcher, 1990). They function as highly efficient carbon uptake and recycling systems based on intimate interactions between resident organisms. Although this web of interlocking physical, chemical, and biological processes is quite dynamic (Sargent and Austin, 1949), significant amounts of organic material are normally not accumulated over time periods longer than a single diurnal cycle (Buddemeier and Keinzie, 1976; Hatcher, 1990). A rigorous, holistic approach to the assessment of entire reef systems is needed for the timely appraisal of pollution effects or generalized endemic stress (Grigg and Dollar, 1990; Wilkinson et al., 1999). Excess organic-carbon production (E), the integration of net community production over time, and net calcification (G), are all key rate variables that describe the status of coral reef ecosystems (Hatcher, 1997a,b). The integration of metabolic process measurements and remote sensing within a spatially distributed model can enable estimation of E and G scaled up over time and space (Andréfouët and Payri, 2001). Multiple-source remote sensing of reef system structure and environment may now be integrated with field determinations of benthic metabolism to evaluate the carbon and carbonate biogeochemistry of reef systems across a range of space and time scales. As a result, holistic analyses of reef community, zone, or system metabolic function can be used to provide a fundamental context for ecological studies, and can inform ecosystem managers on the efficacy of alternate monitoring and remediation strategies.

7. Acknowledgements

The senior author is grateful to I. Kuffner, P. Hallock-Muller, B. Lidz, and an anonymous reviewer for their thoughtful reviews of an earlier version of this manuscript. The authors acknowledge the U.S. Geological Survey's Coastal and Marine Geology Program for funding our investigations of the integration of coral reef metabolic measurements with remote sensing.

8. References

Acker, J.G., A. Vasilkov, D. Nadeau and N. Kuring. 2004. Use of SeaWiFS ocean color data to estimate neritic sediment mass transport from carbonate platforms for two hurricane-forced events. Coral Reefs. 23:39-47.

Ahmad, W. and D.T. Neil. 1994. An evaluation of Landsat Thematic Mapper (TM) digital data for discriminating coral reef zonation: Heron Reef (GBR). International Journal of Remote Sensing. 15:2583-2597.

Andréfouët, S. and C. Payri. 2001. Scaling-up carbon and carbonate metabolism of coral reefs using in-situ data and remote sensing. Coral Reefs. 19:259-269.

Andréfouët, S., F.E. Muller-Karger, E.J. Hochberg, C. Hu and K.L. Carder. 2001a. Change detection in shallow coral reef environments using Landsat 7 ETM+ data. Remote Sensing of Environment. 78:150-162.

Andréfouët, S., J. Pages and B. Tartinville. 2001b. Water renewal time for classification of atoll lagoons in the Tuamotu Archipelago (French Polynesia). Coral Reefs. 20:399-408.

Andréfouët, S., C. Payri, E.J. Hochberg, L.M. Che and M.J. Atkinson. 2003a. Airborne hyperspectral detection of microbial mat pigmentation in Rangiroa atoll (French Polynesia). Limnology and Oceanography. 48:426-430.

Andréfouët, S., J.A. Robinson, C.M. Hu, G.C. Feldman, B. Salvat, C. Payri and F.E. Muller-Karger. 2003b. Influence of the spatial resolution of SeaWiFS, Landsat-7, SPOT, and International Space Station data on estimates of landscape parameters of Pacific Ocean atolls. Canadian Journal of Remote Sensing. 29: 210-218.

Andréfouët, S., E.J. Hochberg, C. Chevillon and F.E. Muller-Karger. 2005. Multi-scale integration of remote sensing tools to understand physical and biological processes in coastal environments: examples on coral reefs In: R.L. Miller, C.E. Del Castillo and B.A. McKee (Eds), Remote Sensing of Aquatic Environments, Springer, pp. 297-315.

Arias-Gonzalez, J.E., B. Delesalle, B. Salvat and R. Galzin. 1997. Trophic functioning of the Tiahura reef sector, Moorea Island, French Polynesia. Coral Reefs. 16:231-246.

Atkinson, M. 1981. Phosphate flux as a measure of net coral reef flat productivity. Proceedings of the Fourth International Coral Reef Symposium, 1:417-418.

Atkinson, M.J. 1988. Are coral reefs nutrient-limited? Proceedings of the Sixth International Coral Reef Symposium, 1:157-166.

Atkinson, M.J. and R.W. Grigg. 1984. Model of a Coral-Reef Ecosystem 2. Gross and net benthic primary production at French Frigate Shoals, Hawaii. Coral Reefs. 3:13-22.

Bainbridge, S.J. and R.E. Reichelt. 1988. An assessment of ground truth methods for coral reef remote sensing data. Proceedings of the Sixth International Coral Reef Symposium, 2:439-444.

Baker, K.S. and R.C. Smith. 1982. Bio-optical classification and model of natural waters. Limnology and Oceanography. 27:500-509.

Barnes, R.A., D.K. Clark, W.E. Esaias, G.S. Fargion, G.C. Feldman and C.R. McClain. 2003. Development of a consistent multi-sensor global ocean colour time series. International Journal of Remote Sensing. 24:4047-4064.

Birkeland, C.E. 1988. Geographic comparisons of coral-reef community processes. Proceedings of the Sixth International Coral Reef Symposium, 1:211-220.

Bishop, J.K.B. and W.B. Rossow. 1991. Spatial and temporal variability of global surface solar irradiance. Journal of Geophysical Research. 96:16, 839-816, 858.

Bohnsack, J.A. and J.S. Ault. 1996. Management strategies to conserve marine biodiversity. Oceanography. 9:73-82.

Bouvet, G., J. Ferraris and S. Andréfouët. 2003. Evaluation of large-scale unsupervised classification of New Caledonia reef ecosystems using Landsat 7 ETM+ imagery. Oceanologica Acta. 26:281-290.

Brock, J.C., C.W. Wright, T.D. Clayton and A. Nayegandhi. 2004. LIDAR optical rugosity of coral reefs in Biscayne National Park, Florida. Coral Reefs. 23:48-59.

Brown, B.E. 1988. Assessing environmental impacts on coral reefs. Proceedings of the Sixth International Coral Reef Symposium, 1:71-79.

Buddemeier, R.W. and R.A.I. Keinzie. 1976. Coral growth. Oceanography and Marine Biology - An Annual Review. 14:83-225.

Catt, P.C. and D. Hopley. 1988. Assessment of large scale photographic imagery for management and monitoring of the Great Barrier Reef. Remote Sensing of the Coastal Zone: International Symposia, Gold Coast, pp. 1-14.

Chauvaud, S., C. Bouchon and R. Maniere. 1998. Remote sensing techniques adapted to high resolution mapping of tropical coastal marine ecosystems (coral reefs, seagrass beds and mangrove). International Journal of Remote Sensing. 19:3625-3639.

Chertock, B., R. Frouin and R.C.J. Somerville. 1991. Global monitoring of net solar irradiance at the ocean surface - climatological variability and the 1982-1983 El niño. Journal of Climate. 4:639-650.

Christensen, N.L., A.M. Bartuska, J.H. Brown, S. Carpenter, C. D'Antonio, R. Francis, J.F. Franklin, J.A. MacMahon, R.F. Noss, D.J. Parsons, C.H. Peterson, M.G. Turner and R.G. Woodmansee. 1996. The report of the Ecological Society of America committee on the scientific basis for ecosystem management. Ecological Applications. 6:665-691.

Clark, C.D., H.T. Ripley, E.P. Green, A.J. Edwards and P.J. Mumby. 1997. Mapping and measurement of tropical coastal environments with hyperspectral and high spatial resolution data. International Journal of Remote Sensing. 18:237-242.

Crossland, C.J., B.G. Hatcher and S.V. Smith. 1991. Role of coral reefs in global ocean production. Coral Reefs. 10:55-64.

Darnell, W.L., W.F. Staylor, S.K. Gupta and F.M. Denn. 1988. Estimation of surface insolation using sun-synchronous satellite data. Journal of Climate. 1:820-835.

Darwin, C. 1897. The structure and distribution of coral reefs, 3rd edn. D. Appleton and Co., New York.

Dedieu, G., P.Y. Deschamps and Y.H. Kerr. 1987. Satellite estimation of solar irradiance at the surface of the earth and of surface albedo using a physical model applied to Metcosat Data. Journal of Applied Meteorology. 26:79-87.

Done, T.J. 1983. Coral zonation: its nature and significance, In: D.J. Barnes (Ed.) Perspectives on Coral Reefs. Australian Institute of Marine Science, Brian Clouston Publisher, Manuka, ACT, Australia, pp. 107-139.

Done, T.J. 1992. Phase shifts in coral reef communities and their ecological significance. Hydrobiologia. 247:121-132.

Done, T.J. 1999. Coral community adaptability to environmental change at the scales of regions, reefs and reef zones. American Zoologist. 39:66-79.

Dustan, P., S. Chakrabarti and A. Alling. 2000. Mapping and monitoring the health and vitality of coral reefs from satellite: a biospheric approach. Life Support & Biosphere Science. 7:149-159.

Frouin, R., C. Gautier, K.B. Katsaros and R.J. Lind. 1988. A comparison of satellite and empirical formula techniques for estimating insolation over the oceans. Journal of Applied Meteorology. 27:1016-1023.

Gattuso, J.P. and J. Jaubert. 1985. Photosynthesis and respiration of caulerpa-racemosa (chlorophyceae, caulerpales) grown in aquaria - effects of light and temperature. Botanica Marina. 28:327-332.

Gattuso, J.P., M. Pichon, B. Delesalle and M. Frankignoulle. 1993. Community metabolism and air-sea CO_2 fluxes in a coral reef ecosystem (Moorea, French Polynesia). Marine Ecology Progress Series. 96:259-267.

Gattuso, J.P., M. Frankignoulle and R. Wollast. 1998. Carbon and carbonate metabolism in coastal aquatic ecosystems. Annual Review of Ecology and Systematics. 29:405-434.

Ginsburg, R.N. and E.A. Shinn. 1993. Preferential distribution of reefs in the Florida reef tract: The past is the key to the present. In: R.N. Ginsburg (Ed.) Proceedings of the Colloquium on Global Aspects of Coral Reefs: Health, hazards, and history, Coral Gables, pp. H21-H26.

Gordon, H.R. 1978. Removal of atmospheric effects from satellite imagery of oceans. Applied Optics. 17:1631-1636.

Gordon, H.R. and D.K. Clark. 1981. Clear water radiances for atmospheric correction of coastal zone color scanner imagery. Applied Optics. 20:4175-4180.

Gordon, H.R. and M. Wang. 1994. Retrieval of water-leaving radiance and aerosol optical thickness over the oceans with SeaWiFS: a preliminary algorithm. Applied Optics. 33:443-452.

Green, E.P. and A.J. Edwards. 2000. Remote sensing handbook for tropical coastal management. Coastal management sourcebooks 3. UNESCO, Paris.

Green, E.P., P.J. Mumby, A.J. Edwards and C.D. Clark. 1996. A review of remote sensing for the assessment and management of tropical coastal resources. Coastal Management. 24:1-40.

Grigg, R.W. and S.J. Dollar. 1990. Natural and anthropogenic disturbance on coral reefs. In: Z. Dubinsky (Ed.) Coral Reefs, Ecosystems of the World, Vol 25. Elsevier, Amsterdam, pp. 439-452.

Grigg, R.W., J.J. Polovina and M.J. Atkinson. 1984. Model of a coral reef ecosystem 3. Resource limitation, community regulation, fisheries yield and resource management. Coral Reefs. 3:23-27.

Gu, J.J. and E.A. Smith. 1997. High-resolution estimates of total solar and PAR surface fluxes over large-scale BOREAS study area from GOES measurements. Journal of Geophysical Research-Atmospheres. 102: 29,685-29,705.

Gurney, R.J., J.L. Foster, and C.L. Parkinson (Eds). 1993. Atlas of satellite observations related to global change. Cambridge University Press, Cambridge.

Hatcher, B.G. 1988. Coral reef primary productivity: A beggars banquet. Trends in Ecology and Evolution. 3:106-111.

Hatcher, B.G. 1990. Coral reef primary productivity: a hierarchy of pattern and process. Trends in Ecology and Evolution. 5:149-155.

Hatcher, B.G. 1997a. Coral reef ecosystems: How much greater is the whole than the sum of the parts? Coral Reefs. 16:S77-S9.

Hatcher, B.G. 1997b. Organic production and decomposition. In: C. Birkeland (Ed.) Life and Death of Coral Reefs. Chapman and Hall, New York, pp. 140-174.

Hatcher, B.G., J. Imberger and S.V. Smith. 1987. Scaling analysis of coral reef systems: An approach to problems of scale. Coral Reefs. 5:171-181.

Hedley, J.D. and P.J. Mumby. 2002. Biological and remote sensing perspectives of pigmentation in coral reef organisms. Advances in Marine Biology. 43:277-317.

Hochberg, E.J. and M.J. Atkinson. 2000. Spectral discrimination of coral reef benthic communities. Coral Reefs. 19:164-171.

Hochberg, E.J. and M.J. Atkinson. 2003. Capabilities of remote sensors to classify coral, algae, and sand as pure and mixed spectra. Remote Sensing of Environment. 85:174-189.

Hochberg, E.J., M.J. Atkinson and S. Andréfouët. 2003. Spectral reflectance of coral reef bottom-types worldwide and implications for coral reef remote sensing. Remote Sensing of Environment. 85:159-173.

Holden, H. and E. LeDrew. 1998. The scientific issues surrounding remote detection of submerged coral ecosystems. Progress in Physical Geography. 22:190-221.

Holden, H. and E. LeDrew. 1999. Hyperspectral identification of coral reef features. International Journal of Remote Sensing. 20:2545-2563.

Holden, H. and E. LeDrew. 2001. Effects of the water column on hyperspectral reflectance of submerged coral reef features. Bulletin of Marine Science. 69:685-699.

Holden, H. and E. LeDrew. 2002. Measuring and modeling water column effects on hyperspectral reflectance in a coral reef environment. Remote Sensing of Environment. 81:300-308.

Holling, C.S. 1992. Cross-scale morphology, geometry, and dynamics of ecosystems. Ecological Monographs. 62:447-502.

Hu, C.M. and K.L. Carder . 2002. Atmospheric correction for airborne sensors: Comment on a scheme used for CASI. Remote Sensing of Environment. 79:134-137.

Hughes, T.P. 1994. Catastrophes, phase-shifts, and large-scale degradation of a Caribbean coral reef. Science. 265:1547-1551.

Joyce, K.E. and S.R. Phinn. 2003. Hyperspectral analysis of chlorophyll content and photosynthetic capacity of coral reef substrates. Limnology and Oceanography. 48:489-496.

Khan, M.A., Y.H. Fadlallah and K.G. Al-Hinai. 1992. Thematic mapping of subtidal coastal habitats in the western Arabian Gulf using Landsat TM data (Abu Ali Bay, Saudi Arabia). International Journal of Remote Sensing. 13:605-614.

Kinsey, D.W. 1977. Seasonality and zonation in coral reef productivity and calcification. Proceedings of the Third International Coral Reef Symposium, 2:383-388.

Kinsey, D.W. 1978. Productivity and calcification estimates using slack-water periods and field enclosures. In: D.R. Stoddart and R.E. Johannes (Eds.) Coral Reefs: Research Methods. UNESCO, Paris, pp. 439-468.

Kinsey, D.W. 1983. Standards of performance in coral reef primary production and carbon turnover. In: D.J. Barnes (Ed.) Perspectives on Coral Reefs. Australian Institute of Marine Science, Brian Clouston Publisher, Manuka, ACT, Australia, pp. 209-220.

Kinsey, D.W. 1985. Metabolism, calcification and carbon production I. Systems level studies. Proceedings of the Fifth International Coral Reef Congress, 4:505-526.

Knight, D., E. LeDrew and H. Holden. 1997. Mapping submerged corals in Fiji from remote sensing and in situ measurements: Applications for integrated coastal management. Ocean and Coastal Management. 34:153-170.

Knowlton, N. 1992. Thresholds and multiple stable states in coral reef community dynamics. American Zoologist. 32:674-682.

Kutser, T., A.G. Dekker and W. Skirving. 2003. Modeling spectral discrimination of Great Barrier Reef benthic communities by remote sensing instruments. Limnology and Oceanography. 48:497-510.

Larkum, A.W.D. 1983. The primary productivity of plant communities on coral reefs In: D.J. Barnes (Ed.) Perspectives on Coral Reefs. Australian Institute of Marine Science, Brian Clouston Publisher, Manuka, ACT, Australia, pp. 221-230.

LeDrew, E.F., H. Holden, D. Peddle and J. Morrow. 1995. Towards a procedure for mapping coral stress rom SPOT imagery with in situ optical correction. Proceedings of the Third Thematic Conference on Remote Sensing for Marine and Coastal Environments, 1:211-219.

Lee, Z.P., K.L. Carder, S.K. Hawes, R.G. Steward, T.G. Peacock and C.O. Davis. 1994. Model for the interpretation of hyperspectral remote sensing reflectance. Applied Optics. 33:5721-5732.

Levin, S.A. 1992. The problem of pattern and scale in ecology. Ecology. 73:1943-1967.

Lewis, J.B. 1977. Processes of organic production on coral reefs. Biological Reviews of the Cambridge Philosophical Society. 52:305-347.

Lewis, J.B., E.H. Gladfelter and D.W. Kinsey. 1985. Metabolism, calcification and carbon production. Proceedings of the Fifth International Coral Reef Congress, 4:540-542.

Li, X., W. Pichel, P. Clemente-Colón, V. Krasnopolsky and J. Sapper. 2001. Validation of coastal sea and lake surface temperature measurements derived from NOAA/AVHRR data. International Journal of Remote Sensing. 22:1285-1303.

Lubin, D., W. Li, P. Dustan, C.H. Mazel and K. Stamnes. 2001. Spectral signatures of coral reefs: Features from space. Remote Sensing of Environment. 75:127-137.

Maeder, J., S. Narumalani, D.C. Rundquist, R.L. Perk, J. Schalles, K. Hutchins and J. Keck. 2002. Classifying and mapping general coral-reef structure using Ikonos data. Photogrammetric Engineering and Remote Sensing. 68:1297-1305.

Malthus, T.J. and P.J. Mumby. 2003. Remote sensing of the coastal zone: an overview and priorities for future research. International Journal of Remote Sensing. 24:2805-2815.

Matsunaga, T. and H. Kayanne. 1997. Observations of coral reefs on Ishigaki Island, Japan, using Landsat TM images and aerial photographs. Proceedings of the Fourth International Conference on Remote Sensing for Marine and Coastal Environments, pp. 657-666.

Mazel, C.R. 1999. Technical issues related to coral reef remote sensing and workshop recommendations. Center for Marine Conservation, Washington.

McClain, C.R., G.C. Feldman and S.B. Hooker. 2004. An overview of the SeaWiFS project and strategies for producing a climate research quality global ocean bio-optical time series. Deep-Sea Research Part II. 51:5-42.

McPherson, B.F. and R.B. Halley. 1996. The south Florida environment: a region under stress. US Geological Survey Circular 1134.

Michalek, J.L., T.W. Wagner, J.J. Luczkovich and R.W. Stoffle. 1993. Multispectral change vector analysis for monitoring coastal marine environments. Photogrammetric Engineering and Remote Sensing. 59:381-384.

Morel, A. 1988. Optical modeling of the upper ocean in relation to its biogenous matter content (case-I waters). Journal of Geophysical Research-Oceans. 93:10749-10768.

Mumby, P.J. and A.J. Edwards. 2002. Mapping marine environments with IKONOS imagery: enhanced spatial resolution can deliver greater thematic accuracy. Remote Sensing of Environment. 82:248-257.

Mumby, P.J., E.P. Green, A.J. Edwards and C.D. Clark. 1997a. Coral reef habitat mapping: how much detail can remote sensing provide? Marine Biology. 130:193-202.

Mumby, P.J., E.P. Green, A.J. Edwards and C.D. Clark. 1997b. Measurement of seagrass standing crop using satellite and digital airborne remote sensing. Marine Ecology Progress Series. 159:51-60.

Mumby, P.J., J.R.M. Chisholm, C.D. Clark, J.D. Hedley and J. Jaubert. 2001. Spectrographic imaging - A bird's-eye view of the health of coral reefs. Nature. 413:36.

Myers, M.R., J.T. Hardy, C.H. Mazel and P. Dustan. 1999. Optical spectra and pigmentation of Caribbean reef corals and macroalgae. Coral Reefs. 18:179-186.

Ouillon, S., P. Douillet and S. Andréfouët. 2004. Coupling satellite data with *in situ* measurements and numerical modeling to study fine suspended-sediment transport: a study for the lagoon of New Caledonia. Coral Reefs. 23:109-122.

Pages, J. and S. Andréfouët. 2001. A reconnaissance approach for hydrology of atoll lagoons. Coral Reefs. 20:409-414.

Peddle, D.E., E. LeDrew and H. Holden. 1995. Spectral mixture analysis of coral reef abundance from satellite imagery and *in situ* ocean spectra, Savusava Bay, Fiji. Proceedings of the Third Thematic Conference on Remote Sensing for Marine and Coastal Environments, 2:563-575.

Pinker, R.T. and J.A. Ewing. 1985. Modeling surface solar radiation: model formulation and validation. Journal of Applied Meteorology. 24:389-401.

Pinker, R.T., R. Frouin and Z. Li. 1995. A review of satellite methods to derive surface shortwave irradiance. Remote Sensing of Environment. 51:108-124.

Purkis, S.J. and R. Pasterkamp. 2004. Integrating *in situ* reef-top reflectance spectra with Landsat TM imagery to aid shallow-tropical benthic habitat mapping. Coral Reefs. 23:5-20.

Sargent, M.C. and T.S. Austin. 1949. Organic productivity of an atoll. Transactions of the American Geophysical Union. 30:245-249.

Sathyendranath, S. and T. Platt. 1988. The spectral irradiance field at the surface and in the interior of the ocean: A model for applications in oceanography and remote sensing. Journal of Geophysical Research-Oceans. 93:9270-9280.

Sathyendranath, S. and T. Platt. 1993. Underwater light field and primary production: Application to remote sensing. In: V. Barale and P.M. Schlittenhardt (Eds.) Ocean Colour : Theory and Applications in a Decade of CZCS Experience. Kluwer Academic, Brussels, pp. 79-93.

Schalles, J., D.C. Rundquist, A. Gitelson and J. Keck. 2000. Close range hyperspectral reflectance measurements of healthy Indo-Pacific and Caribbean corals. Proceedings of the Sixth International Conference on Remote Sensing for Marine and Coastal Environments, 1:431-439.

Shinn, E.A., B.H. Lidz, J.L. Kindinger, J.H. Hudson and R.B. Halley. 1989. Reefs of Florida and the Dry Tortugas: A guide to the modern carbonate environments of the Florida Keys and the Dry Tortugas. International Geological Congress Field Trip Guidebook T176, American Geophysical Union, Washington.

Smith, S.V. 1983. Net production of coral reef ecosystems. In: M.L. Reaka (Ed.) The Ecology of Deep and Shallow Coral Reefs, Symposia Series for Undersea Research. National Oceanic and Atmospheric Administration's Undersea Research Program, Washington, 1:127-131.

Smith, S.V. 1984. Phosphorus versus nitrogen limitation in the marine-environment. Limnology and Oceanography 29:1149-1160

Smith, S.V. 1985. Some thoughts on the past, present, and future of studies on coral reef community metabolism. In: M.L. Reaka (Ed.), The Ecology of Coral Reefs, Symposia Series for Undersea Research. National Oceanic and Atmospheric Administration's Undersea Research Program, Washington, 3(1): 33-36.

Smith, S.V. 1988. Mass balance in coral reef dominated areas. In: B.O. Jansson (Ed.) Coastal-offshore Ecosystem Interactions. Lecture Notes on Coastal and Estuarine Studies. Springer, New York, 22: 209-226.

Smith, S.V. and R.W. Buddemeier. 1992. Global change and coral reef ecosystems. Annual Review of Ecology and Systematics. 23:89-118.

Smith, S.V. and R.W. Buddemeier. 1995. Reflections on the measurement and significance of carbon metabolism on coral reefs. Open File Report Series 95-96, Kansas Geological Survey, Lawrence.

Smith, S.V. and G.S. Key. 1975. Carbon dioxide and metabolism in marine environments. Limnology and Oceanography. 20:493-495.

Smith, S.V. and D.W. Kinsey. 1978. Calcification and organic carbon metabolism as indicated by carbon dioxide. In: D.R. Stoddart and R.E. Johannes (Eds.) Coral Reefs: Research Methods. UNESCO, Paris, pp. 469-484.

Strong, A.E., E.J. Kearns and K.K. Gjovig. 2000. Sea surface temperature signals from satellites: An update. Geophysical Research Letters. 27:1667-1670.

Wilkinson, C. 1992. Coral reefs of the world are facing widespread devastation: Can we prevent this through sustainable management practices? Proceedings of the Seventh International Coral Reef Symposium, 1:11-21.

Wilkinson, C., O. Linden, H. Cesar, G. Hodgson, J. Rubens and A.E. Strong. 1999. Ecological and socioeconomic impacts of 1998 coral mortality in the Indian Ocean: An ENSO impact and a warning of future change? Ambio. 28:188-196.

Wright, W.C. and J.C. Brock. 2002. EAARL: A lidar for mapping shallow coral reefs and other coastal environments. *Seventh International Conference on Remote Sensing for Marine and Coastal Environments,* (CD-Rom).

Yates, K.K. and R.B. Halley. 2003. Measuring coral reef community metabolism using new benthic chamber technology. Coral Reefs. 22:247-255.

Young, I.R., K.P. Black and M.L. Heron. 1994. Circulation in the Ribbon Reef region of the Great Barrier Reef. Continental Shelf Research. 14:117-142.

Section II

Monitoring Applications

Chapter 6

INFRASTRUCTURE AND CAPABILITIES OF A NEAR REAL-TIME METEOROLOGICAL AND OCEANOGRAPHIC *IN SITU* INSTRUMENTED ARRAY, AND ITS ROLE IN MARINE ENVIRONMENTAL DECISION SUPPORT

JAMES C. HENDEE, ERIK STABENAU, LOUIS FLORIT, DEREK MANZELLO, AND CLARKE JEFFRIS

Atlantic Oceanographic and Meteorological Laboratory, National Oceanic and Atmospheric Administration, 4301 Rickenbacker Causeway
Miami, Florida 33149-1026 USA

1. Introduction

The mission of the U.S. National Oceanic and Atmospheric Administration (NOAA) is to understand, predict, and monitor the oceans, coasts, fisheries, and atmosphere. NOAA also participates in the U.S. Coral Reef Task Force created under U.S. Presidential Executive Order 13089 (June 11, 1998, Office of the White House Press Secretary), whose mission is to utilize the combined U.S. governmental agencies to preserve and protect U.S. coral reefs. Although there are many efforts afoot across the globe to increase the breadth of meteorological and oceanographic monitoring networks (e.g., see National Research Council, 2003; U.S. Commission on Ocean Policy, 2005; NOAA Observing System Architecture, 2005; Global Observing Systems Information Center, 2005), few are deployed specifically to monitor coral reef areas, and none contain the information synthesis capabilities and instrumentation designed to monitor coral health directly in near real-time, except as described in this chapter.

1.1 OVERVIEW OF MONITORING STATIONS

Most meteorological and oceanographic monitoring stations have certain common characteristics:

- Measurements are made for air temperature, wind direction, wind speed, wind gusts and barometric pressure. Additional instruments may measure sea temperature, dew point, rainfall, photosynthetically available radiation (PAR) or wave height and period.

- The measurements are usually averaged over a period of one hour and broadcast via a satellite relay (e.g., GOES, Argos or Iridium) or high-frequency radio, and then to a World-WideWeb ("Web") site where the data can be viewed, often in near real-time (see below for further explanation).

- The stations in the network are designed to receive very little maintenance, as the sites are remote and labor is expensive. Instruments requiring high maintenance are not included for these reasons.

L.L. Richardson and E.F. LeDrew (eds.), Remote Sensing of Aquatic Coastal Ecosystem Processes: Science and Management Applications, 135-156.

- There are usually Web-based databases or other mechanisms for retrieving the historical raw data, or quality-controlled data, or both.

- The data are primarily used by weather-predicting agencies to form local and long-range forecasts, and to assist in advising mariners of existing or predicted sea state.

A relatively new application of the technology that led to the development and support of meteorological monitoring stations is deployment of oceanographic monitoring stations that include in-water measurements in addition to the above parameters. Such stations have also been developed for near-shore coastal zones. The newest generation of in-water monitoring stations now provides specific data related to biological processes and stress responses, thus expanding this technology from monitoring of the environment to scientifically based predictive and diagnostic capabilities.

We have been actively involved in the design, construction, deployment, and expansion of near shore (coastal) marine diagnostic and predictive monitoring arrays. This chapter will provide an overview of the current capabilities of our system, and an example of the use of this capability to study aquatic ecosystem processes in a coastal (coral reef) environment.

2. Challenges in Setting up a Network

A number of specific issues and challenges must be addressed before setting up an *in situ* meteorological and oceanographic instrument array. Some of the common and significant problems to setting up a data collection network include the following:

- The financial outlay to construct just one station can be over US $100,000 (but price depends upon many factors). This includes the cost of the instruments, replacements of those instruments, instrument calibrations, travel and transportation costs, diving support, etc.

- In the U.S.A., permission must be received from the U.S. Coast Guard, the Army Corps of Engineers, the local Fish and Wildlife Service, and possibly the local Marine Protected Area (MPA) to construct the station. Such permission may be extremely difficult to come by and may require numerous permits, with long time lags in between application and award of the permit.

- If the site is very remote, the data must be sent via satellite. This requires subscribing to an available and appropriate satellite, then implementing a data retrieval system.

- The cost of field support for technicians is appreciable when you consider salaries, boats, trailers, fuel, insurance, supplies, and other unforeseen costs.

One of the chief disadvantages these stations and networks have is that instruments that malfunction or start to exhibit drift (i.e., begin to record an increasing disparity between true and measured values) cannot be attended to in a timely (i.e., days or weeks) fashion. Another problem is that incoming data in essence "stack up" and are not reviewed in a timely fashion; hence, features of interest may be missed by scientists or MPA managers. We describe below the way in which we have addressed such problems for one particular network. Included is a summary of our current effort to make use of the data collection capabilities of our system to provide a new capability for understanding coral health from afar.

3. The CREWS Network

The Coral Reef Early Warning System (CREWS) network is being developed through the Coral Reef Conservation Program (CRCP) of the National Oceanic and Atmospheric Administration (NOAA) in response to a U.S. Coral Reef Task Force (established through Executive Order 13089) recommendation to install a network of meteorological and oceanographic monitoring stations at all major U.S. coral reef areas (e.g., the Florida Keys, the Bahamas, Hawaii, U.S. Virgin Islands, Puerto Rico, American Samoa, etc.) by 2010. The stations are being constructed and deployed by the Atlantic Oceanographic and Meteorological Laboratory (AOML) in Miami, Florida, and the Coral Reef Ecosystem Division (CRED) in Honolulu, Hawaii. AOML constructs and installs fixed pylon-type stations, and CRED deploys buoys. Other stations providing data to the network are the SEAKEYS Network of seven stations in the Florida Keys (Ogden et al 1994) and the Australian Institute of Marine Science (AIMS) Weather Station Network (http://www.aims.gov.au).

The basic purpose of the CREWS network is to compile long-term data sets upon which MPA managers can be aided in their management decisions, and upon which researchers can determine yearly patterns and trends. CREWS stations (Figure 1) measure wind speeds and gusts, wind direction, air temperature, sea temperature, salinity, PAR, and discrete or broadband ultraviolet radiation (UVR). Oceanographic instruments normally require frequent (every ten days to two weeks) maintenance to prevent biofouling and consequent drift or failure of the instrument. However, because of this level of maintenance, together with the data monitoring software described below, high confidence is attained in the quality of data, and very timely use can be made of the data in the way of an inference engine or expert system (artificial intelligence tool) shell.

Figure 1. A CREWS Station near the Caribbean Marine Research Center, Lee Stocking Island, Exuma Cays, Bahamas.

We describe below the various operational and scientific aspects of the AOML fixed pylon-type stations ("CREWS stations").

4. Station Construction and Deployment

The construction and deployment of CREWS stations involves some formidable and oftentimes lengthy tasks. Table 1 presents a basic list of phases and tasks, while Figure 2 shows a Gantt chart detailing a typical sequence of events involved in the entire process. Although this is not the place to describe each of the phases in depth, a brief discussion of some of the considerations involved in site selection will provide an insight into the follow-up activities that are required in setting up the station.

Table 1. Phases and tasks in CREWS station construction and deployment.

Site Selection	**Pylon Construction**	**Pylon Delivery**
Host Site Discussions	Custom Fabrications	Receive Pylon at Site
Organize Travel	Custom Assemblies	Move to Close Harbor
Dive Plan Instituted	**Electronics Assembly**	Transfer to Boat or Ship
Travel to Host Region	Data Logger	Pylon Pre-Assembly
Boat Trip to Candidate Sites	GOES Transmitter	Electronics Packaging
Return to AOML	Power Supply Assembly	Pack for Shipping
Permitting	Custom Cable Assembly	**Electronics Delivery**
Discover Permitting Agency	Instrument Interfacing	Move to Shipper
Acquire Permit Application	**Testing at AOML**	Station Pre-Installation Assembly
Submit Permit Application	Electronics Tests - AOML	**Pylon Installation**
Receive All Permits	Electronics Tests - Hatchery Tanks	Arrange Travel
Purchasing	**Bottom Plate Installation**	Organize Employee Schedules
Permit Fee(s)	Organize Travel	Travel to Host City
Electronics Infrastructure	Rental or Arrange Boat	Meet with Host Support
Instrumentation	Travel to Site	Travel to Installation Site
Pylon Materials	Deliver Plate and Drilling to Boat	**Install Pylon**
Tools	Install Plate	Electronics Pre-Installation Assembly
Contractor	Review Installation	Electronic Infrastructure
Rentals	Return to AOML	**Electronics Installation**
Other Fees	**Pylon Packaging**	Install Electronics at Site
Freight	Rent or Arrange Lifting Gear	**Field Electronics Testing**
Boat Rentals	Package for Shipment	Test Configuration
Diving Gear	Deliver to Shipper	Feedback from AOML
Drilling Gear		Finalize and Return Home
Software		
Computer Hardware		
Bottom Plate Materials		
Purchase Delivery Trailer		

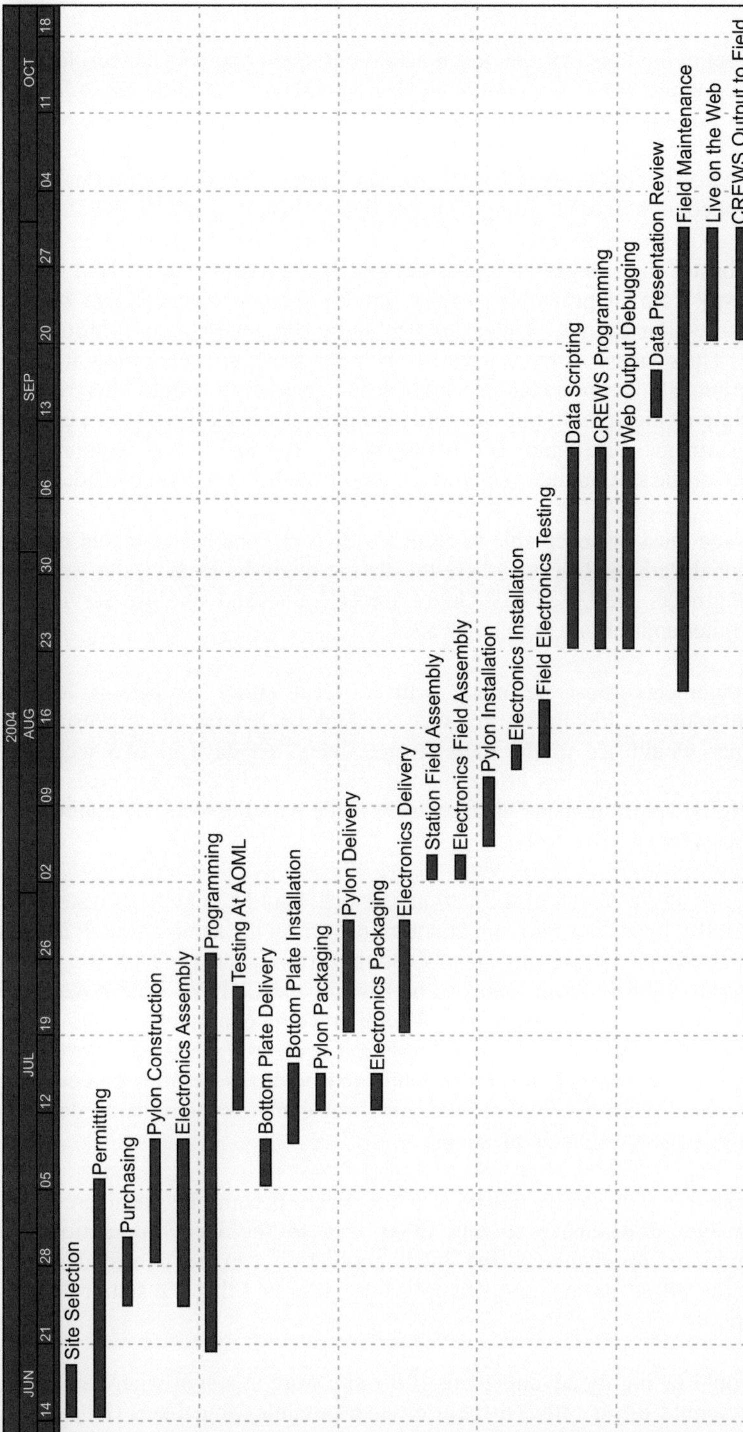

Figure 2. Sequence of events in the planning of installation of a CREWS station.

4.1 SITE SELECTION

It would seem obvious that the best place to put up a coral reef meteorological and oceanographic monitoring station would be near a coral reef, but there are in fact many considerations that must be taken into account. Adhering to the same principals of site selection for each station allows for comparison among the stations, which provides for greater confidence in conclusions. Following are some of the considerations that must be taken into account before choosing a proper site for installing a CREWS station.

- It is best if the site can be on the lee side of the island or local land mass (that is, away from the prevailing winds), but not so close to land so that wind speed and direction cannot be measured to show the general trend throughout the day. The concept is that it would be best not to set up where the station would continuously be pounded by high seas. Not only would this reduce the lifetime of the station, but it would also make it very difficult for personnel to carry out the installation and maintenance. It should be paramount to keep safety of the station maintainer at the top of the list of station considerations.

- The site must be acceptable to NOAA's Coastal Zone Management office, any National Park or Reserve where the station might be located, the Army Corps of Engineers, the U.S. Coast Guard, the U.S. Fish and Wildlife Service, and/or any other appropriate regulatory agency.

- Arrangements must be made with a local entity or person for station maintenance. Ideally, a graduate student or federal or territorial agency partner would see to the maintenance every ten days to two weeks. The upkeep is generally low effort, but maintenance and calibration costs should be carefully determined and an agreement made with the local maintainer on who will pay for specific costs.

- Because of the design of the station, the site should be in 6 m of water. This is especially important for comparison of data from the same research instrument suite deployed at different sites. "Colonized pavement" (NOAA Biogeography Program, 2005) bottom seems to be best for the CREWS station deployment, as it provides for comparisons of similar coral community types at all the regions of installation. Placing the station at an area where extensive research (e.g., coral bleaching) has been conducted before is helpful, especially since the extended research instrument platform contains instruments designed to answer research-related questions.

- The station will ideally not be at a site where it could be considered visually obtrusive or offensive in any other way to the local population. With community involvement, the local population can be educated as to the benefits of the station, and, if applicable, informed that the station location is temporary.

- It would be highly advantageous if the site were at a spot where an interested party could see it often, to help obviate possible vandalism. (Hopefully this and the above conditions go together.)

- In cases where a Web camera is operating (see Figure 3), it would be advantageous if the station were within line-of-sight of an area on shore where the microwave receiver dish could be located. The local receiving dish would relay real-time viewing of the reef below on the observation monitor, and, if desired, that signal could go out as an Internet image stream.

- The station would preferably not be in an area experiencing high tidal range and/or high tidal or other currents. This is for diver and maintainer safety, as well as for the integrity of the station.

- If the station can serve as a navigational aid, this would help bring extra value of the station to the local population.

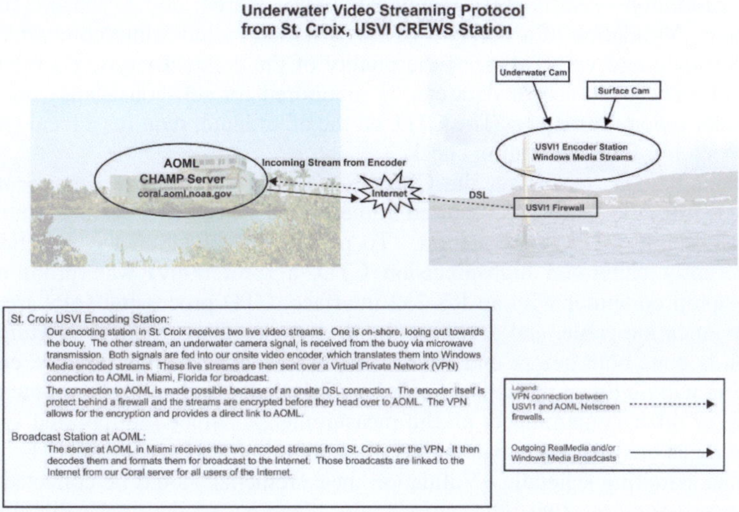

Figure 3. CREWS Web underwater camera architecture. Real-time video is sent from an underwater camera through a microwave transmitter on the CREWS station to a receiver on land, then through an encoder, then finally over a secure service-net to a central Web server where they are broadcast over the Internet.

5. Station Maintenance

The goals of station maintenance are to assure that the highest quality data are collected and delivered, and to prevent the CREWS pylon from significantly altering the environment it is monitoring. Because of the high probability of biofouling and drift of oceanographic instruments, it is essential that the stations receive at least some attention every ten days to two weeks, primarily to reduce biofouling. Thus, it is critical for the CREWS Network to have local cooperating field technicians or scientists at each locale where the stations are deployed. In many cases, utilizing a graduate student to perform the maintenance operations has benefits for both parties: the student receives a stipend, or other support, and gets what he or she knows first hand is good quality data for their thesis/dissertation research; and the station owner gets the required attention to the immediacies of station operation as required.

6. Data Validation

Successful interpretation of the incoming near-real time data by either experts or expert system software (see Information Systems, below) is dependent on the quality of the incoming data. In particular, in remote locations, with sensors chosen for long-term stability and dependability, recorded values for many parameters may experience drift. Automated expert system evaluation can identify suspect data after it has been produced; however, it is necessary for establishing quality of the long-term data set to periodically validate the data. These validations are performed by comparison of the *in situ* sensors that have been exposed to the environment for extended periods against recently calibrated instruments maintained and stored in a clean, controlled environment. Validation of a standard CREWS station includes intercomparison of the wind sensor data and verification of the quality of the conductivity-temperature-depth (CTD) data. The wind sensor data can be monitored for self-consistency and against local weather reports remotely. The CTD, on the other hand, requires a local technician to visit the station for a calibration check.

Validating CTD sensors on the CREWS stations is made by comparison of the CTD values from the station, as reported to the CREWS Web site, with readings from a calibrated portable CTD at the station. To perform a validation the maintainer will need a recently calibrated high-precision CTD, a (preferably) waterproof or water resistant laptop computer with an RS-232 interface, CTD processing software, an RS-232 communication cable, and access to the Web to see the currently reporting values. Validation is done both before and after a simple in-water cleaning process, consisting of including wiping down the sensor surfaces with a soft cloth, to ascertain the effect of biofouling or other contaminants on the measurements. Since the reported values are hourly averages of six discrete data points, the calibration CTD will need to be operated on the same sampling schedule. Validation measurements should be collected for two time periods before and two time periods after cleaning, so the total calibration time, including the cleaning dive, is approximately five hours. It is possible for approved divers to use the calibration time to enter the water to clean and inspect the mooring and supporting hardware for signs of wear. If the CREWS station CTD is found to be out of calibration, it is removed and replaced with the calibrated CTD. The field CTD that was removed would then be returned for cleaning and calibration. Maintainers should follow this validation process at least once every two weeks.

7. Information Systems

7.1 PRESENTATION OF REAL-TIME RAW DATA

At the station each instrument sends data (or is polled by a data logger), which averages the number of readings over the hour and then transmits a stream of numbers ("datastream") up to a GOES satellite. The datastream is then relayed to Wallups Island, Virginia, and then to AOML (see Figure 4). Individual sensor values are then parsed from the datastream and presented as near real-time values on the CREWS Web Page (http://www.coral.noaa.gov/crews), with the disclaimer that the data should only be considered as provisional since they have not yet been reviewed. The disclaimer is necessary because mariners and others who are trying to make decisions based on the data should keep in mind that the data might be wrong (e.g., through data drift or other mechanical problem), and to thus make their decisions carefully. However, because of the CREWS suite of expert systems, notice is usually given automatically when data look suspect, and the maintainer is usually able to attend to a failing or drifting instrument in a timely fashion.

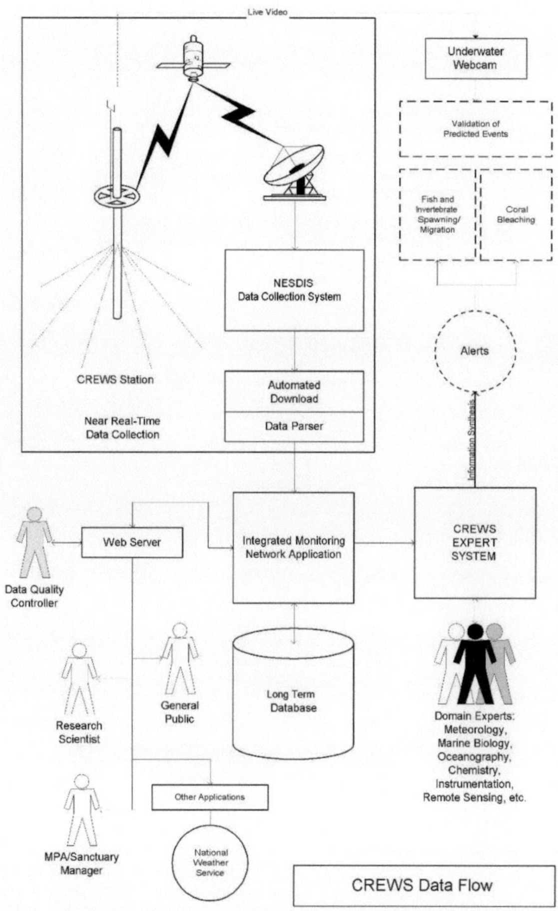

Figure 4. Flow of data from the CREWS station to the various destinations.

7.2 DATA QUALITY CONTROL

After the data have been parsed from the raw data stream, data quality control personnel receive files of the raw data. The files are reviewed and edited in an attempt to remove unreliable numbers. Generally speaking, data are accepted if they fall within a range of expected, ambient values that are determined from trends of previously received data, personal experience, and the published literature. For example, sea surface temperatures (SST) in the Caribbean are expected to be in the range of 26-32° C during summer months, whereas SST values of 18-24° C are normal for winter months. Care must be exercised when determining ranges, given that these vary from location to location. The CREWS expert system software reviews data for appropriate ranges as they arrive (see below), but follow-up screening by a data quality specialist ensures the quality of the data before archiving in the CREWS Integrated Monitoring Network (see Figure 5).

Figure 5. An example interface from Integrated Monitoring Network, which collects data from the CREWS and SEAKEYS networks.

Completeness reports are compiled for each station for each year and are included in metadata reports. Completeness reports list the availability of each parameter for that year. For example, if salinity readings are not available this is noted indicating over the range (in days) for which values do not exist. Once the completeness report is finished, metadata files are then emailed to and stored by NOAA's Coral Reef Information System (CoRIS). Annual data files are then uploaded to the CREWS Web site where they are accessed by the general public.

7.3 EXPERT SYSTEM ANALYSIS

The underlying structure of the expert system has been previously described in Hendee (1998, 2000). Production rules (basically, if/then-type heuristics) utilized in CREWS are drawn from published data, field observations, and from discussions in the literature.

The approach presented here reflects first-order laboratory and field-based testing of instruments, in conjunction with programming of the expert system. The expert system acts as a model, which will eventually help to further elucidate the role of the physical environment in biological events which are influenced by the physical environment, and which can be measured with a robust sensor.

Table 2 presents the terms used for the inference engine within the expert system. Basically, these subjective interpretations help to reduce complexity in modeling the environment and in assessing its role in influencing a marine behavioral event (e.g., coral bleaching). Data from each sensor are categorized according to the table into *subjective data ranges* (e.g., drastically low, very low, etc.), as described by experts who use the sensors or work with the parameters in question. The terms "unbelievably low" or "unbelievably high" represent thresholds beyond which the measurements would be considered unrealistic in nature. The *subjective periods of the day* explained in Table 2 are those perceived by humans, and which also quite often correspond to periods of biotic behavior (e.g., crepuscular feeding behavior at "dawn" or "sunset"). When an observed condition (e.g., high sea temperature) holds to the same subjective data range beyond one of the basic periods, which are three hours each, the condition is reassigned to the next larger category. For instance, if sea temperature is "very high" for "dawn" and "morning" (each of which is a three hour period) it becomes reassigned as "dawn-morning," a six hour period. Similarly, if the high sea temperatures persist for all daylight hours, the condition is reassigned as occurring for "daylight-hours."

8. Research Application

The CREWS stations have been designed to provide an extensible architecture so that instrumentation may be added relatively easily to provide answers to selected research questions. One of the primary goals of the CREWS stations has been to elucidate the role of light and temperature in the phenomenon of coral bleaching. Another research question of value is determining what role carbon dioxide flux plays in coral growth, bleaching, and coral larval settlement. In select situations, where sediment resuspension and turbid outflow may have deleterious effects on the coral reef, turbidity sensors might also be added. The list of possible research questions that can be addressed by quality long term investigation of specific parameters includes, in addition to the above mentioned parameters, monitoring of nutrients, dissolved oxygen, pH (potential hydrogen), Eh (electrochemical potential), and video-based ecosystem analysis. For the present discussion, research into the role of light and high sea temperature in coral bleaching events will serve as an example of the effectiveness of the CREWS station design and expert system analysis.

Subjective Data Ranges:			
Abbreviation	**Descriptor**	**Abbreviation**	**Descriptor**
ul	unbelievable low	sh	somewhat high
dl	drastically low	hi	high
vl	very low	vh	very high
lo	low	dh	drastically high
sl	somewhat low	uh	unbelievably high
aw	average		
Subjective Periods of the Day:			
Abbreviation	**Period**	**Local Time**	
(Basic Periods)			
midn	midnight	2200 - 0100	
pdaw	pre-dawn	0100 - 0400	
dawn	dawn	0400 - 0700	
morn	morning	0700 - 1000	
midd	mid-day	1300 - 1600	
psun	pre-sunset	1600 - 1900	
suns	sunset	1900 - 2200	
even	evening		
(Larger Groupings)			
all	all-day	2200 - 2200	
dayl	daylight-hours	0400 - 1900	
nite	night-hours	1900 - 0400	
dayb	dawn-morning	0400 - 1000	
aftn	afternoon	1300 - 1900	

Table 2. Local time periods are subjectively grouped for use by the expert system software to allow the automated interpretation of data with respect to easily understood diurnal periods. If the conditions characterized in the subjective data ranges (e.g., "somewhat low") hold for beyond one or more of the Basic Periods, they are recategorized into the next Longer Period (e.g., "daylight-hours"). See Table 3 for the next step in processing.

8.1 CORAL BLEACHING

Coral bleaching may be described as the general whitening of coral colonies due to the loss of symbiotic zooxanthellae from coral tissues and/or a reduction in the densities of zooxanthellae and photosynthetic pigment concentrations within the zooxanthellae (Glynn 1993). Typically, bleaching is a response to environmental stressors including extreme water temperature (Fitt, et al. 2001, Saxby, et al. 2003), intense light (Anderson, et al. 2001), or biological factors such as infection by bleaching-inducing bacteria (Sutherland, et al. 2004). Intense coral bleaching may result in extensive mortality of reef-building corals and a reduction in the growth rate of surviving colonies. A further consequence is the loss of production of calcareous substrate (the reef framework) necessary to support the abundance and variety of reef-dwelling and reef-dependent aquatic life. Interestingly, scientists have discovered that corals live in environments that are very close to the limit in which they can survive with respect to temperature, insolation, and even pH. A water temperature increase above a selected coral's upper thermal tolerance of 1 to 2°C for two weeks has been accepted as a rough heuristic for coral bleaching. This temperature range is quite close to the upper limit that the coral normally experience (Coles and Jokiel 1976).

It is known that not all corals, even of the same species, will exhibit signs of coral bleaching over their entire surface under the conditions of thermal stress described above. In addition to temperature stress, increases in the duration and intensity of light exposure beyond the range of photoacclimatization has shown a strong correlation with the coral bleaching response (Shick et al. 1996, Hoegh-Guldberg 1999, Lesser and Farrell 2004). It is believed that the combination of high temperature combined with increased insolation in surface waters leads to coral bleaching, and that the areas of corals of the same species and symbiotic (zooxanthellate) clade that do not bleach are perhaps not exposed to equivalent amounts of sunlight. The intensity of light on the surface could be limited locally by physical shading, as in the bottom surfaces of corals as opposed to the tops, or due to UVR screening substances. These include chromophoric dissolved organic matter (CDOM) in the water column or mycosporine-like amino acids within the coral itself (Lesser and Farrell 2004, Otis, et al. 2004).

In order to gain an understanding of the role of light in coral bleaching, it is necessary to determine the spectra and intensity of light at the coral surface. Providing individual light sensors at each coral in a study region is prohibitively expensive and may disrupt the environment. Therefore, an approach taken by the CREWS research team has been to calculate coral light exposure by extrapolating from CREWS station irradiance data.

8.1.1 *Remote verification of coral bleaching alerts and predictions*

One method we have developed to verify our coral bleaching alerts and predictions has been to install an underwater camera that can transmit images locally to a shore-based station and then to the Web. The underwater Coral Camera infrastructure at St. Croix (Figure 3) is composed of two cameras (above and below water) and a Niagara™ streaming video computer server (which encodes the signal from analog to digital) located on shore, and also at AOML. The above-water camera, located on shore and pointing directly at the CREWS station, is connected to a video server inside a building. This camera is a manual zoom camera mounted in an industrial-strength all-weather housing within direct line-of-sight of the station. The second camera is located underwater and tethered to the CREWS station in such a manner that a diver is able to move the camera to different viewing locations around the CREWS station, to as far away as approximately 150 m, and down to a depth of 20 m. The camera is powered by the central station's solar-powered batteries and is in communication with the onshore video server via a microwave link, consisting of a transmitter on the station and a receiver onshore. Once the video signal is processed and converted into Windows Media™ by the encoder, the signal is then prepared for streaming over the Internet. This is possible only via an Internet Virtual Private Network (VPN) link between St. Croix and AOML, since security is of paramount concern for all federal installations. When an Internet user wishes to view the video stream, a request is sent to the AOML based server. The server in turn creates a link over the VPN to the video server in St. Croix. The server then packages the video stream and provides it to the Internet requestor. In this way, after CREWS initiates bleaching alerts for St. Croix, a user can look to the underwater Coral Camera to see if, indeed, bleaching is taking place. Since different species bleach before others, however, it is of course important to position the camera to point at the species of interest.

8.1.2 *The Underwater Light Field*

The wavelength dependence of biological responses to light is well established. The relationship varies from the induction of photosynthetic activity by PAR (400 – 700 nm) to the induction of repair enzymes (or production of UVR-screening pigments) during exposure to component bands of ultraviolet–A radiation (UV-A, 315 – 400 nm) (Corredor, et al. 2000) or to direct photochemical damage to DNA by ultraviolet–B radiation (UV-B, 280 – 315 nm) (Lyons, et al. 1998). Due to the complexity of coral responses to light, a number of studies aimed at investigating the relationship between different spectral regions and coral bleaching have been undertaken (Gleason and Wellington 1993, Fitt and Warner 1995). In these studies, both the density of zooxanthellae cells and the concentration of chlorophyll in zooxanthellae showed significant reductions with increasing UVR and visible light intensity.

UV and blue light penetration in shallow oligotrophic environments, with chlorophyll *a* concentrations of less than 0.5 mg/m^3, is largely controlled by CDOM (Markager and Vincent 2000, Nelson and Siegel 2002). This macromolecular mixture of organic molecules contains a complex array of unidentified chromophores with overlapping absorbance spectra (Stabenau and Zika 2004). The mixture is often characterized by its exponential increase in absorbance with decreasing wavelength (Green and Blough 1994, Kuwahara, et al. 2000). A simple exponential equation:

$$a_\lambda = a_{\lambda o} \exp(-S(\lambda - \lambda_o))$$

where $a_{\lambda o}$ is the absorption coefficient at λ_o (i.e., 290 nm) and S is the spectral slope coefficient, is used to fit measured light absorbance data, allowing differentiation between classes of CDOM by differences in S (Blough and Green 1994). It has been shown (Otis, et al. 2004) that $a_{\lambda,\mathrm{CDOM}}$ dominates the total attenuation of light below 500 nm near a CREWS station at Lee Stocking Island, Bahamas. Since the diffuse attenuation coefficient (Kd) is dominated by absorbance from CDOM at these wavelengths, it is expected that the wavelength dependence of Kd will show an exponential increase with decreasing wavelength, similar to that observed for CDOM. The diffuse attenuation spectral slope coefficient (S_{Kd}), can be used to describe this behavior and subsequently to predict the spectra and intensity at any depth within the well mixed waters found near the CREWS station.

One complexity in the determination of Kd from the CREWS data is the necessity for two in-water irradiance values from different depths. The CREWS station typically employs only a single in-water and a single above-water irradiance sensor. To determine Kd, the above water sensor data is corrected for reflective losses and refractive differences at the sea surface to produce theoretical irradiance values for a subsurface, z = -0 m, depth. Once the corrections have been performed, the subsurface values and measured in-water values are used to calculate Kd at the measured wavelengths (305, 330, 380 nm and a broadband measure of PAR). The S_{Kd} value is then calculated to allow interpolation between these measured wavelengths in order to determine the total wavelength dependent Kd. This value is applied to standard surface spectra to correct for in-water attenuation, valid for CDOM and other absorptive features for the specific hour the data were collected, and used to predict in near real-time the spectra and intensity of light at the coral surface.

An inherent feature of this approach is that variations in S_{Kd} are related to variations in the type and processing history of CDOM, which in coastal zones is closely coupled to variations in biota, including seagrass communities and surface water run-off (Stabenau, et al. 2004). Variations in CDOM type or concentration may indicate normal seasonal changes in these communities or may indicate long term ecosystem changes, due to either natural or anthropogenic causes. Thus, one achievement of CREWS station long term monitoring via the methods described here may be the elucidation of the causes and implications of variation in the shading properties of CDOM on coral health.

8.1.3. *Fluorescence Efficiency*

Previous studies have shown reduction in coral fluorescent yield as a coral stress response to, for example, high sea temperature (Jones, et al. 2000) or exposure to intense UVR (Jones and Hoegh-Guldberg 2001). This reduction in fluorescent yield typically precedes either mild or extensive coral bleaching, including the expulsion of zooxanthellae from the coral host. Variation in fluorescent yield during a diurnal cycle is larger than the night-to-night, dark-adapted variation in efficiency observed over time. Monitoring of the nighttime fluorescence yield of corals provides an early indication of the onset of coral bleaching, prior to when subjective determination of changes in the corals' color can be observed. For this reason, select CREWS stations have been designed to incorporate a direct measure of fluorescent yield to provide better coral bleaching predictions. Long term monitoring of fluorescence yield provides an additional baseline from which the expert system software can calculate deviation and determine significance.

The typical method for determining fluorescent yield is via pulse amplitude modulation (PAM) fluorometry. PAM fluorometers have been used at individual locations and with variable sampling intervals to determine photosynthetic health, both for terrestrial plants and, more recently, for submerged vegetation and corals (for a review, see Fitt et al. 2001). The next generation PAM fluorometer, a multi-sensor "monitoring PAM" designed for long-term underwater deployments, has been incorporated into the CREWS station architecture (see Figures 6-8). Typical PAM fluorometry output data (Figure 9) provide the dark adapted background fluorescence (F_0) and the maximum fluorescence observed when the coral is exposed briefly to a saturating pulse of broad spectrum light (F_m). The instrument then calculates the variation in fluorescence ($F_v = (F_m - F_0)/F_m$) which is the fluorescent yield. In a monitoring PAM deployment, the time series data will show very low values of F_v during the day since the corals undergo dynamic photoinhibition, essentially eliminating a portion of the excess light energy from photosystem II for protection from cellular damage (Lesser and Gorbunov 2001). However, in the nighttime hours (a subjective period recognized by the expert system software, cf. Table 2), the F_v values quickly rise to a stable range, with healthy corals achieving values as high as 0.65. Higher values are exhibited in conditions favorable for coral growth, while lower values are observed when corals are stressed. Subjective data ranges are applied to these values to allow expert system-based early warnings of coral bleaching events.

Figure 6. Power from the central CREWS station is supplied to the central canister of the PAM fluorometer (see text for further explanation). Here a diver unravels the four cables from the central canister which lead to the individual monitoring PAM sensors, which will be applied to four different species of corals.

Figure 7. Divers position one of the monitoring PAM sensors at one of the species (*Agaricia agaricites*) to be monitored.

Figure 8. A diver must position the monitoring PAM sensor head precisely over the coral species in question (here, *Siderastrea siderea*) using a pre-measured spacing tool.

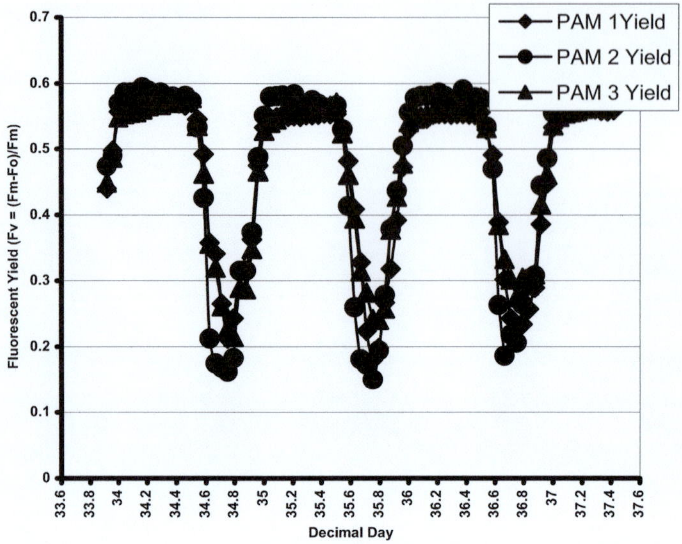

Figure 9. Diurnal cycle observed from multiple detectors of a monitoring pulse amplitude modulating (PAM) fluorometer deployed at Lee Stocking Island, Bahamas, 2005. Decrease in nighttime fluorescent yield would be expected when corals are experiencing thermal or photochemical stress (not indicated here). Separation of the signal into subjective day and night periods is necessary to interpret long term trends.

The decision table utilizing the abbreviations and subjective ranges shown in Table 2 may be seen in Table 3 as an example for use of these data to predict coral bleaching. The monitoring and information architecture, therefore, not only models conditions conducive to coral bleaching, but also reports when the coral is actually undergoing physiological stress consistent with coral bleaching. Such output is of value not only to coral researchers seeking to understand the environmental stressors and physiological mechanisms associated with bleaching, but also to MPA managers who wish to directly assess the status of a species of coral being monitored.

Deployment of a monitoring PAM fluorometer requires special consideration. Since it is an optical instrument utilizing a light source, it needs to be frequently cleaned because of the detrimental affect of biofouling organisms. Unfortunately, anti-fouling options are potentially harmful to the corals being monitored and, at the time of this publication, a successful mechanism for the automated cleaning or protection of optical interfaces has yet to be developed. Here again, high quality and intense *in situ* monitoring requires an attentive station maintenance plan attuned to output from the CREWS software and a regular station maintenance schedule. Fortunately, the lens itself is a polycarbonate material and apparently has similar resistance to biofouling as has been observed with Teflon™ coatings on irradiance sensors. These coatings exhibit fairly low rates of biofouling because the non-porous material isn't easily colonized, or adhered to, by biofouling organisms in the marine environment.

In certain applications, the target corals may be some distance from the CREWS station. However, utilization of the monitoring PAM fluorometer is still possible, with communication to the tower achieved through use of acoustic modem technology (Figure 10), currently under development at AOML. Acoustic modems transfer data underwater acoustically, rather than through the use of cables, satellite or radio transmission, as do aerial data transmission modes. AOML's Telemetered Instrument Array (TIA) architecture consists of a receiving acoustic modem on the CREWS station, and a sending acoustic modem on a special remote instrumented array. The TIA is capable of physically supporting and combining the signals from a monitoring PAM fluorometer, a Conductivity-Temperature-Depth (CTD) instrument, an irradiance spectrophotometer, and a transmissometer. Data are acquired from each of these instruments, averaged over a period of one hour, and then sent as a composite data stream to the receiving modem on the CREWS station. All data parameters from the station are then combined for transmission to the GOES satellite in the usual fashion. The flexibility of combining multiple instruments at great ranges from the station, with data collection and delivery in near-real time at the station, greatly expands the possibilities for support of *in situ* coral research since different needs and different researchers may be accommodated.

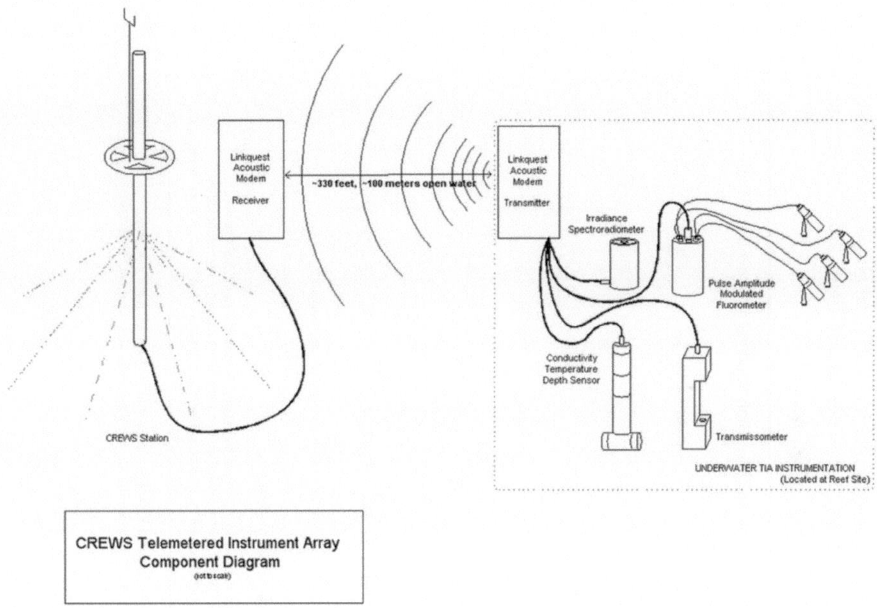

Figure 10. Envisioned deployment of a Telemetered Instrument Array, containing a monitoring PAM fluorometer, a CTD instrument, a transmissometer, an irradiance spectrophotometer, and a transmitting modem, in the vicinity of a CREWS station with a receiving acoustic modem for acquiring data remote from the station. See text for details.

Table 3. Expert software system decision table, set at initial conditions based on the study herein, used to provide reports on the probability of a coral bleaching event based on combinations of environmental conditions and coral photosynthetic health indicators in specific time intervals. Conditions monitored include Fv (unitless ratio of coral fluorescence at rest and under intense light), Ed_λ (wavelength specific downwelling irradiance, $\mu W/cm^2/Sr^1$) at the coral surface, and sea surface temperature.

Rule: High Sea Temp + Low Fluoro Yield (night) + High Noon Irradiance + Low Winds
(Julian Day : 172 to 264... Season: Summer)

		ul	dl	vl	lo	sl	av	sh	hi	vh	dh	uh
IF	sea temp									all (24) dayl (15) nite (9) dayb (6) aft (6) basic (3)	all (48) dayl (30) nite (18) dayb (12) aft (12) basic (6)	
and:	fluor yield		nite (18)	nite (9)	nite (6)							
and:	irradiance									mid (6)	mid (12)	
and:	wind speed		all (24) dayl (15) nite (9) dayb (6) aft (6) basic (3)	all (24) dayl (15) nite (9) dayb (6) aft (6) basic (3)								

THEN

Conditions are (probably/possibly) conducive to mass coral bleaching.

See Table 2 for day-period designations.

Note: The expert system adds up all the points in parentheses for conditions that hold, then outputs an "alert" which shows totals for each parameter, as well as a combined total. The total number of points acts as an environmental index of stress that results in coral bleaching.

KEY: Subjective assignments for fluorescence yield and maximum noon insolation.
(Note: These are starting levels yet to be verified through lab and field experimentation.)

	FV		ED
dl	.30 - .40	uh	> 1000
vl	.40 - .55	dh	800 - 1000
lo	.55 - .65	vh	600 - 800
uh	> .80		

9. Acknowledgements

The CREWS project has evolved through the efforts of many private and government agencies (both foreign and domestic). The individual contributors are too numerous to list here. However, we would like to acknowledge our co-workers in the core CREWS crew who daily strive to make the program successful. They are listed here alphabetically: Jeff Absten, Jules Craynock, John Halas, Mike Jankulak, Jeff Judas, Chris Langdon, Emy Roque-Rodriguez, Michael Shoemaker, and Scott Stolz.

The CREWS project has been funded through NOAA's Coral Reef Conservation Program, and the NOAA High Performance Computing & Communications office. The findings and conclusions in this report are those of the authors and do not necessarily represent the views of NOAA.

10. References

Anderson, S., R. Zepp, J. Machula, D. Santavy, L. Hansen and E. Mueller. 2001. Indicators of UV exposure in corals and their relevance to global climate change and coral bleaching. Human and Ecological Risk Assessment, 7:271-1282.

Blough, N.V. and S.A. Green. 1994. Spectroscopic characterization and remote sensing of non-living organic matter. In: Zepp RG, Sonntag C (eds.) Role of Non-Living Organic Matter in the Earth's Carbon Cycle. Wiley, New York. 42-57 pp.

Blough, N.V. and R. Del Vecchio. 2002. Distribution and dynamics of chromophoric dissolved organic mater (CDOM) in the coastal environment, p. 503-541. In D. Hansell and C. Carlson [eds.], Biogeochemistry of Marine Dissolved Organic Matter. Academic Press, 209 pp.

Coles, S.L., P.L. Jokiel and C.R. Lewis. 1976. Thermal tolerance in tropical versus subtropical Pacific reef corals. Pacific Science, 30:156-166.

Corredor, J.E., A.W.Bruckner, F.Z. Muszynski, R.A. Armstrong, R. Garcia and J.M. Morell. 2000. UV-absorbing compounds in three species of Caribbean zooxanthellate corals: depth distribution and spectral response. Bulletin of Marine Science, 67:821-830.

Fitt, W.K., B.W.Brown, M.E. Warner and R.P. Dunne. 2001. Coral bleaching: interpretation of thermal tolerance limits and thermal thresholds in tropical corals. Coral Reefs, 20:51-65.

Fitt, W.K. and M.E. Warner. 1995. Bleaching patterns of four species of Caribbean reef corals. Biological Bulletin (Woods Hole), 189:298-307.

Gleason, D.F. and G.M. Wellington. 1993. Ultraviolet-radiation and coral bleaching. Nature, 365:836-838.

Global Observing Systems Information Center. 2005. http://www.gosic.org/ios/GOOS_ios.htm, observed March 31, 2005.

Glynn, P.W. 1993. Coral bleaching: ecological perspectives. Coral Reefs, 12:1-18.

Green, S.A. and N.V. Blough. 1994. Optical-absorption and fluorescence properties of chromophoric dissolved organic matter in natural waters. Limnology and Oceanography, 39:1903-1916.

Hendee, J.C. 1998. An expert system for marine environmental monitoring in the Florida Keys National Marine Sanctuary and Florida Bay. 2nd International Conference on Environmental Coastal Regions. 1:57-66.

Hendee, J.C. 2000. A data-driven soft real-time expert system for producing coral bleaching alerts. Ph.D. Thesis, Nova Southeastern University, 131 pp.

Hoegh-Guldberg, O. 1999. Climate change, coral bleaching and the future of the world's coral reefs. Marine and Freshwater Research, 50:839-866.

Jerlov, N.G. 1968. Optical Oceanography. Elesevier,194 pp.

Jones, R.J. and O. Hoegh-Guldberg. 2001. Diurnal changes in the photochemical efficiency of the symbiotic dinoflagellates (Dinophyceae) of corals: photoprotection, photoinactivation and the relationship to coral bleaching. Plant Cell and Environment, 24:89-99.

Jones, R.J., S. Ward, A.Y. Amri and O. Hoegh-Guldberg. 2000. Changes in quantum efficiency of Photosystem II of symbiotic dinoflagellates of corals after heat stress, and of bleached corals sampled after the 1998 Great Barrier Reef mass bleaching event. Marine and Freshwater Research, 51:63-71.

Kieber, D.J. and N.V. Blough. 1990. Determination of Carbon-Centered Radicals in Aqueous Solution by Liquid Chromatography with Fluorescence Detection. Analytical Chemistry, 62:2275-2283.

Kuwahara, V.S., H. Ogawa, T. Toda, T. Kikuchi and S. Taguchi. 2000. Variability of bio-optical factors influencing the seasonal attenuation of ultraviolet radiation in temperate coastal waters of Japan. Photochemistry and Photobiology, 72:193-199.

Lesser, M.P. and J.H. Farrell. 2004. Exposure to solar radiation increases damage to both host tissues and algal symbionts of corals during thermal stress. Coral Reefs, 23:367-377.

Lesser, M.P. and M.Y. Gorbunov. 2001. Diurnal and bathymetric changes in chlorophyll fluorescence yields of reef corals measured *in situ* with a fast repetition rate fluorometer. Marine Ecology-Progress Series, 212:69-77.

Lyons, M.M., P. Aas, J.D. Pakulski, L. Van Waasbergen, R.V. Miller, D.L. Mitchell and W.H. Jeffrey. 1998. DNA damage induced by ultraviolet radiation in coral-reef microbial communities. Marine Biology, 130: 537-543.

Markager, S. and W.F. Vincent. 2000. Spectral light attenuation and the absorption of UV and blue light in natural waters. Limnology and Oceanography, 45:642-650.

Miller, R.L. and B.F. McPherson. 1995. Modeling photosynthetically active radiation in water of Tampa Bay, Florida, with emphasis on the geometry of incident irradiance. Estuarine Coastal and Shelf Science, 40:359-377.

Miller, W.L., M.A. Moran, W.M. Sheldon, R.G. Zepp and S. Opsahl. 2002. Determination of apparent quantum yield spectra for the formation of biologically labile photoproducts. Limnology and Oceanography, 47:343-352.

Moran, M.A. and R.G. Zepp. 1997. Role of photoreactions in the formation of biologically labile compounds from dissolved organic matter. Limnology and Oceanography, 42:1307-1316.

National Research Council. 2003. Enabling ocean research in the 21st century. The National Academic Press; Washington, D.C., 220 pp.

Nelson, N.B. and D.A Siegel. 2002. Chromophoric DOM in the open ocean. In: Hansell DA, Carlson CA (eds.) Biogeochemistry of Marine Dissolved Organic Matter. Academic Press, San Diego, pp 547-578).

NOAA Biogeography Program. 2005. Benthic Habitats of Puerto Rico and the U.S. Virgin Islands Habitats; Habitats. http://biogeo.nos.noaa.gov/products/benthic/ htm/frames_h.htm. (Viewed March 28, 2005).

NOAA Observing System Architecture. 2005. http://www.nosa.noaa.gov/observing_systems.html, observed March 31, 2005.

Ogden, J., J. Porter, N. Smith, A. Szmant, W. Jaap and D. Forcucci. 1994. A long-term interdisciplinary study of the Florida Keys seascape. Bulletin of Marine Science, 54(3):1059-1071.

Otis, D.B., K.L. Carder, D.C. English and J.E. Ivey. 2004. CDOM transport from the Bahamas Banks. Coral Reefs, 23:152-160.

Sandvik, S.L.H., P. Bilski, J.D. Pakulski, C.F. Chignell and R.B. Coffin. 2000. Photogeneration of singlet oxygen and free radicals in dissolved organic matter isolated from the Mississippi and Atchafalaya River plumes. Marine Chemistry, 69:139-152.

Saxby T, Dennison WC, Hoegh-Guldberg O (2003) Photosynthetic responses of the coral *Montipora digitata* to cold temperature stress. Marine Ecology-Progress Series 248: 85-97

Shick, J.M., M.P. Lesser and P.L. Jokiel. 1996. Ultraviolet radiation and coral stress. Global Change Biology, 2:527-545.

Stabenau, E.R., R.G. Zepp, E. Bartels and R.G. Zika. 2004. Role of the seagrass *Thalassia testudinum* as a source of chromophoric dissolved organic matter in coastal south Florida. Marine Ecology Progress Series, 282:59-72.

Stabenau, E.R. and R.G. Zika. 2004. Correlation of the absorption coefficient with a reduction in mean mass for dissolved organic matter in Southwest Florida River Plumes. Marine Chemistry, 89:55-67.

Sutherland, K.P., J.W. Porter and C. Torres. 2004.) Disease and immunity in Caribbean and Indo-Pacific zooxanthellate corals. Marine Ecology Progress Series, 266:273-302.

U.S. Commission on Ocean Policy. 2004. An ocean blueprint for the 21st century. Final Report. Washington, D.C.

Vaughan, P.P. and N.V. Blough. 1998. Photochemical formation of hydroxyl radical by constituents of natural waters. Environmental Science and Technology, 32:2947-2953.

Chapter 7

AIRBORNE LASER ALTIMETRY FOR PREDICTIVE MODELING OF COASTAL STORM-SURGE FLOODING

TIM L. WEBSTER[1] AND DONALD L. FORBES[2]
[1]*Applied Geomatics Research Group, Center of Geographic Sciences, Nova Scotia Community College, 50 Elliot Road, RR# 1 Lawrencetown, Nova Scotia, Canada B0S 1M0*
[2] *Geological Survey of Canada, Bedford Institute of Oceanography, P.O. Box 1006, Dartmouth, Nova Scotia, Canada B2Y 4A2*

1. Introduction

1.1 THE CHALLENGE

Next to fire, floods are the most common and widespread natural disasters. For coastal communities, the risk of flooding associated with sea-level rise and storm surges is of particular concern. Global mean sea-level has been rising at a rate between 0.1 and 0.2 m per century over the past 100 years (McLean et al., 2001). The Intergovernmental Panel on Climate Change (IPCC) has predicted a further increase of 0.09 to 0.88 m, with a central value of 0.48 m, from 1990 to 2100 (Church et al., 2001). Adjusted to a 100 year base, the central value (0.44 m) is between 2.2 and 4.4 times the global mean rate of sea-level rise during the 20th century, implying an acceleration in response to global warming. This will result in more frequent extreme high water levels (storm-surge flood levels) and the return interval for a given flood level may be further reduced by potential increases in storm intensity (e.g. Walsh and Katzfey, 2000).

Relative sea-level rise at any one place is a combination of changes in regional sea-level and vertical ground motion. Subsidence in many coastal areas will result in more rapid relative sea-level rise. Projections of the number of people potentially subject to storm-surge flooding on an annual basis increase from about 10 million today to 50-80 million in the 2080s, depending on adaptive response and population growth, and assuming a sea-level rise comparable to the central value of the IPCC projections (Nicholls et al., 1999). This points to the need for mapping of flood risk hazard zones in areas of low relief where existing maps are rarely adequate. A high-resolution digital elevation model (DEM) produced using airborne laser altimetry (LIDAR) provides a cost-effective and efficient method to achieve this end. These types of maps and information products provide the background required for coastal zone managers to make policy decisions related to future developments, as well as adaptation possibilities to mitigate damage and loss of existing coastal resources.

Superimposed on sea-level rise, coastal erosion and flooding associated with large waves and runup riding on high storm-surge water levels pose additional hazards. Other factors being equal, coastal erosion will increase with a rise in mean sea-level (Bruun, 1962; Cowell and Thom, 1994; Leatherman et al., 2003). On dune-backed coastal beaches and barriers, erosion often takes the form of dune-front scarping, while backshore flooding may occur if runup and overwash exploit low points in the dune crest to create breaches and washover channels. The potential for breaching is difficult to assess, but availability of a high-resolution DEM from LIDAR surveys would enable

L.L. Richardson and E.F. LeDrew (eds.), Remote Sensing of Aquatic Coastal Ecosystem Processes: Science and Management Applications, 157-182.

rapid pinpointing of risk sites, while repetitive LIDAR can be used to assess storm impacts on coastal geomorphology and sand storage (Stockdon et al., 2002).

Surges along the east coast of Canada can add up to 2 m or more of water to the predicted water level along the coast (Parkes et al., 1997). A storm surge is defined by the difference between observed water level and the level predicted for the astronomical tide. Surges are caused by high winds and atmospheric low pressure systems associated with storms. Coastal communities are most vulnerable if a storm surge makes landfall during a high-tide event, especially during large spring high tides. With accelerated sea-level rise as described above, the extent and frequency of flooding will increase in the future, as will the impact of related factors such as shoreline erosion. Thus, there will be a growing demand for information products to predict areas at risk from such events, at present and into the future, as a basis for sustainable coastal zone management, effective planning, and the development of appropriate adaptation strategies. In this chapter, following the introductory discussion of coastal flooding associated with sea level rise and storm surges, we consider remote-sensing technologies available to produce information products suitable for defining susceptible coastal areas. We then describe a case study from Atlantic Canada, where a multi-disciplinary scientific team has produced information products to aid local resource managers in identifying areas prone to coastal flooding and erosion. The Planning Department for the City of Charlottetown was involved in the project and has incorporated the flood risk and flood depth maps into their GIS system to allow them to develop a policy to minimize damage from future flooding events (Webster et al., 2003).

1.2 REMOTE SENSING TECHNOLOGIES FOR FLOOD RISK MAPPING

Remotely sensed data can be used in flood risk applications in two main ways: 1) using remote sensing to map the extent of flooding (active, cloud-penetrating, sensors such as synthetic aperture radar (SAR) on Radarsat are ideal for this application); and 2) using remote sensing to obtain high-resolution elevation information that can be used in predicting flood-risk areas. Because storm surges are typically 0.6 to 2 m in height, technologies with vertical precision significantly finer than these values must be employed to generate flood risk maps of sufficient resolution. Airborne LIDAR (light detection and ranging) is an emerging technology that offers the vertical accuracy and high spatial sampling density required for this purpose. Many LIDAR systems have vertical accuracies on the order of 15 cm or better. LIDAR technology has been employed for a number of years in atmospheric studies (e.g. Post et al., 1996; Mayor and Eloranta, 2001) and as an airborne technique for shallow bathymetric charting (e.g. Guenther et al., 2000), although cost remains an impediment to widespread acceptance for the latter purpose. The technology can also be used to image the land and water surface (Hwang et al., 2000), as in the case study presented here. A general overview of airborne laser scanning technology and principles is provided by Wehr and Lohr (1999). Applications have been demonstrated in forestry (Maclean and Krabill, 1986), sea-ice studies (Wadhams et al., 1992), and glacier mass balance investigations (Krabill et al., 1995, 2000; Abdalati and Krabill, 1999). Use of LIDAR in coastal process studies in the USA have been reported by Sallenger et al. (1999), Krabill et al. (1999), and Stockdon et al. (2002), among others. Preliminary trials in Atlantic Canada were reported by O'Reilly (2000) and subsequent efforts described by Webster et al. (2002). Most of the coast of the conterminous USA has now been mapped using this technology (Brock et al., 2002).

Another applicable remote-sensing technique, with resolution on the order of 1 m, is interferometric SAR (inSAR). This technology has been used for flood risk mapping in the United Kingdom (Galy and Sanders, 2002). A complete review of LIDAR, photogrammetry, inSAR and other technologies used in the production of digital elevation models (DEMs) is provided by Maune (2001).

1.3 CASE STUDY: FLOOD RISK MAPPING IN PRINCE EDWARD ISLAND, CANADA

Storm-surge flood risk mapping was one of the major objectives of a recent project (McCulloch et al., 2002) to evaluate coastal impacts of climate change and sea-level rise on Prince Edward Island in southeastern Canada (Figure 1). Airborne LIDAR surveys were employed in this project and the resulting data sets provided the essential foundation for flood risk mapping in the urban centre of Charlottetown and in a representative rural area in the vicinity of North Rustico (Webster et al., 2002). It was recognized at the outset that a high-resolution representation of the coastal topography would be essential for predicting areas at risk of storm-surge flooding (Webster et al., 2001, 2003). A multi-disciplinary scientific team was involved in this project and contributed related analyses of sea-level change, storm-surge climatology, wave and sea-ice climatology, statistics of flood probability, coastal erosion, socio-economic impacts, and adaptation options (Chagnon, 2002; Forbes and Manson, 2002; Forbes et al., 2002; Manson et al., 2002; Milloy and MacDonald, 2002; Parkes and Ketch, 2002; Parkes et al., 2002; Thompson et al., 2002). The City of Charlottetown Planning Department also participated in the project and incorporated the results into their information system. The data have wide applicability, beyond climate-change impacts assessment, among other fields, including geological and ecological research, urban and regional planning, coastal management, and agriculture. The remainder of this chapter describes the flood risk mapping component of this project utilizing LIDAR for coastal areas on Prince Edward Island.

1.3.1 *LIDAR mapping*

LIDAR mapping involves an aircraft emitting laser pulses toward the ground and measuring the return time of the pulse (see Webster et al., 2004). The laser scan is acquired by rapid repetition of the laser pulse transmitter and cross-track deflection of the beam using an oscillating mirror to produce a zigzag pattern of laser hits on exposed surfaces below the aircraft (Figure 2). A Time Interval Meter (TIM) records the mirror scan angle, the time when the pulse is transmitted from the sensor, the time of the returning reflected pulse, and in some cases the intensity of the return. The configuration of the TIM is what determines if the sensor captures the first or last reflected returns. New generation LIDARs are capable of capturing first, last, and intermediate returns, along with multiple intensities. The data volume with such sensors is a potential problem and the information content of the intermediate returns is an area of active research. Using precise differential Global Positioning System (GPS) technology to determine the location of the aircraft (Krabill and Martin, 1987) and an Inertial Measurement Unit (IMU) to measure the aircraft attitude (pitch, yaw, and roll), the location of individual laser returns measured by the TIM can be determined (Figure 2).

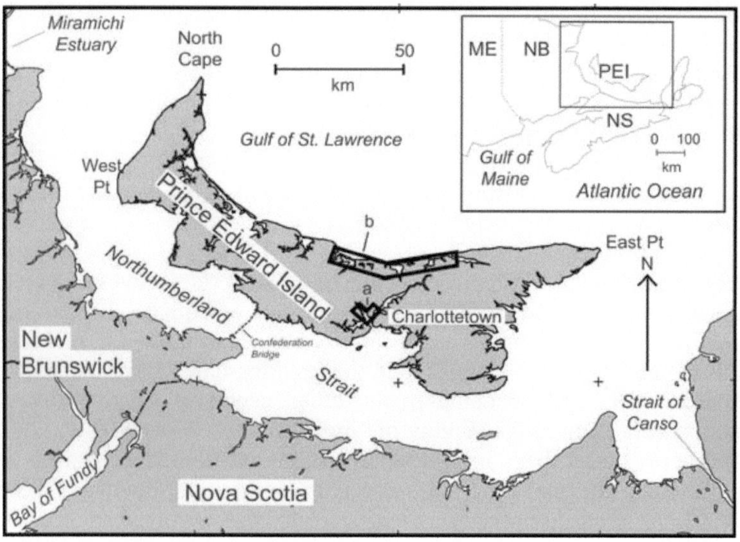

Figure 1. Study area locations for LIDAR acquisition, a) Charlottetown, b) North Rustico (from Webster et al., 2004).

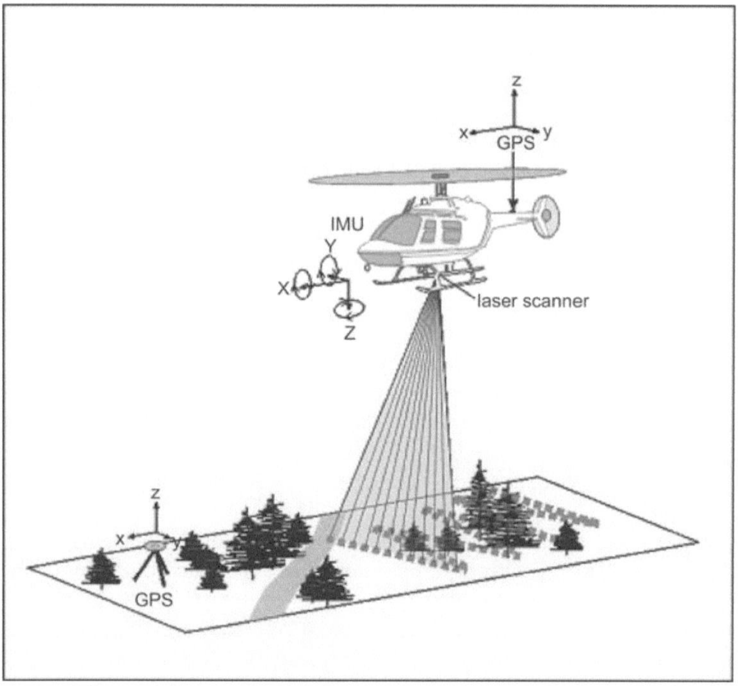

Figure 2. LIDAR helicopter configuration. GPS control for sensor location, Inertial Measurement Unit (IMU) for attitude angles to position each laser shot. Note the zig-zag pattern of LIDAR ground shots as a result of the forward motion of the aircraft and the across-track beam displacement by the oscillating mirror (from Webster et al., 2004).

The nominal accuracy of the system used in this study is ±30 cm both horizontally and in the vertical. The preliminary data output includes geographic coordinates (longitude and latitude) and elevation (in meters) for each laser point reflection, all referenced to the World Geodetic System ellipsoid of 1984 (WGS 84), the ellipsoid employed in the GPS system. If the data are to be used for GIS applications such as flood risk mapping, the horizontal coordinates are converted to an appropriate map projection, in this case using the Universal Transverse Mercator (UTM) grid. Elevations on most land-based topographic maps are measured relative to a geodetic vertical datum. For Canada this is known as the Canadian Geodetic Vertical Datum of 1928 (CGVD28). Thus, for many applications, including flood risk mapping, the LIDAR elevations are transformed from ellipsoid heights to orthometric heights. Orthometric heights are based on the geoid, an equipotential surface defined by the earth's gravity field, approximately equal to mean sea-level (Figure 3). To obtain orthometric heights, an adjustment must be made for the local vertical separation between the ellipsoid and the geoid. The difference between the WGS84 ellipsoid and the CGVD28 geoid is obtained by using the HT1_01E model, since replaced by HTv2.0, with an accuracy of ±5 cm with 95% confidence in southern Canada (Geodetic Survey Division, Natural Resources Canada at the following website: www.geod.nrcan.ca/index_e/products_e/software_e/gpsht_e.html).

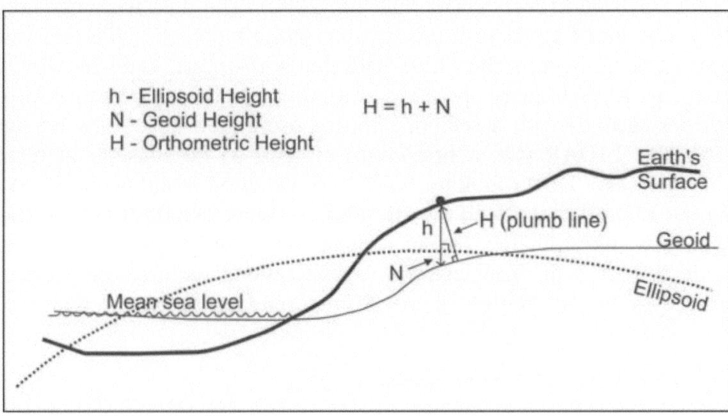

Figure 3. Relationship between ellipsoidal height and orthometric height. The ellipsoid is a smooth mathematical surface. The geoid is an equipotential surface defined by the earth's gravity field. Orthometric heights are measured from the earth's surface normal to the geoid. The HT1_01 model is used to determine the separation between the ellipsoid and geoid for this study.

When the LIDAR system scans the ground, several targets are potentially illuminated on each laser pulse, including the ground, tree canopy, and building tops. After initial processing, the suite of reflections forms the LIDAR *point cloud* and must be further classified. The most common scheme is to classify the LIDAR points into two categories: ground points and non-ground points. This is required to enable creation of a *bald earth* representation, without vegetation or buildings, when generating a DEM for flooding analysis.

The standard delivery product for many LIDAR vendors at the time of this survey was an ASCII file of (x,y,z) data. Currently, there is a proposal from the American

Society of Photogrammetric Engineering and Remote Sensing for a new binary data format standard for LIDAR data (Schuckman, 2003). Both ellipsoidal and orthometric heights were requested for the final delivery of the LIDAR data in this study. The study area was split into 1 km by 1 km tiles and two ASCII files were delivered for each tile, one for ground points and one for non-ground points. Each ASCII file consisted of the following fields:

- UTM easting (m)
- UTM northing (m)
- height above WGS84 ellipsoid (m)
- orthometric height above CGVD28 (m)
- GPS time (s) from the start of each GPS week.

The addition of the GPS time stamp allowed us to analyze the data based on flight lines, made it possible to extract data for times when the GPS constellation was poor, and provided a time stamp for water level (from the tide gauge). Without times, it is difficult to separate LIDAR returns between flight lines.

Terra Remote Sensing Inc. of Sidney, British Columbia, Canada, was contracted to acquire the LIDAR survey data for two study areas: low-lying parts of the City of Charlottetown and a coastal strip extending about 50 km along the central North Shore of Prince Edward Island (Forbes and Manson, 2002). The data were acquired on 1-2 August 2000. The aircraft was positioned using phase kinematic GPS, referenced to a geodetic ground monument north of the Charlottetown Airport. The LIDAR system was a diode-pumped I/R YAG laser operating at a pulse repetition rate of 10 kHz (10,000 laser pulses per second) with a scanning mirror oscillation rate of 15 Hz and a scan angle of 50°. The LIDAR was a first-return system, so no subsequent returns were measured in this case. Down-looking video was acquired simultaneously to assist in interpreting the LIDAR data. With the aircraft at a flying height of 600 m, the LIDAR ground swath was approximately 600 m wide and the ground spacing between LIDAR points was less than 2 m. The technical specifications required the horizontal and vertical accuracy to be 95% within 30 cm of measured GPS points.

2. Validation of LIDAR Elevation Models

Highly accurate DEMs of the coastal zone are required to predict flooding extent from storm-surges of the order of 1 m. The accuracy of a DEM derived from LIDAR data depends on the successful removal of systematic errors associated with the acquisition system, and on a validation process to confirm that the specifications are met. Filin (2003) provides an overview of systematic error types and treatment of these errors in LIDAR systems. In order to ensure that the LIDAR data meet high vertical accuracy specifications, independent ground validation data are required. However, in the Charlottetown study area, the available topographic control at the time of the survey was inadequate and a variety of approaches were used to test the LIDAR data.

During the initial quality assurance of the LIDAR data, water levels from the Charlottetown tide gauge (Parkes et al., 2002) were used to assess the accuracy of water surface hits in the LIDAR data set (Webster et al., 2002). During the LIDAR survey, the winds were light, the harbour surface had no significant waves (although it may have been rippled), and it was assumed that the water level at the tide gauge was a good measure of water level throughout the harbour. A total of 16,515 LIDAR point hits on the harbour surface in eight areas showed a mean offset of -0.85 m. A bimodal distribution with five areas revealed a mean offset of -0.93 m, while three others gave a mean of -0.66 m (Webster et al., 2002). The standard deviation of the LIDAR elevations on the water surface in the eight sample areas ranged from 0.14 to 0.31 m for sample sizes between 1,013 and 3,634. The disagreement in the LIDAR elevations and water level data was later interpreted as possibly relating to tidal hydraulics in the Yorke and Hillsborough River arms of the harbour (Webster et al., 2004).

A separate comparison was made between a Canadian Hydrographic Service (CHS) benchmark CHTN 1-1963 on the Coast Guard wharf near the tide gauge at the foot of Queen Street in Charlottetown and LIDAR hits on the wharf surface within a radius of 3 m. The ellipsoidal and geoidal (CGVD28) elevations of the benchmark were determined as part of the vertical datum control survey reported in King et al. (2002). Initial comparison with presumed ground (*bald-earth*) points in the LIDAR data gave an offset of -2.2 m, but it turned out that the classification algorithm had erroneously identified water surface points as *ground* points and wharf deck points as *non-ground*. Subsequent identification of points on the wharf deck in the vicinity of the benchmark showed an offset of -0.92 m ellipsoidal. On the basis on these results and consultation with the data acquisition contractor in which no systematic errors were identified, an adjustment of +0.9 m was applied to all data points prior to incorporation in the DEM. This produced realistic flooding levels for a simulation of the 21 January 2000 storm surge (Webster et al., 2002, 2003), whereas the original DEM without adjustment had suggested much more extensive flooding.

There was some question whether the 0.9 m adjustment would be equally applicable in the North Shore study area some distance from Charlottetown. Therefore, validation work was undertaken in that area as well, consisting of surveying cross-shore profiles using real-time kinematic (RTK) differential GPS techniques with horizontal and vertical resolution better than 5 cm. These data were available at eight monitoring sites maintained by the Geological Survey of Canada in the study area (Forbes and Manson, 2002). Comparison of gridded LIDAR points (processed using GRASS as described below) with RTK survey points along one such transect at Brackley Beach (Figure 4) showed good correspondence. This profile crossed two dune ridges and revealed that the crest elevation of the high narrow dune crest was somewhat underestimated in gridded LIDAR topography. Reflection from the tops of rose and bayberry plants was also evident in an overestimation of ground elevations in the depression between the two dune crests. Otherwise, this example showed successful validation and indicated that the 0.9 m offset adjustment was appropriate in the North Shore study area.

To further test the validity of the 0.9 m adjustment, a high-precision GPS campaign was carried out within the Charlottetown survey area in the summer of 2001 (Fraser, 2001). Carrier phase static GPS measurements were collected and processed at 15 sites throughout the city so that a more detailed analysis of the height differences could be done (Figure 5). The GPS sites were selected based on a variety of factors including: spatial distribution throughout the LIDAR study area; GPS satellite geometry to minimize obstructions (e.g. away from large buildings where multipath could be a problem); flat smooth areas of dense LIDAR point coverage such as grass fields in city parks; and critical waterfront features such as wharf decks. Local GPS base stations were established using the provincial geodetic control network. Baselines between the

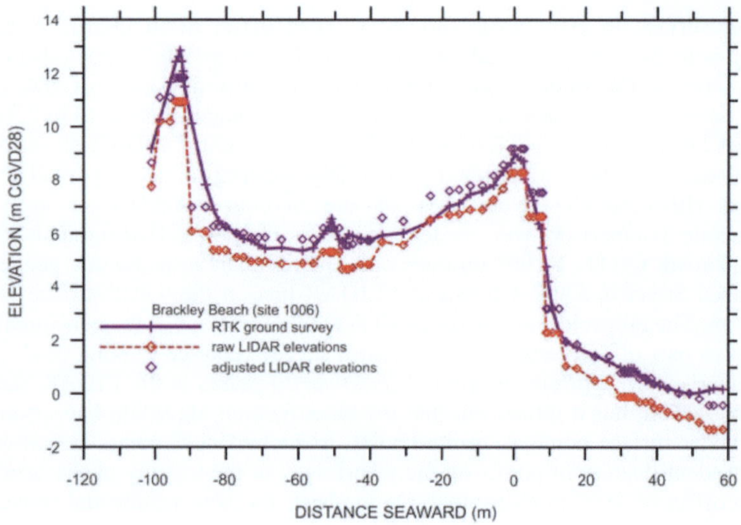

Figure 4. Comparison of raw gridded LIDAR points with RTK survey profile on cross-shore transect over two dune crests and beach at right on central survey line at Brackley Point monitoring site, North Shore of Prince Edward Island. Apparent step functions on steep slopes are artifacts of gridded elevation extraction corresponding to closely spaced RTK survey points. When the 0.9 m adjustment is added to the LIDAR elevations, the LIDAR profile shows good correspondence with the ground survey. Lower elevation at narrow dune crest is related to the grid size. Higher elevations in trough between dune crests are related to the height of rose and bayberry plants growing in that area. Offset at base of beach profile is a function of topographic change between the ground and airborne surveys.

GPS base and rover units were kept to less than 5 km. Combinations of Trimble dual and single frequency P-code receivers were used to collect the GPS observations and post processed in order to maintain centimeter level accuracy of the points.

The processed GPS data were brought into an Arc/Info GIS to be integrated and compared with the LIDAR survey points. The validation procedure consisted of two approaches:

- comparing the GPS points to the interpolated LIDAR DEM surface, and
- comparing the GPS points to LIDAR points within a fixed radius around each GPS point (as in the comparison with the CHS benchmark on the wharf).

In both cases the orthometric heights measured by GPS were compared to those obtained from the LIDAR data set. The HT1-01 model was used to transform the GPS ellipsoidal heights into orthometric heights. As noted above, two validation methods were used because they provided different and complimentary information. The first method was computationally simple and gave details on the accuracy of the DEM produced from the LIDAR data. This method involved comparing each GPS point with the interpolated DEM surface, thus ensuring availability of a LIDAR surface elevation regardless of the original LIDAR point distribution. The drawback of this method was that the details of the original LIDAR points compared to the GPS points were lost. The second method, although more time consuming and complicated in both computation

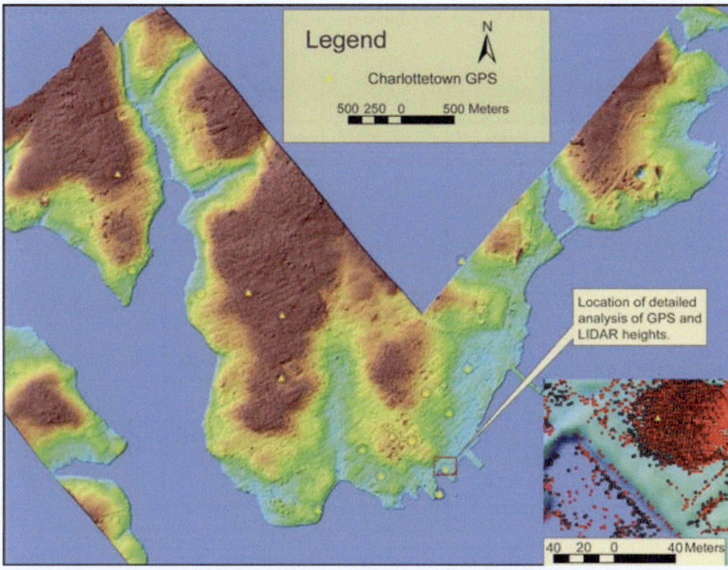

Figure 5. Digital Elevation Model (DEM) derived from LIDAR ground points for Charlottetown with GPS locations (yellow triangles), with the red box denoting the inset map location. Inset map shows LIDAR ground points (red and grey points) and GPS location (yellow triangle) for a city park lawn near the waterfront. The grey and red LIDAR points are colour coded based on the GPS time stamp of the aircraft.

and interpretation, addressed the specific details of individual LIDAR point hits. Because the LIDAR points rarely if ever coincided exactly with the GPS ground validation points, in the second method the cluster of LIDAR points within a specified radius around the GPS point was used to ensure an adequate sample. It was important to consider the radius of the search area to ensure that the LIDAR points selected were representative of the ground feature surveyed by GPS (e.g. if GPS points were collected on a road, a large search radius could include LIDAR points in the ditch and thus indicate an erroneous vertical offset). With those conditions in mind, the second method provided more details on the raw LIDAR data and systematic errors could be more readily detected.

The first validation procedure involved overlaying the GPS points on the LIDAR DEM in order to obtain the cell elevation value (Figure 5). Fifteen GPS points were compared to the adjusted DEM surface (Table 1). The average difference between GPS measurements and the surface was -4.1 cm, thus confirming that the +0.9 m offset was appropriate throughout the study area (Figure 5). However, the standard deviation of the differences between the GPS and LIDAR surface values was 0.54 m and the average magnitude of the height difference was 0.45 m indicating a high degree of variance in the data (Table 1). This simple validation approach gave a sense of the potential resolution and validity of the DEM, but observed offsets could be attributed to the influence of height differences in adjoining cells.

Table 1. Comparison of 15 GPS carrier phase measurements to adjusted LIDAR DEM surface.

GPS (m)	LIDAR (m)	ΔZ (m)
4.46	5.36	-0.90
4.60	4.02	0.57
4.22	4.10	0.13
10.42	9.77	0.65
5.26	4.27	0.98
9.07	8.84	0.22
2.34	2.92	-0.59
14.99	15.54	-0.55
19.30	19.81	-0.52
6.20	6.19	0.00
18.64	18.93	-0.29
19.48	19.86	-0.37
6.90	7.03	-0.13
2.09	2.44	-0.35
2.78	2.28	0.50
Mean offset		-0.04
Standard deviation		0.54
mean absolute difference		0.45
standard deviation		0.28

The second validation approach involved an analysis of all LIDAR ground points within 5 m of each of the GPS ground validation points. Many of the latter were located in city parks and sports fields, with short grass (typically less than 5 cm), and level terrain. Two outputs were generated to assess the height differences. The first considered all LIDAR ground points in the vicinity of each GPS point and computed the following values:

- the number of LIDAR points within 5 m,
- the minimum height difference (GPS-LIDAR),
- the maximum height difference (GPS-LIDAR),
- the mean height difference (GPS-LIDAR), and
- the standard deviation of the height difference (GPS-LIDAR).

The second output represented the LIDAR points within 5 m of a GPS point and computed the following values:

- the distance to the nearest GPS point,
- the GPS orthometric height, and
- the height difference.

For all 15 GPS control points, a total of 342 LIDAR points fell within the 5 m radius limits. The average difference in orthometric heights between the LIDAR and GPS points was -0.37 m, with a standard deviation of 0.25 m. Because the majority of the GPS validation points were on extensive flat surfaces within city parks and sports fields, one would expect the height differences between the locations of individual LIDAR hits and the nearest GPS points to be nearly constant within the 5 m radius. Figure 6 shows a typical site in a city park near the waterfront with an extensive, flat, grass field. In this case, the observed height difference based on the LIDAR elevations was not constant, and the spatial distribution of differences showed a pattern. Because we recorded the GPS time tag, we were able to classify the LIDAR points based on the GPS time to distinguish points from different flight lines.

Taking the site shown in Figure 6 as an example, Figure 7 shows the LIDAR *ground* points coded by flight line. The LIDAR points within 5 m of the GPS point are represented by larger symbols and consist of two flight lines. Figure 8 shows the same LIDAR points within 5 m of the GPS location, in this case coded by height difference. From these figures, it appears that the height difference is related to the LIDAR flight line. When the height differences are plotted against GPS time, the systematic height difference is more apparent (Figure 9). One set of points flown at GPS time 47,600 s, is designated line 1; another at time 62,010 s is termed line 2. The height differences for line 1 range from -0.26 m to +0.03 m, with a mean difference of -0.12 m and a standard deviation of 0.08 m. The height differences for line 2 range from -0.46 m to -0.20 m, with a mean difference of -0.33 m and a standard deviation of 0.07 m. Thus the height differences are distinct for each flight line although they have a similar variance range and overlap slightly. It is clear that if adjustments in the form of a vertical offset are applied to the data, these should preferably be specific to flight lines. Examining the rest of the LIDAR points within 5 m of each of the GPS survey validation points reveals a similar pattern, with distinctive height differences for different flight lines (Figure 10). Another way to look at this relationship is to plot the GPS-LIDAR height differences against GPS time (Figure 11). It is clear from this plot that, while the 0.9 m adjustment to the LIDAR elevations was appropriate in an aggregate sense, the variation in offsets between flight lines add an extra error term to the DEM elevations. In many cases this difference is reduced because the mean elevation for a DEM grid cell is computed from all points lying within the cell.

Another problem with the LIDAR data collected in 2000 involved large variations in the density of returns. A reduction in the laser power resulted in flying the Charlottetown LIDAR survey at lower altitude than planned (Webster et al., 2004). This also resulted in a lack of LIDAR returns from near-infrared targets such as black asphalt pavement and rooftops (Figure 5). This resulted in the clear delineation of the

Figure 6. Example of a typical GPS site. City park near the waterfront with a flat level grass field. This location corresponds to the inset map on figure 5. The GPS unit is a Trimble model 4600 and was used for post processing to determine the height of the ground surface to centimeter level accuracy.

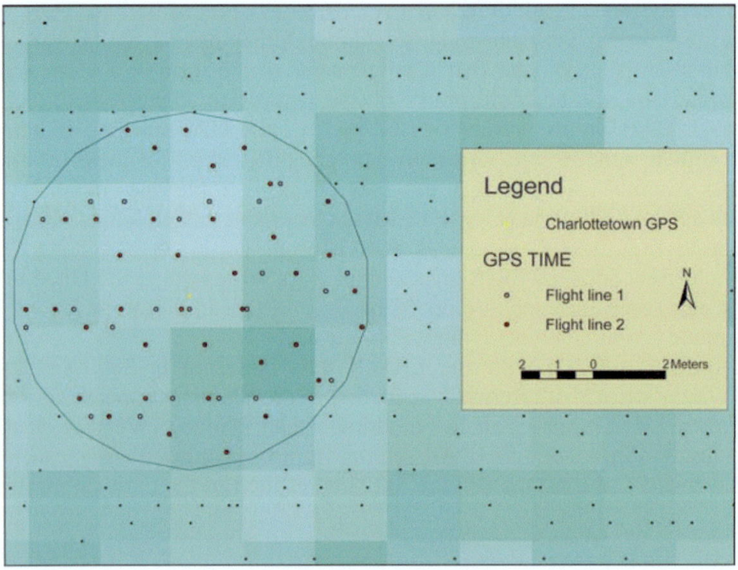

Figure 7. Larger grey and red points are LIDAR ground points within 5 m of the GPS site at the city park in figures 5 and 6. LIDAR points (large and small) are colour coded by GPS time and denoted as flight lines 1 and 2. The raster grid in the background represents the DEM interpolated from the LIDAR ground points as shown in figure 5.

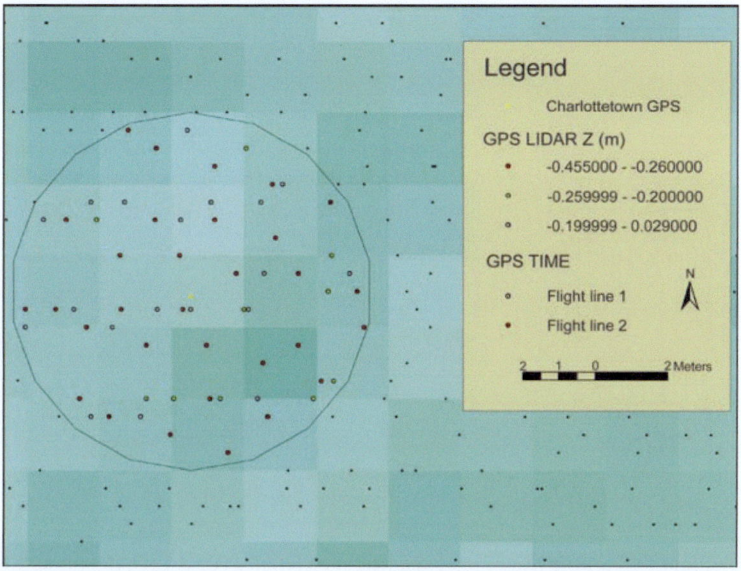

Figure 8. Larger points are LIDAR ground points within 5 m of the GPS point shown in figure 5, 6, and 7. These points are coded into three classes based on the height difference compared to the GPS point (GPS-LIDAR). The smaller grey and red points are colour coded based on GPS time as in figure 7. Since the field is flat, the height difference should be constant. However, as can be seen from the spatial pattern of the height differences there is a range between –0.45 to –0.029 m difference. The pattern of the height differences corresponds to the differences in the flight lines of the LIDAR data.

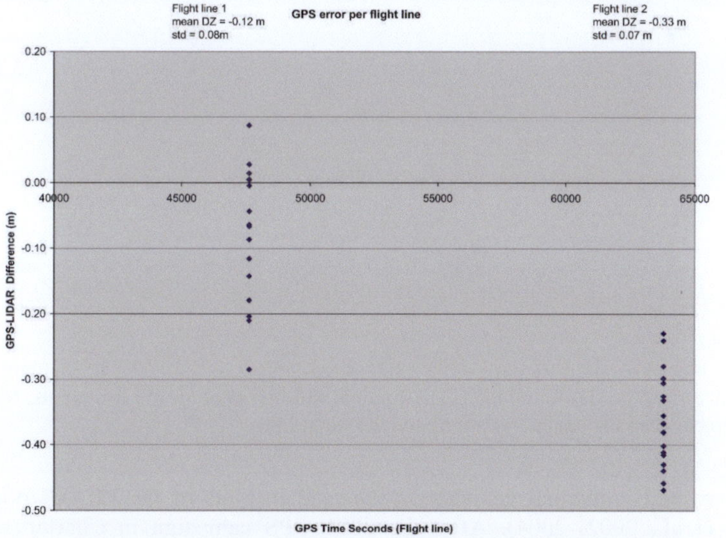

Figure 9. Plot of GPS time (flight line) and height difference (GPS-LIDAR) for the LIDAR points within 5 m of the GPS point within the city park shown in figures 5-8. Note the distinct difference in height differences between flight lines covering the city park site. Flight line 1 is denoted by GPS time 47,600 seconds and flight line 2 is denoted by GPS time 62,010 seconds.

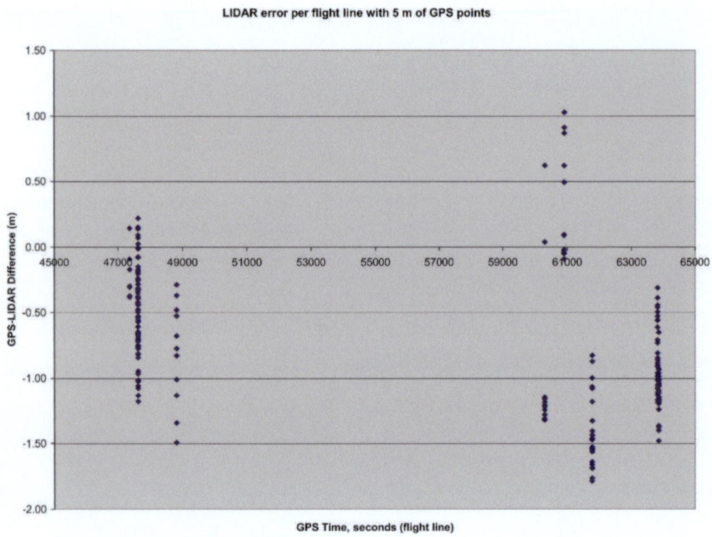

Figure 10. Plot of GPS time (flight line) and height difference (GPS-LIDAR) for all LIDAR ground points within 5 m of all 15 GPS locations. Note the distinct difference in height differences between flight lines for the entire GPS survey area.

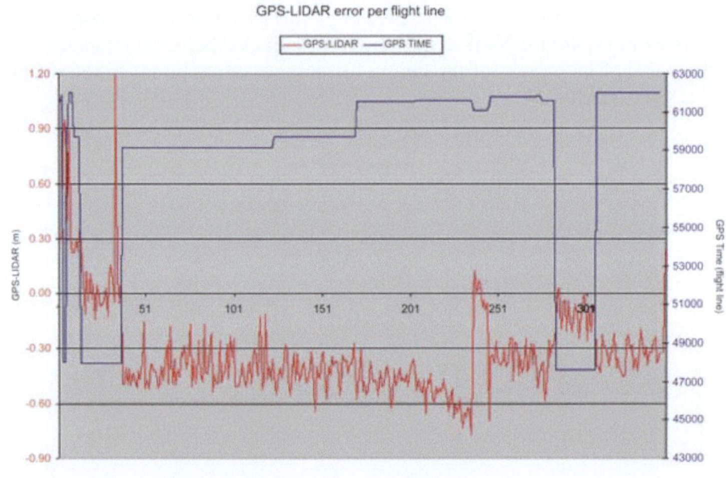

Figure 11. Plot of height difference (GPS-LIDAR red-left y-axis) and GPS time (flight line) (seconds blue-right y-axis) for all the LIDAR points within 5 m of all 15 GPS points. Note the relationship between the height variations and the flight lines.

road pattern and individual vehicles on the road in plots of the LIDAR point cloud (Webster et al., 2002, 2004). After the 2001 GPS campaign in Charlottetown and another more extensive campaign in the Annapolis Valley of Nova Scotia (where another LIDAR survey had been flown during the summer of 2000), the vertical offset problems in the LIDAR system were found to be related to calibration procedures

(Webster et al., 2004). Specifically, they were related to differences in flying altitude between the calibration flights and the survey flights. The laser range calibrations were carried out at the planned survey altitude of 900 m. However, because of the power loss problems, the actual survey was flown at an altitude nearer 600 m, introducing a range bias. The problem was initially difficult to assess, because the LIDAR data showed a good match in the calibration areas used by the data acquisition team (usually near the airport where the GPS base station was established). To confirm this interpretation, several lines of the LIDAR data have been reprocessed by applying a range scale factor and offset, resulting in agreement of the orthometric heights in overlapping lines and a better match to the GPS data.

The above analysis demonstrates the necessity for independent validation of LIDAR data. The expected level of accuracy of this technology is such that most existing data (e.g. 1:10,000 scale topographic maps) are inadequate for this purpose. Therefore carrier-phase GPS data must be collected specifically for validation purposes.

3. DEM Construction from LIDAR

3.1 INTERPOLATION METHODS AND CLASSIFICATION OF THE LIDAR POINT CLOUD

As mentioned above, it is standard practice to classify the LIDAR returns into *ground* and *non-ground* points to enable the production of a *bald earth* DEM. An accurate representation of the ground surface is critical for coastal zone flood risk mapping from storm-surge events. As outlined in Maune (2001), several methods have been developed for constructing DEMs from point data. Most involve interpolation between the LIDAR points to generate a continuous surface. For this project two approaches were used: 1) for both the Charlottetown and North Shore survey areas, a DEM was constructed by direct gridding of the LIDAR points; and 2) for the Charlottetown area, a DEM was constructed through interpolation.

In the first approach, public domain GRASS software was used to build a grid from the LIDAR ground points. This is the method we have been using for a number of years to process multibeam bathymetric data to generate digital seabed models and shaded-relief imagery (cf. Courtney and Shaw, 2000). A 2 m grid was overlaid on the LIDAR points and each grid cell was assigned the mean orthometric height for the point(s) lying in the cell. This produces a DEM without interpolation, but areas of sparse or missing LIDAR points will not be assigned a value in the DEM. Limited interpolation can be used to fill small gaps in the resulting grid. Shaded-relief images derived from the resulting DEMs were reproduced in Forbes and Manson (2002) and Forbes et al. (2004).

In the second approach, an Arc/Info geographic information system (GIS) was used to construct a triangular irregular network (TIN) from the LIDAR ground points. A 2 m grid was then built from the TIN using the quintic interpolation method (5th order polynomial). Although it is computationally more intensive than linear interpolation, this method ensured a smooth surface that honored all data points. The DEM grid was then transferred to the PCI Geomatica suite of image processing tools for visualization and modelling. A colour shaded-relief model was constructed from the DEM and used for qualitative assessment and flood modelling. At this stage, two problems were identified:

- the TIN interpolation crossed areas of no data, such as the inner gap in the 'V' shape of the study area (Figure 1) and sparse data points in the water, thus producing a surface in these areas that was not reliable; and
- wharves and sea walls with vertical faces along the waterfront appeared in the shaded-relief image to have slanted sides (Dickie, 2001).

3.2 GROUND SURFACE REFINEMENT

The problem of inaccurate surface interpolation across areas of no valid data can be resolved by clipping the grid using a mask covering only the areas of interest. This is also required along the shoreline to exclude water surface returns. Complications can develop where the area of interest is large, as in the North Shore survey area, and there are significant variations in tide level during the survey or hydraulic effects causing different water levels inside and outside estuaries (Figure 12).

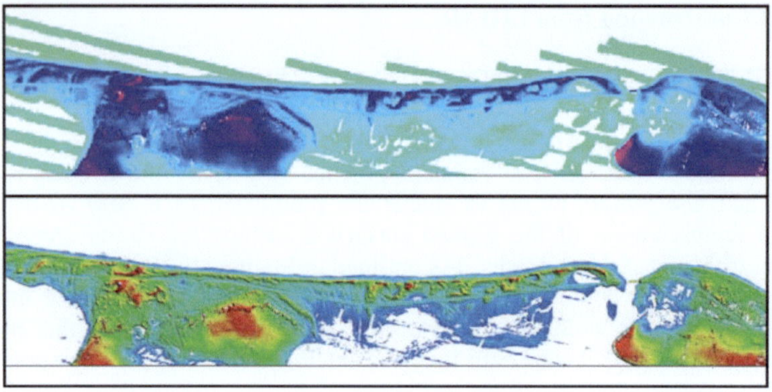

Figure 12. Filtering the LIDAR data at the water line. Upper panel: Initial topographic model (darker blues are higher). Lower panel: Colour shaded-relief image after trimming at an appropriate water line (yellows to reds are higher). Note that minor nadir reflections from the water surface remain inside the bay at bottom centre right.

The problem of the wharves and waterfront structures not being accurately modeled was serious for our intended use of the DEM for flood visualization. Overlaying the *ground* points on the colour shaded-relief DEM, it was immediately clear that there were very few around the edges of the wharves. Numerous presumed *ground* points were located on the water surface and in the central areas of the wharf decks away from the edges (Figure 5 inset). As noted earlier, we discovered that the classification algorithm had coded the wharf edges as *non-ground.* This problem has been encountered elsewhere with LIDAR datasets – for example, large flat roofed buildings are often misclassified. The rooftops near the edge will be correctly classified as non-ground, but in the center of the roof the points will often be coded as ground. Similarly, in our Prince Edward Island data set (e.g. in the area shown in Figure 12), dune crests adjacent to dune-face scarps were miscoded as *non-ground* points. In the case of the Charlottetown waterfront, this issue was resolved by manually extracting the correct ground points from the non-ground files. A set of software tools developed by Helical Systems was used to examine the non-ground points in 3-D and select those that represented the wharf-edge and sea-wall ground features. The original ground points were combined with the new points extracted from the non-ground file to generate a new TIN along the waterfront (Dickie, 2001). In this case, a linear interpolation method was used to construct a 2 m DEM. A linear interpolation was chosen to better represent the abrupt vertical changes associated with the waterfront. The final DEM consisted of a mosaic combining the quintic 2 m DEM for the area landward of the waterfront with the linear 2 m DEM for the waterfront itself (Figure 13).

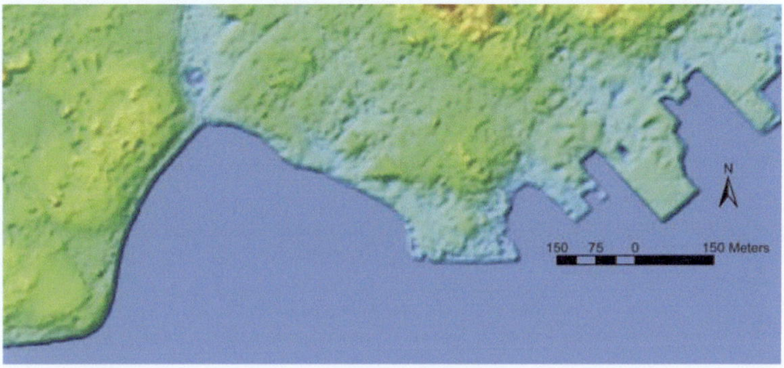

Figure 13. Revised DEM of waterfront, ground and some non-ground points used to construct the surface.

A GIS database of the street network, building footprints, and other infrastructure was overlaid on the DEM to assess the horizontal accuracy of the model. The DEM fit the vector data sufficiently. As mentioned above, the GPS survey data were examined in relation to the DEM to assess the LIDAR adjustment and the final DEM product.

A digital surface model (DSM) was also constructed that incorporated all of the LIDAR points, both *ground* and *non-ground*. Because it represents the top of the canopy in wooded areas (Figure 14), this model is not appropriate for flood risk modeling. The DSM was generated for visualization purposes only, but it does provide useful information on land cover, buildings and other structures, and more realistic visual clues.

4. Flood-risk Mapping Using a LIDAR DEM

4.1 WATER LEVELS FOR THE FLOOD MODELING

Once a reliable DEM is constructed that accurately represents the natural and man-made coastal morphology, flood-risk modeling can begin. The coastal area of Prince Edwards Island was selected for this study due to its vulnerability to flooding and other impacts during storm-surge events with rising relative sea-level (Shaw et al., 1998).

Shortly after the study began, the City of Charlottetown was severely impacted during a storm on 21-22 January 2000 (Forbes et al., 2000, 2001). This event, which caused extraordinary damage, was recorded by the Charlottetown tide gauge (Figure 15). The surge occurred during a run of perigean spring tides (Parkes and Ketch, 2002). The downtown waterfront was flooded by a record high water level of 4.229 m CD (Chart Datum) resulting from a surge of almost 1.5 m superimposed on a large high tide.

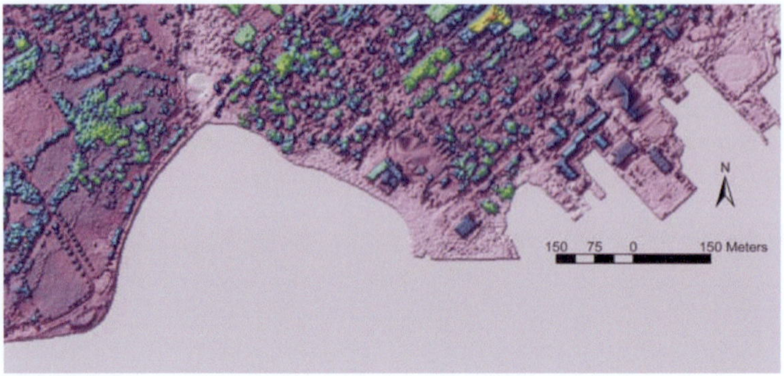

Figure 14. Digital Surface Model (DSM) all LIDAR points used to construct this surface.

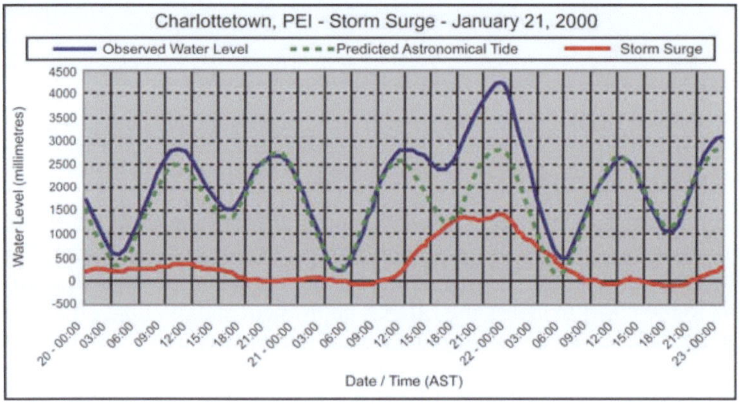

Figure 15. Tide-gauge water level records for Jan. 21, 2000 storm-surge event. The predicted tide is in green and the observed water level is in blue. The difference between the observed and predicted water level represents the storm-surge and is shown by the red line. The largest storm-surge event of 1.5 m occurred at the highest tide level, resulting in significant coastal flooding. The heights are in millimeters above Chart Datum for Charlottetown.

The Charlottetown gauge provides one of the longest continuous records of sea-level in the region (1911 to present) as documented by Parkes et al. (2002). That study showed a rising trend of relative sea-level amounting to 32 cm per century over the length of the tide-gauge record. This rise in sea-level relative to a fixed reference point on land results from a combination of climate-induced sea-level rise and regional crustal subsidence. As relative sea-level rises, the probability of flooding to a given level increases and the maximum potential flood level rises. Thus, the combined effect of subsidence and sea-level rise results in an increased vulnerability to coastal damage by storm-surges.

The 21-22 January 2000 storm provided an opportunity to test our flood-risk modeling efforts and to validate the results by comparing the areas flooded in the model to those observed during the event. After the flood risk maps were constructed by modeling the flood level of the January 2000 storm, the results were shared with city engineers, planners and other municipal officials for a qualitative validation. The maps of flooding extent were in agreement with flood limits observed by officials during the event. Three water levels were used to generate flood risk maps from the LIDAR DEM: 1) the peak water level for the 21-22 January 2000 storm event (4.23 m CD); 2) the same storm event superimposed on a moderate sea-level rise (4.73 m CD); and 3) the same storm event superimposed on a realistic estimate of relative sea-level rise over the coming 100 years (4.93 m CD).

In order to map flood limits in this project, it was also necessary to relate the LIDAR elevation data to hydrographic Chart Datum (CD), the vertical datum used in the tide-gauge records from which storm-surge flood levels are obtained. This datum is approximately equivalent to the lowest astronomical tide and varies from place to place around the coast. The local Chart Datum at each secondary port has typically been related to a local benchmark near the tide gauge, often a permanent marker on the wharf or nearby structures. A separate component of the Prince Edward Island project focused on determining the vertical differences between Chart Datum, the ellipsoid, and CGVD28 throughout the study area (King et al., 2002). In the case of Charlottetown, Chart Datum was determined to be 1.685 m below CGVD28, the vertical reference for the LIDAR DEM. Thus, a simple translation of the 21-22 January 2000 water level and future water levels to the DEM was made by subtracting 1.685 m from the water levels referred to Chart Datum.

4.2 GIS FLOOD MODELING OF STORM-SURGE WATER LEVELS

Many sophisticated numerical models have been developed for simulating tidal hydraulics and these can be particularly useful for flooding of tidal reaches in rivers. In the present study of coastal flooding, it was decided to use existing GIS and image processing capabilities combined with the LIDAR DEM to visualize the potential areas of flooding. Galy and Sanders (2002) recently used a similar approach where they used a DEM derived from a SAR data to map flood risk along the River Thames in the United Kingdom.

In our study, the initial DEM was built using tools within Arc/Info and the data were transferred to PCI Geomatics image processing software for visualization and the generation of raster flood risk maps. The maps were then transferred back to Arc/Info for final vector processing and overlay analysis. It was assumed that a given water level from the storm-surge event would form a horizontal flood plane extending landward from the open harbour. Thus hydraulic effects, associated time lags, or flood expansion or dampening were not considered in the modeling effort.

With the water levels now referenced to CGVD28, a model was written to threshold the DEM into two classes for a given water level, one wet and one dry. This initial threshold procedure did not include any connectivity checks with the source harbour area. The resultant raster image was converted to vector polygons. With the flood extent data in this form, it was quite simple to select only those polygons that were connected to the open harbour, thereby excluding low-lying areas landward of barriers that would check the spread of the flood. Specific conditions such as bridges and causeways with culverts had to be dealt with individually. Areas separated from the harbour in the LIDAR DEM by an apparent barrier such as a bridge were included in the flooded areas. Other less obvious situations, such as causeways with culverts, were more problematic and local municipal officials were consulted in these cases. Many of the culverts are equipped with one-way valves to allow water to flow toward the harbour but stop it from flowing upstream. In these cases the low-lying areas beyond the barrier were not included. Presently there are no built in tools within the commercial GIS and image processing systems we used in this project to automate this type of connectivity while mapping flood risk. Therefore, a software program was developed by M. Gould at the Applied Geomatics Research Group, Centre of Geographic Sciences to automate the procedure. The program allows the user to enter a water level and a starting location, and the application will use a DEM to determine all of the areas below that water level that are connected to the starting point (a location in the harbour was used in this case).

As mentioned earlier, the 21-22 January 2000 storm-surge flood provided an ideal validation test. Areas flooded during that storm were compared to those generated from the DEM model and a very good match was observed. One circumstance that the model did not account for was the backup of seawater through the storm drain system with water flowing up and out of street drains near the waterfront. Many of these areas were in low-lying areas that were predicted to be flooded in the model.

Having obtained good agreement between the observed and predicted water levels of the 21-22 January 2000 storm, we proceeded to model two additional future water levels using a similar approach (Figure 16). The flood risk vectors show the variation of slope within the coastal zone. Many areas have steep slopes and are not vulnerable to flooding, although erosion may be a problem. The waterfront of Charlottetown is vulnerable to flooding, as is the residential area to the west of the downtown core (Figure 17).

4.3 FLOOD-DEPTH MAPS

Another set of layers from our analysis of the flood risk was the depth of the water within the flooded areas. This was calculated by subtracting a constant value of the flood water level from the DEM only for the areas at flood risk. This produced a grid of flood depth values and provides more information than just the flood risk vector that shows the area of inundation from flooding. The amount of damage will be related to the depth of the floodwater, where more damage is expected with deeper flood depths. This is especially true for residential areas where the floodwater must reach a high enough point to enter the basement through windows that are low to the ground or doorway entrances. An example of the flood depth map is shown on Figure 18 for the 21-22 January 2000 flood level. A revised economic impact assessment has not yet been implemented using these maps. However, such maps can add significantly to the information regarding flood risk, impact mitigation, and adaptation planning required to minimize the effects of flooding.

4.4 FLOOD-RISK IMPACT ANALYSIS AND ADAPTATION

An additional component of the project focused on the potential economic impacts of flooding events (Milloy and MacDonald, 2002). Much of the analysis involved the use of GIS to summarize the areas potentially affected by flooding. This involved compilation of the municipal GIS database, including property boundaries and building footprints, linked to other databases such as property and building assessment information, in relation to flood extent at various water levels (Dickie, 2001). Initially a simple overlay of the flood-risk areas and the property boundaries was used to determine that the flood affected more than 460 properties. By summarizing the tax

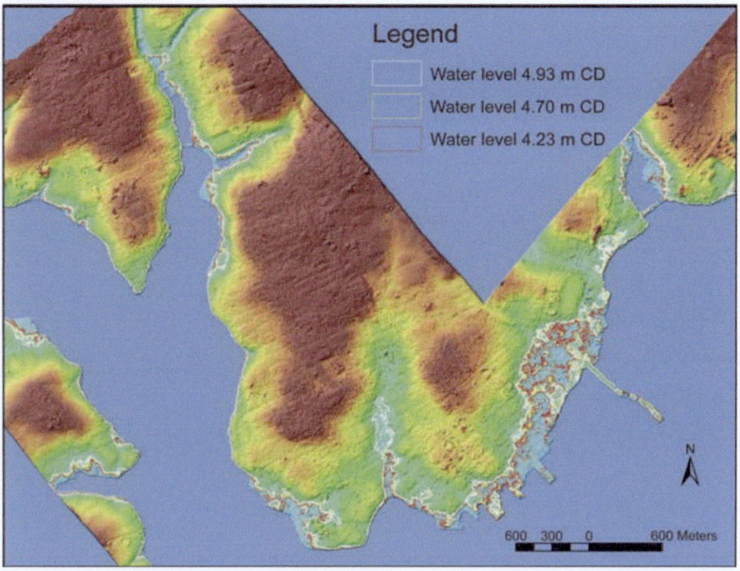

Figure 16. Flood risk areas, flood water levels are referenced to chart datum (CD). The 4.23 m water level corresponds to the Jan. 21, 2000 storm-surge event. The DEM in the background was constructed from the "ground" LIDAR points and refined along the waterfront using some "non-ground" points to ensure an accurate representation.

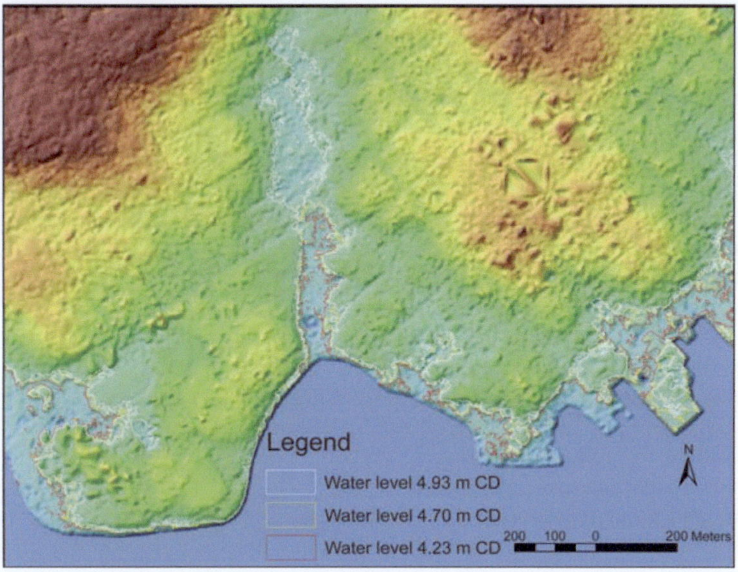

Figure 17. Close up of flood risk areas. The downtown waterfront is on the right and a dense residential area is at the center of the map. The large area in the center of the map that would be flooded at the 4.93 m level corresponds to an ancient stream channel.

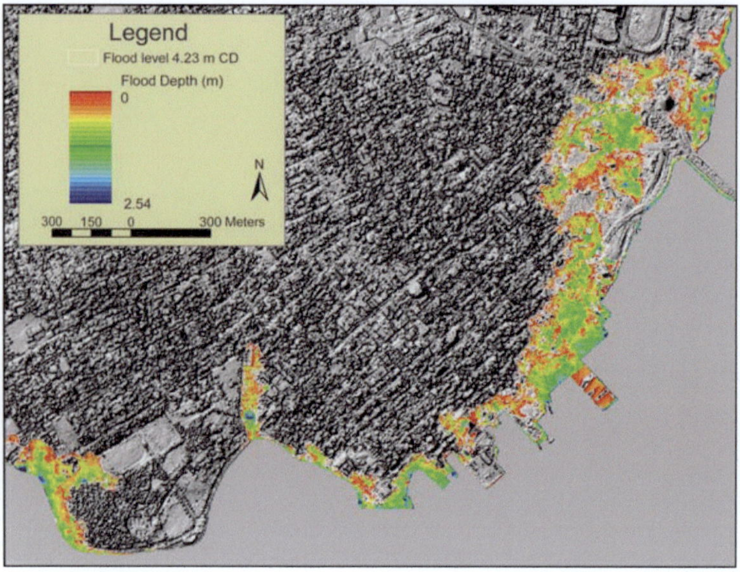

Figure 18. Floodwater depth associated with the Jan. 21, 2000 storm-surge event overlaid on a shaded relief of the Digital Surface Model (DSM). The DSM was constructed from the combined "ground" and "non-ground" LIDAR points. Water depth can have a significant impact on the amount of damage an area sustains from flooding.

assessment values of these properties, an economic impact was calculated (Milloy and MacDonald, 2002). It was also shown that public lands, critical infrastructure, and transportation links (e.g. access to the hospital) were affected during the surge event.

The flood-risk vectors were passed to the City of Charlottetown Planning Department to assist in defining future development plans for the waterfront. On a municipal basis, some planning strategies can be pursued to limit vulnerability. These include measures such as appropriate zoning, acquisition of flood-prone properties, flood-proofing, or taking advantage of replacement schedules to move key infrastructure to less vulnerable locations (Forbes et al., 2002).

Retreat from flooding may not be a broadly viable adaptation option in urban settings such as Charlottetown, so alternative accommodation and protection strategies may need to be considered. Furthermore, horizontal setbacks may not be appropriate in areas where the risk is primarily related to flooding. In such cases, vertical thresholds or setbacks in which the location is defined by elevation and flood probability may be more effective in reducing exposure. Accommodation options are already being considered in Charlottetown as a result of this study. These include flood-proofing of basements and other measures to reduce damage in the event of flooding, as well as more stringent assessment of building proposals in potentially flood-prone areas. These are progressive moves. Other straightforward accommodation measures may include raising foundation heights or the heights of protection structures, wharves, and other coastal infrastructure (O'Reilly et al., 2003).

5. Conclusions

This work demonstrates the application of LIDAR technology to the mapping of storm-surge flood risk for coastal areas. The high-resolution of the LIDAR data allowed a DEM to be constructed that can now be used to model the inundation effect of water levels of 2 m or more higher than usual. However, the results of this study also demonstrate the need for independent validation data to ensure the reliability of such high-resolution topographic mapping.

Careful analysis of validation data using several different approaches revealed the presence of an altitude calibration bias in the LIDAR elevations for the present study. A bulk adjustment of the elevations by 0.9 m provided a reasonable representation of flood levels, but variation in the flying altitude between flight lines resulted in failure to meet the intended 0.3 m vertical error specification. This was partially mitigated by the density of survey points, so that the mean elevation determined for each small grid cell was more often within specification. New LIDAR surveys recently undertaken in another coastal area of eastern Canada are building on the lessons learned in the Prince Edward Island study to provide more accurate and precise DEM data. Nevertheless, the LIDAR data obtained in Prince Edward Island provided unprecedented topographic detail and enabled highly detailed delineation of flood hazard zones. The flood risk maps and information products have made it to the hands of the coastal resource managers, who have to deal with these risks on an annual basis, and have been incorporated into their GIS system. The maps have provided a tool to allow the local planning officials to begin a long-term adaptation process, and to initiate community discussions. For the short term, planning officials now use the maps to inform local developers of the future predictions of flooding events and possible water depths associated with different areas of the waterfront. The planning department has used the information to identify areas where development may be restricted along the waterfront because of the risk of flooding.

6. Acknowledgements

We are happy to acknowledge the efforts of the data acquisition contractors, including Rick Quinn, Roger Shreenan and others (Terra Remote Sensing Inc.), Herb Ripley, Andrew Cameron, and Laura Roy (Hyperspectral Data International), the pilots and other support staff. We thank Paul Fraser and Dan Deneau (Applied Geomatics Research Group, Centre of Geographic Sciences, NSCC) for the Charlottetown GPS campaign and preliminary analysis. We would like to acknowledge George Dias and the other members of the AGRG class (2002) for coding the ARC AML that does the comparison between GPS points and LIDAR points within a fixed radius. We are also grateful to Steve Dickie for his work on the LIDAR processing and report for the PEI project. We acknowledge the contributions of Mike Butler and Brent Rowley for helping in the coordination of data acquisition during the summer of 2000. Gavin Manson (Geological Survey of Canada) provided critical field and office support. Glen King (CHS) assisted with vertical control data, as did Charles O'Reilly (CHS), who was the prime inspiration behind the initiative to acquire laser altimetry for this project. Don Poole (Planning and Development Officer, City of Charlottetown) provided invaluable assistance. This study was funded in large part by the Climate Change Action Fund (CCAF) of the Government of Canada and we are grateful to the entire CCAF project team for their ideas and support on the project. Additional funding to the

AGRG was through the Canada Foundation for Innovation, Industry Canada. The manuscript benefited from internal reviews by Bob Maher of the AGRG and Gavin Manson (GSC). This paper is a contribution to the Natural Resources Canada Program on Reducing Canada's Vulnerability to Climate Change and is Geological Survey of Canada contribution 2004000.

7. References

Abdalati, W. and W.B. Krabill. 1999. Calculation of ice velocities in the Jakobshavn Isbrae area using airborne laser altimetry. Remote Sensing of the Environment, 67:194-204.

Brock, J.C., C.W. Wright, A.H. Sallenger, W.B. Krabill, and R.N. Swift. 2002. Basis and methods of NASA airborne topographic mapper LIDAR surveys for coastal studies. Journal of Coastal Research, 18:1-13.

Bruun, P. 1962. Sea level rise as a cause of shore erosion. Journal of Waterways and Harbors Division, American Society of Civil Engineers, 88:117-130.

Chagnon, R. 2002. Sea-ice climatology. In Coastal Impacts of Climate Change and Sea-Level Rise on Prince Edward Island. Edited by Forbes, D.L. and Shaw, R.W. Geological Survey of Canada, Open File 4261, Supporting Document 5, 24 pp. (on CD-ROM).

Church, J.A., J.M. Gregory, P. Huybrechts, M. Kuhn, K. Lambeck, M.T. Nhuan, D. Qin, and P.L. Woodworth. 2001. Changes in sea level. Chapter 11 in *Climate change 2001: the scientific basis.* Contribution of Working Group I to the Third Assessment Report of the Intergovernmental Panel on Climate Change. Cambridge University Press, Cambridge and New York, 639-693 pp.

Courtney, R.C. and J. Shaw. 2000. Multibeam bathymetry and backscatter imaging of the Canadian continental shelf. Geoscience Canada, 27:31-42.

Cowell, P.J. and B.G. Thom. 1994. Morphodynamics of coastal evolution. In Coastal Evolution: Late Quaternary Shoreline Morphodynamics. Edited by Carter, R.W.G. and Woodroffe, C.D. Cambridge University Press, Cambridge and New York, pp. 33-86.

Dickie, S. 2001. Storm surge innundation mapping for Charlottetown, Prince Edward Island: Analysing flood risk using high-resolution imagery. Unpublished Applied Geomatics Research Group report, NSCC, Middleton NS.

FGDC ([US] Federal Geographic Data Committee). 1998. Geospatial accuracy standards, Parts 1 to 3 (www.fgdc.gov/standards/status/textstatus.html, documents FGDC-STD-007.1, FGDC-STD-007.2, FGDC-STD-007.3).

Filin, S. 2003. Recovery of Systematic Biases in Laser Altimetry Data Using Natural Surfaces. Photogrammetric Engineering and Remote Sensing, 69(11):1235-1242.

Forbes, D.L. and G.K. Manson. 2002. Coastal geology and shore-zone processes. In Coastal Impacts of Climate Change and Sea-Level Rise on Prince Edward Island. Edited by Forbes, D.L. and Shaw, R.W. Geological Survey of Canada, Open File 4261, Supporting Document 9, 84 p. (on CD-ROM).

Forbes, D.L., G. Parkes, C. O'Reilly, R. Daigle, R. Taylor, and N. Catto. 2000. Storm-surge, sea-ice, and wave impacts of the 21-22 January 2000 storm in coastal communities of Atlantic Canada. Abstracts, Canadian Meteorological and Oceanographic Society, 34th Congress, Victoria BC.

Forbes, D.L., R.W. Shaw, and G.K. Manson. 2002. Adaptation. In Coastal Impacts of Climate Change and Sea-Level Rise on Prince Edward Island. Edited by Forbes, D.L. and Shaw, R.W. Geological Survey of Canada, Open File 4261, Supporting Document 11, 18 p. (on CD-ROM).

Forbes, D.L., G.S. Parkes, G.K. Manson, and L.A. Ketch. 2004. Storms and shoreline retreat in the southern Gulf of St. Lawrence. Marine Geology, 210:169-204.

Fraser, P. 2001. LIDAR validation using GPS, Charlottetown, Prince Edward Island. Unpublished Applied Geomatics Research Group report, NSCC, Middleton NS.

Galy, H.M. and Sanders, R.A., 2002. Using synthetic aperture radar imagery for flood modelling. *Transactions in GIS*, 6(1):31-42.

Gould, M., 2001. Flood risk modeling program. Unpublished Applied Geomatics Research Group report, NSCC, Middleton NS.

Guenther, G.C., M.W. Brooks, and P.E. LaRocque. 2000. New capabilities of the SHOALS airborne LIDAR bathymeter. Remote sensing of the Environment, 73:247-255.

Hwang, P.A., W.B. Krabill, W. Wright, R.N. Swift, and E.J. Walsh. 2000. Airborne scanning LIDAR measurement of ocean waves. Remote sensing of the Environment, 73:236-246.

King, G., C. O'Reilly, and H. Varma, H. 2002. High-precision three-dimensional mapping of tidal datums in the southwest Gulf of St. Lawrence. In Coastal Impacts of Climate Change and Sea-Level Rise on Prince Edward Island. Edited by Forbes, D.L. and Shaw, R.W. Geological Survey of Canada, Open File 4261, Supporting Document 7, 16 p. (on CD-ROM).

Krabill, W.B. and C.F. Martin. 1987. Aircraft positioning using global positioning system carrier phase data. Navigation, 34:1-21.

Krabill, W., W.Abdalati, E. Frederick, S. Manizade, C. Martin, J. Sonntag, R. Swift, R. Thomas, W. Wright, and J. Yungel. 2000. Greenland Ice Sheet: high-elevation balance and peripheral thinning. Science, 289: 428-430.

Krabill, W.B., R.H. Thomas, C.F. Martin, R.N. Swift, and E.B. Frederick, E.B. 1995. Accuracy of airborne laser altimetry over the Greenland ice sheet. International Journal of Remote Sensing, 16:1211-1222.

Krabill, W.B., C. Wright, R. Swift, E. Frederick, S. Manizade, J. Yungel, C. Martin, J. Sonntag, M. Duffy, and J. Brock, J. 1999. Airborne laser mapping of Assateague National Seashore Beach. Photogrammetric Engineering and Remote Sensing, 66:65-71.

Leatherman, S.P., B.C. Douglas, and J.L. LaBrecque. 2003. Sea level and coastal erosion require large-scale monitoring. Eos, Transactions American Geophysical Union, 83 (2):13 and 16.

Maclean, G.A. and W.B. Krabill. 1986. Gross-merchantable timber volume estimation using an airborne LIDAR system. Canadian Journal of Remote Sensing, 12:7-18.

Manson, G.K., D.L. Forbes, and G.S. Parkes. 2002. Wave climatology. In Coastal Impacts of Climate Change and Sea-Level Rise on Prince Edward Island. Edited by Forbes, D.L. and Shaw, R.W. Geological Survey of Canada, Open File 4261, Supporting Document 4, 31 p. and 1 attachment (on CD-ROM).

Maune, D. F. 2001. Digital elevation model technologies and applications: The DEM Users Manual. Edited by Maune, D.F. American Society of Photogrammetry and Remote Sensing.

Mayor, S.D. and E.W. Eloranta. 2001. Two-dimensional vector wind fields from volume imaging LIDAR data. Journal of Applied Meteorology, 40:1331-1346.

McCulloch, M.M., D.L. Forbes, R.W. Shaw, and the CCAF A041 Scientific Team. 2002. Coastal Impacts of Climate Change and Sea-Level Rise on Prince Edward Island. Edited by Forbes, D.L. and Shaw, R.W. Geological Survey of Canada, Open File 4261, xxxiv + 62 p. and 11 supporting documents (on CD-ROM).

McLean, R.F., A. Tsyban, V. Burkett, J.O. Codignotto, D.L. Forbes, N. Mimura, R.J. Beamish, V. and Ittekkot. 2001. Coastal zones and marine ecosystems. In Climate Change 2001: Impacts, Adaptations and Vulnerability. Contribution of Working Group II to the Third Assessment Report of the Intergovernmental Panel on Climate Change. Cambridge University Press, Cambridge and New York, pp. 343-379.

Milloy, M. and K. MacDonald, K. 2002. Evaluating the socio-economic impacts of climate change and sea-level rise. In Coastal Impacts of Climate Change and Sea-Level Rise on Prince Edward Island.Edited by Forbes, D.L. and Shaw, R.W. Geological Survey of Canada, Open File 4261, Supporting Document 10, 90 p. (on CD-ROM).

Nicholls, R.J., F.M.J. Hoozemans, M. and Marchand. 1999. Increasing flood risk and wetland losses due to global sea-level rise: regional and global analyses. Global Environmental Change, 9:S69-S87.

O'Reilly, C. 2000. Defining the coastal zone from a hydrographic perspective. Proceedings, Workshop on risk assessment and disaster mitigation: Enhanced use of risk management in integrated coastal management. International Ocean Institute, Bermuda. Backscatter, April 2000, pp. 20-24

O'Reilly, C.T., D.L. Forbes, and G.S. Parkes. 2003. Mitigation of coastal hazards: adaptation to rising sea levels, storm surges, and shoreline erosion. Proceedings, 1st Coastal, Estuary and Offshore Engineering Specialty Conference, Canadian Society of Civil Engineering, Moncton, NB, CSN-410-(1-10).

Parkes, G.S. and L.A. Ketch. 2002. Storm-surge climatology. In Coastal Impacts of Climate Change and Sea-Level Rise on Prince Edward Island.Edited by Forbes, D.L. and Shaw, R.W. Geological Survey of Canada, Open File 4261, Supporting Document 2, 87 p. (on CD-ROM).

Parkes, G.S., D.L. Forbes, and L.A. Ketch, L.A. 2002. Sea-level rise. In Coastal Impacts of Climate Change and Sea-Level Rise on Prince Edward Island.Edited by Forbes, D.L. and Shaw, R.W. Geological Survey of Canada, Open File 4261, Supporting Document 1, 33 p. and 5 attachments (on CD-ROM).

Parkes, G.S., L.A. Ketch, and C.T. O'Reilly. 1997. Storm surge events in the Maritimes. Proceedings, Canadian Coastal Conference 1997. Canadian Coastal Science and Engineering Association, Guelph, Ontario, pp.115-129.

Post, M.J., C.J. Grund, A.M. Weickmann, K.R. Healy, and R.J. Willis. 1996. A comparison of the Mt. Pinatubo and El Chichon volcanic events: LIDAR observations at 10.6 and 0.69 μm. Journal of Geophysical Research, 101(D2):3929-3940.

Sallenger, A.B., Jr., W. Krabill, J. Brock, R. Swift, M. Jansen, S. Manizade, B. Richmond, M. Hampton, and D. Eslinger. 1999. Airborne laser study quantifies El Niño- induced coastal change. Eos, Transactions, American Geophysical Union, 80:89, 92.

Schuckman, K. 2003. Announcement of the Proposed ASPRS Binary LIDAR Data File Format Standard. Photogrammetric Engineering and Remote Sensing, 69(1):13-19.

Shaw, J., R.B. Taylor, D.L. Forbes, M.H. Ruz, and S. Solomon. 1998. Sensitivity of the Coasts of Canada to Sea-Level Rise. Geological Survey of Canada Bulletin 505.

Stockdon, H.F., A.H. Sallenger, J.H. List, and R.A. Holman. 2002. Estimation of shoreline position and change using airborne topographic LIDAR data. Journal of Coastal Research, 18(3):502-513.

Thompson, K., H. Ritchie, N.B. Bernier, J. Bobanovic, S. Desjardins, P. Pellerin, W. Blanchard, B. Smith, and G. Parkes, G. 2002. Modelling storm surges and flooding risk at Charlottetown. In Coastal Impacts of Climate Change and Sea-Level Rise on Prince Edward Island.Edited by Forbes, D.L. and Shaw, R.W. Geological Survey of Canada, Open File 4261, Supporting Document 6, 48 p. (on CD-ROM).

Wadhams, P., W.B. Tucker III, W.B. Krabill, R.N. Swift, J.C. Comiso, and N.R. Davis. 1992. Relationship between sea ice freeboard and draft in the Arctic Basin and implications for ice thickness monitoring. Journal of Geophysical Research, Vol. 97 (C12):20,325-20,334.

Walsh, K.J.E. and J.J. Katzfey. 2000. The impact of climate change on the poleward movement of tropical cyclone-like vortices in a regional climate model. Journal of Climate, 13:1116-1132.

Webster T.L., S. Dickie, C. O'Reilly, D.L. Forbes, K. Thompson, and G. Parkes, G. 2001. Integration of diverse datasets and knowledge to produce high resolution elevation flood risk maps for Charlottetown, Prince Edward Island, Canada. In CoastGIS2001, Halifax, Nova Scotia, Canada.

Webster, T.L., D.L. Forbes, S. Dickie, R. Covill, and G.S. Parkes. 2002. Airborne imaging, digital elevation models and flood maps. In Coastal Impacts of Climate Change and Sea-Level Rise on Prince Edward Island. Edited by Forbes, D.L. and Shaw, R.W. Geological Survey of Canada Open File 4261, Supporting Document 8, 36 p. (on CD-ROM).

Webster T.L., S. Dickie, C. O'Reilly, D.L. Forbes, G. Parkes, D. Poole, and R. Quinn. 2003. Mapping storm surge flood risk using a LIDAR-derived DEM. In Elevation, A Supplement to Geospatial Solutions, May, pp 4-9.

Webster, T.L., D.L. Forbes, S. Dickie, and R. Shreenan. 2004. Using topographic LIDAR to map flood risk from storm-surge events for Charlottetown, Prince Edward Island, Canada. Canadian Journal of Remote Sensing, 30(1):1-13.

Wehr, A. and U. Lohr, U. 1999. Airborne laser scanning – An introduction and overview. *Journal of* Photogrammetry and Remote Sensing, 54:68-82.

Chapter 8

INTEGRATION OF NEW DATA TYPES WITH HISTORICAL ARCHIVES TO PROVIDE INSIGHT INTO COASTAL ECOSYSTEM CHANGE AND VARIABILITY

JENNIFER GEBELEIN
Department of International Relations, Florida International University, Miami, Florida, 33199, USA. Jennifer.Gebelein@fiu.edu

1. Introduction

As populations increase around the world, anthropogenic influences are changing the environment with unforeseen rapidity and unknown consequences (Lunetta, 1998). The changes that are occurring result in environmental transformations at local, regional, and global scales. Earth observing technologies now allow us to monitor landscape changes at multitemporal (as well as regional and global spatial) scales. However, the speed that technology has advanced has not been matched by parallel progress in conservation, environmental management, or developmental regulations. We can lessen this gap by making scientists and managers aware of historical data archives as well as current baseline datasets and products.

Recently developed technologies allow scientists to integrate large and varied datasets over many different spatial and temporal scales. This type of integration was not possible 40 years ago (Chang, 2002). Since that time there has been an evolution of remote sensing and Geographic Information Systems (GIS) technology which has allowed for many types of data integration. The importance of data integration is directly linked to current multidisciplinary research that incorporates aspects of both human and physical science data sets (Konecny, 2003).

When current and historical data are combined, new patterns, relationships, and results emerge (Maynard, 2003). Utilizing historical datasets for time series analyses allows scientists to model what may happen in the future, based on observed change in the past (Longely, 1996; DeMers, 2002). These types of historical, integrative analyses allow researchers to ask new questions such as: Where are the most threatened ecosystems? What types of anthropogenic stresses are influencing an ecosystem? How do we identify indicators of ecosystem health at different scales? Is it possible to predict what the effect of long-term environmental impacts will be (Lunetta, 1998)? Since anthropogenic activities are the catalyst for most of the environmental changes we are observing, data creation, integration, and analyses should address issues of not only physical change, but also the social impetus for *why* such changes occurred. Understanding the reasons driving such change allow scientists to better analyze data and help managers develop better and creative conservation strategies (Walsh et al., 2002).

1.1 ECOSYSTEM CHANGE

Many ecosystems are vulnerable to degradation through human misuse and mismanagement. However, some are at a higher risk than others. According to the World Resource Institute's (WRI) Pilot Analysis of Global Ecosystems (PAGE) report

L.L. Richardson and E.F. LeDrew (eds.), Remote Sensing of Aquatic Coastal Ecosystem Processes: Science and Management Applications, 183-197. © 2006 *Springer. Printed in the Netherlands.*

(2000), there were just over 2.2 billion people living within 100 km (62 miles) of a coast in 1995. The coastal zone as a land cover class only accounts for twenty percent of all land area. Most lands within the zone have been permanently altered. Apart from Antarctica, 19% of all landmass within 100 km of the coast are classified as agricultural or urban areas; 10% are classified as a mosaic of natural and altered vegetation; and 71% are classified in the least modified category. A significant percentage of this least modified class incorporates many uninhabited areas in northern latitudes. Within this coastal zone, numerous coastal habitats are quickly vanishing, such as wetlands, mangroves, coral reefs, and seagrasses (WRI 2000, Edinger et al., 1998, Mumby et al., 1999). Approximately 5-80% of original mangrove area in countries where such data are available is believed to have been lost. Many coastal ecosystem communities have been significantly altered by anthropogenic changes to the landscape. These changes, in turn, can negatively impact coastal habitats such as coral reef and wetlands. An example of use of remote sensing to detect such considerable change over time is shown in Figure 1. Figure 1 also positively links the Reefs at Risk indicator with such changes. Such widespread changes and losses have increased remarkably in the last 50 years (WRI, 2000). This is of direct concern for coastal managers since the different ecosystems within the coastal zone include a myriad of habitats supporting juvenile and mature aquatic species, and indigenous and migratory bird species. The coastal zone is also increasingly under pressure by a growing population, and urban expansion and development, all of which result in a loss of original habitat. Reports documenting the extreme loss of coastal habitats have been produced sporadically. Herein lies the importance of identifying archival datasets, data products, and sites of ongoing data acquisition of remotely sensed data. Identifying sources of data, data products, and ongoing efforts can lessen the cost of data and value-added product attainment. This chapter provides a case study of an organization dedicated to coastal (and other) data dissemination, a description of current archival data, and insight into the issues of data integration. This chapter also discusses the integration of new and archival data and how the integration of these datasets can provide insight into determining current and future ecosystem change, variability, and management decisions. Since coastal ecosystems are identified here as one of the fastest changing, highly altered ecosystems on the planet, data sources that would benefit studies in this area will be the major focus of this chapter. However, many of the data sources discussed could also be used in support of research in other ecosystems.

2. Historical Satellite Imagery and Related Data Products

2.1 GLOBAL LAND COVER FACILITY (GLCF) – A CASE STUDY

In this section the most notable archive of available imagery in the United States, the University of Maryland's Global Land Cover Facility (GLCF), is discussed in detail. Since this is the most comprehensive, free data distribution facility at the time, other image archives will be mentioned (but not discussed at length) at the end of this section for the reader's reference. Not all of the mapping and data products available are discussed, only those which could potentially be utilized in conjunction with coastal or aquatic study.

The mission of the GLCF is "*To create and communicate improved understanding of the nature and causes of land cover change and its impact on the Earth System through the use of remote sensing.*" At the University of Maryland in 1993, the

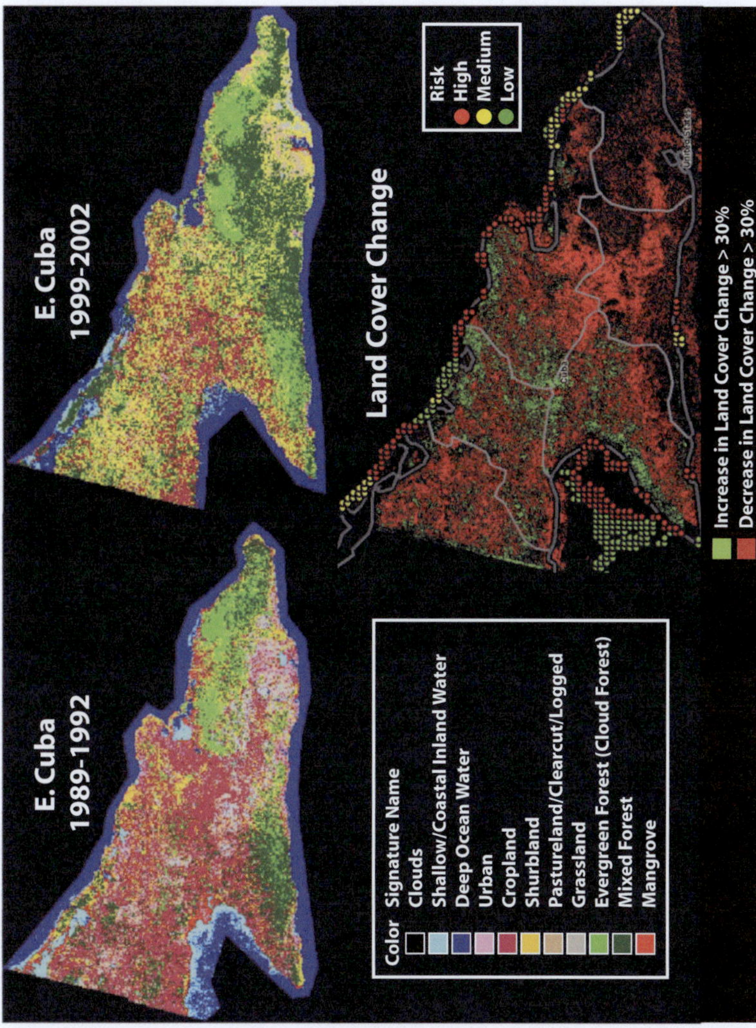

Figure 1. Land Cover Change and Mangrove and Reef Health.

Department of Geography began a joint project with the Institute for Advanced Computer Studies (UMIACS) to employ advanced computational methodologies to tackle issues of data volume, data processing, and data analyses in global and regional-scale studies (Davis and Townshend, 1993). Because of this joint effort, a National Science Foundation Grand Challenge grant was awarded to UMIACS and the Department of Geography. This award challenged the University of Maryland team to resolve the data volume issue involved in global and regional-scale data analyses as well as addressing the user community's need for faster processing time (Townshend et al., 2000). The results of this research were the basis for the current GLCF structure. Currently, the GLCF is funded by NASA's Earth Science Information Partnership (ESIP).

2.2 PRODUCTS GENERATED USING GLCF DATA AND DISSEMINATED ON THE GLCF WEBSITE

The GLCF develops, disseminates, and archives not only satellite imagery but earth science products as well. The type of data the GLCF provide includes (but is not limited to): 1) Landsat imagery (Enhanced Thematic Mapper (ETM), Thematic Mapper (TM) and Multispectral Scanner (MSS); 2) Landsat derived products (e.g. mosaics); 3) Moderate Resolution Imaging Spectro-radiometer (MODIS) products (e.g. 32-day composites); 4) AVHRR products; 5) satellite-derived calculations of radiative flux; 6) NOAA's Geostationary Operational Environmental Satellite (GOES) data; and 7) urban growth of major United States metropolitan centers. The following sections briefly describe each of the major data products applicable to coastal ecosystem research.

2.2.1 *High Resolution Data Products*

Deforestation Mapping Product. The purpose of the Deforestation Mapping Project (DMP) was to generate digital forest cover maps and offer them to other researchers for use as baseline data. Deforestation can have extremely negative affects on aquatic communities (fish species, aquatic invertebrates) due to increased runoff and sediment deposition, as well as coastal land structure (Duarte, 1996). The DMP dataset used multiple dates of Landsat TM and ETM+ to classify change in forest extent over various time periods. There are six major classes in these datasets: forest, degraded forest, nonforest, water, cloud, and shadow. The (DMP) generated several major products: 1) 1980s and 1990s country map products of Bolivia, Peru, Colombia, Ecuador (1990s only), and the Democratic Republic of Congo; and 2) 1990s and 1980s time series products of Pan-Amazon Deforestation Hotspots, the Central African region, other Pan-Amazon areas of interest, and a Bolivia (1970s, 1980s and 1990s) land cover change map (1999, UMD).

Coastal Marsh Project. The objective of the Coastal Marsh Project (CMP) was to analyze Landsat TM images in combination with data from the USDA National Wetlands Inventory. The results of this analysis allowed researchers to locate varied surface conditions of coastal marshes and classify areas according to percentage of water as follows: "up to 10% water is healthy marsh; up to 20% is slightly deteriorated marsh; up to 30% is moderately deteriorated marsh; and above 50% indicates complete deterioration" (Kearney et al., 1995). This classification helped determine the mechanism for total marsh loss. This dataset is available for most of the east coast of the United States. The final products include the development of a classified health dataset of coastal marshes and an improved methodology of coastal marsh health assessment. This project was managed and completed in the Department of Geography at the University of Maryland, College Park from 1993 to 1998 (Kearney et al., 1995; Rizzo et al., 1996).

NASA EOS Land Validation Products. The NASA Earth Observing System (EOS) has multiple core sites around the globe. At each of these core sites, *in situ* data have been collected to compare with remote sensing data products from space and airborne sensors including the Sea-viewing Wide Field-of-view Sensor (SeaWifs), Multi-angle Imaging Spectroradiometer (MISR), Advanced Very High Resolution Radiometer (AVHRR), Advanced Spaceborne Thermal Emission and Reflection Radiometer (ASTER), and Landsat Thematic and Enhanced Thematic Mapper (TM/ETM+). These core site datasets are intended for land product validation over a variety of ecosystem types. Most of the core sites continue to build on the current program of long-term measurements and have the infrastructure for continuous *in situ* data collection. This project evolved from a Science Working Group meeting for the AM Platform Land Validation Coordination meeting in 1997 (Justice et al., 1998).

The GLCF Landsat Image Archive. The GLCF is perhaps most famous for its Landsat imagery archive. Landsat provides a relatively high resolution counterpart to NASA's Earth Observing System (EOS) sensors such as MODIS and MISR. In terms of time series analyses, the Landsat program offers one of the longest continuous records in Earth observations. Landsat provides multispectral imagery with 30-90 meter spatial resolution of the Earth's land and coastal areas. There have been several Landsat missions, beginning in 1972 with the Landsat I MSS sensor, and concluding most recently with Landsat 7-ETM+ sensor (Jensen, 2003). The GLCF offers the ability to search by path and row or by global map. The GLCF also has a function called "workspace" that allows a user to preview images and image metadata, decrease download time, organize download files, and save image query results. Users may refine their search by date, sensor, path/row, data format, level of processing, validation (yes/no), or if the image has been orthorectified. The map and path/row search, data preview and download, and workspace areas are all located on the GLCF Earth Science Data Interface website: http://glcfapp.umiacs.umd.edu:8080/esdi/index.jsp.

Landsat historical data sets have filled, and continue to fill, an important niche in many research communities. Because Landsat has one of the longest continuous records of Earth observations, it has been analyzed and compared with almost every sensor in orbit as well as newer GIS spatial data types. Examples of aquatic coastal multitemporal Landsat data analysis combined with GIS data include studies such as: 1) watershed modeling using Landsat and micro-computer based GIS system (Berich and Smith, 2000); 2) decision analysis in administration using Landsat and GIS modeling of a Southwestern watershed (Kepner and Edmonds, 2002); 3) water quality monitoring (Erkkilä, 2004); 4) assessment of variation in coastal sea surface temperature (Thomas, 2002); 5) mapping of mangrove extent (Sulong et al., 2002); and 6) determination of concentrations of chlorophyll and suspended sediment in surface waters for monitoring coastal water quality (Keiner, 1998).

Utilization of the historical Landsat dataset has allowed researchers to conduct time series analyses in a number of studies. These include: 1) changes in coastal sediment transport processes (El-Asmar, 2002); 2) monitoring changing coastline shape (White, 1999); 3) water body detection and delineation with Landsat TM data (Frazier and

Page, 2000); 4) coral reef bleaching (Yamano, 2004); 5) water quality monitoring in estuarine waters (Lavery, 1993); 6) and changes in coral reef communities over multiple years (Dustan et al., 2001).

Because of the significant length of the Landsat dataset, there have been many studies in which Landsat images are combined with other air- or spaceborne sensor types to yield an enhanced data product. Some examples of research that use this approach are: 1) use of remote sensors to classify coral, algae, and sand as pure and mixed spectra (Hochberg and Atkinson, 2003); 2) merging Landsat TM and SPOT via wavelet transformation (Zhou and Civco, 1998); 3) coral reef habitat mapping with Ikonos and Landsat (Capolsini, 2003); 4) change detection in coral reef environments using Landsat and SeaWiFS data (Andréfouët et al., 2001); and 5) mapping shallow-water marine environments (Mumby and Edwards, 2002)

2.2.2 *Moderate Resolution Data Products*

The GLCF also has moderate resolution products and data available, including Terra (Latin for "land") MODIS data, United States vegetation index product, Global MODIS-derived 500 meter tree cover product, MODIS-derived 500 meter 32-day composite products, and MODIS-derived vegetation product for Central Brazil and Idaho/Montana at 250 meter resolution. These GLCF products are generally more appropriate for land based research, and will not be discussed here. However, the NASA MODIS-derived products can provide researchers with useful information concerning coastal and aquatic areas. Thus, those aspects of the MODIS instrument are the focus of this section

MODIS. The MODIS sensor has 36 spectral bands, most of which have 1 km spatial resolution and a wavelength range of 0.4 μm to 14.4 μm. MODIS is mounted on two platforms: Aqua and Terra. Terra's orbit is from North to South in the morning and Aqua's orbit is South to North in the early evening. The two MODIS sensors yield global coverage once every 1-2 days (Townshend et al.,2005). NASA-derived data products from MODIS can be utilized to detect, monitor, and analyze ecosystem processes at the local, regional, and global scales. These data products include, but are not limited to, chlorophyll concentration, organic matter concentration, sea surface temperature, ocean primary productivity, ocean aerosol properties, and normalized water-leaving radiance (http://modis.gsfc.nasa.gov/data/dataproducts.html).

MODIS historical data sets remain an important asset to the coastal (global) research community. Research based on a combination of MODIS with newer datasets such as Landsat 7 imagery, serve to enhance and validate derived products. Based on MODIS data, scientists have conducted static and time series analyses focused on global monitoring of air pollution (Chu et al., 2003), Landsat-derived training data for MODIS classifiers (DeFries et al., 1998), validation of ocean color imagery (Chomko et al., 2003), and mapping concentrations of total suspended matter in coastal waters (Miller and McKee, 2004).

MODIS data have also been combined with other data types to enhance and validate its detection, monitoring and analysis capabilities. Several studies that exemplify this combination include *in situ* measurements compared to MODIS-derived spectral reflectances of snow and sea ice (Zhou and Li 2003), ocean-color observations of the tropical Pacific Ocean (McClain et al., 2002), measurements of sea-skin temperature for validation of satellite data (Minnett, 2003), and an investigation of product accuracy as a function of inputs including Sea Wifs and MODIS data (Wang et al., 2001).

2.2.3. *Coarse (Global and Regional-Scale) Resolution Data Products*

In response to a demand for large-scale land cover products to assess change, several global datasets are now available. These include the AVHRR-derived global 1 km vegetation product; the 1 km, 8 km and 1 degree global land cover classified maps; the Large-Scale Biosphere-Atmosphere Experiment in Amazonia (LBA); and NOAA's GOES archival data. With the exception GOES archival data, most of these data products are only of peripheral interest to coastal and aquatic researchers. A much larger collection (archived at the Jet Propulsion Laboratory's Physical Oceanography DAAC) of regional and global resolution data is discussed in the following section.

NOAA GOES archival data. The GOES archival data are distributed through the UMD's department of meteorology in association with the GLCF. GOES data are primarily used for basic weather monitoring, forecasting operations, sea surface temperature, sea surface height, and climate modeling. Associated coastal land surface conditions, winds, and cloud cover can also be derived from GOES data. GOES derived products can be generated near real-time. Examples of these products include atmospheric profiles (total precipitable water in cloud column), sea surface temperature, fire detection, and biomass burning (SSEC, 2001).

2.2.4 *Other Available Historical Image Datasets*

There are several other noteworthy archives of satellite imagery. Two are briefly discussed below.

The Jet Propulsion Laboratory's Physical Oceanography DAAC (PO.DAAC) The PO.DAAC houses an extensive online collection of data and data products applicable to research pertaining to the physical state of the ocean. Data regarding ocean surface topography, ocean wind vectors, sea surface temperature, ocean currents, and wave height are provided. With respect to sea surface temperature, there is also *in situ* data available to validate remotely sensed measurements. There are 19 sensors that provide data at varying regional to global spatial resolutions. The spectral resolution is also quite broad, considering all 19 instruments. The data available on PO.DAAC include, for example, data from TOPEX/POSEIDON, ATSR, AVHRR, CZCS, GOES-3, GEOSAT, MODIS, and SEASAT. Most datasets have at least three years of temporal coverage, making time-series analysis possible. The site also contains viewing and download tools allowing users to access large datasets easily and efficiently. The web address is http://podaac.jpl.nasa.gov/index.html.

Land Processes Distributed Active Archive Center (DAAC) This center provides online access to all available data in NASA's Earth Observing System. The data vary in their scale, size, temporal, and global coverage. There is a large variety of atmospheric, land, and oceanic data acquired by EOS sensors such as the Tropical Rainfall Measuring Mission (TRMM), MISR, Landsat ETM+, Terra ASTER, Terra MODIS, Landsat Pathfinder, AVHRR, global elevation data, global land cover dataset, airborne imagery, and SIR-C radar data. DACC is an institute that produces EOS data products and carries out data archival, management, and distribution for NASA. Cost of data products depends on the product and level of processing. These data are distributed by the Land Processes Distributed Active Archive Center (LP DAAC), located at the U.S. Geological Survey's EROS Data Center (http://LPDAAC.usgs.gov).

Users who have downloaded data from the described sources include scientists from varied research communities that include physical oceanography, applied social science, resource management, environmental policy, computer science, and disaster management. Technology has matured to the point where scientists are now routinely combining archival and current data over differing spatial and temporal scales. This type of data amalgamation allows for identification of natural and anthropogenic ecosystem changes not previously possible. The tools are now available to combine not only archival current data, and data at different scales, but also data of different production types. When different data types are combined, such as *in situ* data measurements validating AVHRR sea surface temperature data, or coastal erosion models incorporating decadal census data, this technology becomes a robust, integrated tool that can be used to derive meaningful policy regarding any kind of ecosystem management.

3. Sample Historical GIS Data and Related Data Products

The World Resources Institute The World Resources Institute (WRI) is a non-profit independent institution that conducts and funds research to contribute to the following goals: 1) guarantee public assess to data, information, and decisions concerning natural resources and environment; 2) assess and help reverse degradation of ecosystems while building their capacity to provide humans with goods and services; 3) support climate change research; and 4) support sustainable enterprise research such that markets may expand and succeed without further damaging the environment (http://www.wri.org/). Through these goals, the WRI enables scientists to conduct research utilizing the data that have been generated from many projects dealing with local, regional, and global issues. The WRI also has data products generated from various research studies. There are two major differences between the University of Maryland's GLCF online databank and WRI databank. The first is the WRI data are generally in GIS format and not remotely sensed imagery. Second, the WRI data tend to have primarily more current data than that offered on the University of Maryland site. All of the downloadable datasets available on the WRI Internet site may be found at the following website: http://pubs.wri.org/datasets.cfm?SortBy=1. In the following section, the major datasets and data products will be discussed.

3.1 Data and Data Derived Products Disseminated on the WRI Website

Pilot Analysis of Global Ecosystems (PAGE) The PAGE program is a collaborative effort among many agencies, including WRI, the International Food Policy Research Institute (IFPRI), intergovernmental organizations, research agencies, and individual scientists in more than 25 countries globally. There are five major categories of ecosystems that the PAGE collaboration focuses upon: coastal, forests, freshwater systems, grasslands, and agroecosystems. All are pilot studies. All categories of analysis utilized various physical measurements such as water condition, temporal land use change, and soil conditions. There were two outcomes of these studies. The first was a written publication including an in-depth analysis of all data acquired (Burke et al., 2000). The second result was spatial data. Baseline historic and current data are now available online and may be downloaded from the WRI website. For this section, we will only focus on coastal and freshwater area ecosystems.

Coastal Ecosystems The PAGE Coastal Systems Analysis publication clearly illustrates the declining health of the global coastal ecosystem. The analysis focuses on five major services and goods provided by the coastal zone: tourism, biodiversity, shoreline stabilization, filtering water, and food for human consumption. More specifically, the report includes qualitative and quantitative measures indicating the level of anthropogenic alteration of the global coastal environment. Tourism, urban growth, and industrial expansion have altered the global coastal land area by almost 30%. Along with coastal population increase is a parallel increase in nutrient pollutant loads from sewage, fertilizer, and aquaculture (Burke et al., 2000). Population increase also increases pressure on local marine fisheries, which has left numerous valuable fish stocks depleted, or in serious decline (Botsford et al., 1997; Castilla, 1999).

Freshwater Systems This analysis, completed in 2000 by Revenga et al. (2000), demonstrates an assessment of global freshwater basins and their extent, including Europe, the United States, and inland fisheries of all countries. GIS data are available for European and United States' basins, watersheds, and fisheries. This study developed indicators of the world's freshwater systems to show the level of human intervention in the hydrologic cycle. Global, regional, and local datasets were utilized to demonstrate significant trends, concepts, issues, and ultimately derive indicators of freshwater. Inland and coastal freshwater sources are extremely important in terms of aquatic habitat (Bogan, 1993; Carpenter et al., 1996), health of adjacent vegetation (Sasser and Gosselink, 1984; Osborne and Kovacic, 1993), and bird species (Chapman and Loftus, 1986; Crossley, 1999). This study included analysis of several topics related to freshwater health, including inland fisheries, water quality, water quantity, and biodiversity. The major conclusions of this research included the following: "...although humans have enhanced water availability through dams and reservoirs, over 40 percent of the global population lives in conditions of water stress. This percentage is estimated to grow to almost 50 percent by 2025. Surface and groundwater is being degraded in almost all regions of the world by intensive agriculture and rapid urbanization, aggravating the water scarcity problem. Food production from wild fisheries has been affected by habitat degradation, overexploitation, and pollution to a point where most of these resources are not sustainable without fishery enhancements. Finally, the capacity of freshwater ecosystems to support biodiversity is highly degraded at a global level, with many freshwater species facing rapid depopulation declines or extinction." (Revenga et al., 2000). This dataset is discussed at length in the associated publication for this freshwater study (Revenga et al., ibid), and is one of four technical reports as part of the PAGE research at WRI.

Other Data and Data Products Provided by WRI With reference to coastal resources, there are several past and ongoing datasets and products available from WRI. The first major GIS data product is the Reefs at Risk dataset. There are three complete reef datasets available: a Reefs at Risk Global dataset, Reefs at Risk in Southeast Asia dataset, and Reefs at Risk in the Caribbean. The global dataset is a georeferenced (GIS) point file that gives an assessment of anthropogenic threat to over 55,000 coral reefs worldwide. The four major threat types include: inland pollution and erosion, marine-based pollution, coastal development, and overexploitation. The online data are accompanied by a report (Burke, 1998) describing key findings from dataset analysis, including the findings that most United States' reefs are at risk, with special concern to Puerto Rico, Hawaii and Florida reefs; almost two-thirds of Caribbean reefs are at risk;

and Southeast Asian reefs are the most threatened of any region, with greater than 80% at high risk of destruction (Burke, 1998). This dataset and report were completed due to the efforts of many collaborators, including WRI, the World Conservation Monitoring Centre (WCMC), the International Center for Living Aquatic Resources Management (ICLARM), and the United Nations Environment Programme (UNEP).

Since the Global dataset indicated that the greatest risk to reefs was found in Southeast Asia, this led to the second Reefs at Risk dataset specific to this region. This dataset was completed in 2002 by Burke et al. (2000), and is also a georeferenced (GIS) point database with reefs categorized by threats. Overfishing, tourism, increased fishing pressure, proximity to cities, coral disease, land cover type, proximity to shipping lanes, and inflow from rivers are only a few examples of the threats to reefs in this region. The accompanying report states that although there is widespread knowledge of the reef degradation in this region, specific information for local reefs was typically minimal prior to the Reefs at Risk project. The Reefs at Risk in Southeast Asia project was intended to address this information deficit through an extensive data compilation effort (Burke et al., 2002).

Reefs at Risk in the Caribbean is the most recent WRI publication regarding reefs and threats to reef health (Burke et al., 2004). The available georeferenced datasets incorporate vector and raster data, including but not limited to, reef locations, threats from anthropogenic activities, river mouths, bathymetry, ports, oil/gas wells, airports marine protected areas, population densities, soil type, dive centers, and watershed boundaries. Following the integration and analysis of these datasets, there were several important trends and patterns which emerged. First, there is a clear decline in marine protected area effectiveness. There has been a tremendous increase in tourism and development in areas adjacent to coral reefs. Currently, there are more than 285 MPAs in the Caribbean, but the protection provided by those MPAs is not consistent throughout the Caribbean. According to Burke et al. (2004) only 6% of MPAs were successfully managed. Threats to reef survival can include disease, increase in ocean temperature, fishing pressure, increased tourism, sedimentation, and pollution. The datasets provided by Burke et al. (2004) provide an excellent source of data for further reef research and future data integration.

4. Data Integration Issues

Combining datasets of different data format, scales (temporal, spectral, spatial), and geometric rectification can yield a more clear and sophisticated analyses of any ecosystem. Data integration can also introduce error or uncertain results. There are general integration issues, and there are uncertainty issues that belong solely to imagery or data from acquired for an aquatic environment. The most basic type of uncertainty arises from data acquisition error in terms of geometric aspects (scale, projection, illumination geometry), data acquisition method per different sensor system (SAR, TM, LIDAR), platform stability (aerial versus orbital), ground control corrections, and atmospheric conditions (Lunetta et al., 1991). Merging datasets with inherent errors in the data structure or geometry can lead to unreliable results.

Difficulties unique to aquatic environments include lack of reliable ground control points and the rapid change in ocean, lake, and freshwater ecosystems. First, if imagery does not contain an identifiable set of ground control reference points, accurate georectification cannot occur. This situation is not only true in underwater areas, it also occurs due to tidal changes, particularly in marsh areas. Second, a water-based environment changes at a much faster rate than a land-based environment. Substantial transformations may occur within minutes, rather than weeks, years, or decades, as is the case with land-based changes. This basic potential for swift change makes *in situ* validation data even more important than for land-based analyses. Increased repeat sensor coverage also becomes more important to gain a better understanding of the more minute changes occurring over a shorter period of time. The greater the time period between data acquisitions, the greater the number of assumptions that are typically made, resulting in increased errors that could therefore directly influence final results and conclusions.

Other types of error may be introduced in the data integration stage. In processing multi-sensor, multi-stage satellite images for change detection, for example, there are several factors that must be addressed. The first is geometric processing so that each image matches the other in terms of coverage extent. If the image registration is not done correctly, determining where and how much change has occurred is impossible (Igbokwe, 1999). Another important potential source of error in change detection analysis is correction for atmospheric effects. Since different dates of imagery are required, atmospheric effects will not be the same for each scene. This topic is thoroughly explained and discussed in Conghe et al. (2001).

A fourth element of data integration error occurs when mathematically evaluating and comparing satellite-derived products to one another. Such errors occur, for example, when comparing thematic maps based on map values to decide which map is more appropriate for an application. Several comparative statistics may be employed to help assess whether the derived maps are significantly different, or no better than a random result. These evaluative methods include the error matrix, kappa statistic, and Z statistic. These statistics can be very helpful to the novice remote sensing researcher and are discussed at length in Congalton and Green (1999), Stehman (1999), and Congalton (2001).

The integration of remote sensing and *in situ* data can also lead to serious error. In terms of coastal land cover or land use change, for example, it is important to note whether the researcher starts from survey data and links it to landscape change, or begins from remotely sensed data of the land and links it to survey data (Rindfuss et al., 2001). In either case, the investigator must accurately identify and georeference the parcel of land under investigation. This can be very difficult, especially in less developed countries lacking cadastral surveys. An exceedingly time consuming approach involves going to the field with a GPS unit. However, this will usually result in smaller sample sizes. After the parcel is georeferenced, it must be co-registered with the remotely sensed image. Co-registration in most cases is fairly uncomplicated. However, in some instances registration may be difficult due to lack of ground control points, thereby introducing spatial uncertainty. This spatial uncertainty would potentially increase the errors of any subsequent research using the two datasets, such as overlay analysis in a GIS (Rindfuss and Stern, 1998; Evans and Moran, 2002). Additionally, a spatial mismatch between plot size and pixel size may exist. The remote sensing spatial resolution may be larger than the plot sizes. Higher resolution imagery or aerial photography may be a solution in some instances, but for most studies their purchase will be outside the means of investigators (Evans and Moran, 2002).

There is a progressive trend to integrate remote sensing and social science data. There are two ways of approaching data integration with these two data types. The first is to take social science data and create a grid based on socioeconomic data, thus matching it with the format of earth science data. This approach has been coined "pixelizing the social" in coastal change (Geoghegan et al., 1998). One method to pixelize the social has been developed by the Center for International Earth Science

Information Network's (CIESIN) Socioeconomic Data and Applications Center (SEDAC). SEDAC has created a gridded population dataset called Gridded Population of the World (Deichmann et al., 2000). SEDAC utilized census data at the most basic administrative units available, and converted them to a grid of 2.5' by 2.5' latitude-longitude cells. A comparable method was developed for a global urban-rural dataset - Landscan, created by Oak Ridge National Laboratory. This method uses gridded census data with additional algorithms to categorize population in relation to data on lights at night, land cover classification, elevation, slope, and transportation infrastructure (Dobson et al., 2000). Once the socioeconomic data have been gridded in either way, they can more easily be combined with remote sensing and GIS data (CIESIN, 2004).

A second approach to social and physical science data integration is to convert data in the opposite direction, i.e. translating physical science data within gridded formats and converting them to tabular data formats more useful and familiar to social scientists (CIESIN, 2004). The SEDAC Population, Landscape and Climate Estimates (PLACE) data set is one of the first efforts to do this. It can be found at the following website: http://sedac.ciesin.columbia.edu/plue/nagd/place.html. The methodology in this case is to take remotely sensed data, or data originally derived from remote sensing instruments, including coastlines, elevation, slope, climate zones and biomes, and aggregate human populations within those categories. These datasets can subsequently be joined with additional tabular data (economic, environmental, or trade statistics, for example), and aggregated at different spatial scales to identify patterns via statistical analyses (CIESIN, 2004). It is critical that such efforts increase aimed at the integration of social and physical science data in order to identify coastal change and coastal urban expansion, and to make physical science data more accessible to social scientists for more inclusive and extensive analyses of coastal, global changes.

5. Conclusions

Remote sensing observations and spatial data of coastal ecosystem dynamics have historically been separate from *in situ* monitoring of those same dynamic systems. Increasingly, however, the two data types have been utilized in tandem to determine the nature of the landscape, the change over time, and the implementation of management strategies. The combination of historical data archives with current data sets has strengthened many studies of ecosystem change and variability. There are many sources of free and at-cost archival and contemporary data that can support analyses of change on a global, regional, or local scale. Integrating multiple datasets as well as *in situ* and remotely sensed datasets and products allows researchers to analyze an ecosystem and include all the influences upon and within it. Environmental impacts affecting every ecosystem due to anthropogenic actions have become more evident as technology is better able to link ecosystem changes to their causal source. As ecosystem vulnerability increases worldwide, it is imperative that scientists lend their understanding of technologies to better identify the links between environment, weather, global climate change, and human activities. The principle interest of sustainable development needs to incorporate integrated research including urban growth, environmental education, marine park development, tourism revenue, and marine resource management to avoid a global "Tragedy of the Commons" (Ehler and Basta, 1993; van da Weide, 1993; Christie and White, 1997; Courtney and White, 2000).

6. Acknowledgements

Funding in support of this chapter was partially provided by NASA through "Using Landsat 7 Data in a GIS-based Revision of ReefBase (A Global Database on Coral Reefs and Their Resources): Distributing Information on Land Cover and Shallow Reefs to Resource managers," NASA Office of Earth Science Application, 2001-2004. The author also appreciates the comments of Dr L. Richardson.

7. References

Andréfouët, S., F. Müller-Karger, E. Hochberg, C. Hu and K. Carder. 2001. Change detection in shallow coral reef environments using Landsat 7 ETM+ data. Remote Sensing of Environment, 78:50-162.

Berich, R.H., and M. B. Smith. 2000. Miscellaneous Paper, Purdum and Jeschke, Consulting Engineers, Baltimore, MD.

Bogan, A.E. 1993. Freshwater bivalve extinctions. American Zoologist, 33:599-600.

Botsford, L.W., J.C. Castilla and C.H. Peterson. 1997. The management of fisheries and marine ecosystems. Science, 277:509-515.

Burke, L., Y. Kura, K. Kassem, C. Revenga, M. Spalding and D. McAllister. 2000. Pilot analysis of global ecosystems (PAGE): Coastal ecosystems, World Resources Institute, Washington, D.C.

Capolsini, P., S. Andréfouët, C. Rion and C. Payri. 2003. A comparison of Landsat ETM+, SPOT HRV, Ikonos, ASTER, and airborne MASTER data for coral reef habitat mapping in South Pacific islands. Canadian Journal Remote Sensing, 29:187-200.

Carpenter, S., T. Frost, L. Persson, M. Power and D. Soto. 1996. Freshwater ecosystems: linkages of complexity and processes. Chapter 12 In: H. A. Mooney (Ed.), Functional Roles of Biodiversity: A Global Perspective. John Wiley and Sons, New York, 493 pp.

Castilla, J.C. 1999. Coastal marine communities: trends and perspectives from human exclusion experiments. Trends in Ecology and Evolution, 14:280-283.

Chang, K. 2002. Introduction to Geographic Information Systems. McGraw-Hill Press, 386 pp.

Chapman, J.D. and W.F. Loftus. 1986. American white pelicans feeding in freshwater marshes in Everglades National Park, Florida. Florida Field Naturalist, 15:20-21.

Chomko, R.M., H. R. Gordon, S. Maritorena and D. A. Siegel. 2003. Simultaneous retrieval of oceanic and atmospheric parameters for ocean color imagery by spectral optimization: a validation. Remote Sensing of Environment, 84:208-220.

Christie, P. and A. White.1997. Trends in development of coastal area management in tropical countries: from central to community orientation. Coastal Management, 25:155-181.

Crossley, G.J. 1999. A Guide to Critical Bird Habitat in Pennsylvania, Pennsylvania Audubon Society, Important Bird Areas Program, 219 pp.

Chu, D.A., Y. J. Kaufman, G. Zibordi, J. D. Chern, J. Mao, C. Li and B. Holben. 2003. Global Monitoring of Air Pollution over Land from EOS-Terra MODIS. Journal of Geophysical Research, 108: 4661-4678.

Congalton, R. 2001. Accuracy assessment and validation of remotely sensed and other spatial information. The International Journal of Wildland Fire, 10:321-328.

Congalton, R. and K. Green. 1999. Assessing the Accuracy of Remotely Sensed Data: Principles and Practices. CRC/Lewis Press, 137 pp.

Conghe, S., C.E. Woodcock and K.C. Sero. 2001. Classification and change detection using Landsat TM data: when and how to correct atmospheric effects. Remote Sensing of Environment, 75:230-244.

Courtney, C. and A. White. 2000. Integrated coastal management in the Philippines: testing new paradigms. Coastal Management, 28:39-53.

Davis, L.D. and J. Townshend. 1993. High Performance Computing for Land Cover Dynamics. UMIACS/University of Maryland.

DeFries, R., M. Hansen, J.R. Townshend and R. Sohlberg. 1998. Global land cover classifications at 8 km spatial resolution: The use of training data derived from Landsat imagery in decision tree classifiers. International Journal of Remote Sensing, 19:3141-3168.

Deichmann, U., D. Balk and G. Yetman. 2001. Transforming Population Data for Interdisciplinary Usages: From census to grid. CIESIN. 20 pp.

DeMers, M.N. 2002. GIS Modeling in Raster. John Wiley and Sons Inc., 208 pp.

Dobson, J. E., E.A. Bright, P.R. Coleman, R. C. Durfee and B. A.Worley. 2000. LandScan: A global population database for estimating populations at risk. Photogrammetric Engineering & Remote Sensing, 66:849-57.

Dustan, P., E. Dobson and G. Nelson. 2001. Landsat Thematic Mapper: Detection of Shifts in Community Composition of Coral Reefs. Conservation Biology, 15:1523-1739.

Duarte, C.M. 1996. Annual Report of the Project Responses to Coastal Ecosystems to Deforestation-derived siltation in Southeast Asia (CERDS). European Union, Brussels.

Edinger, E.N., J. Jompa, G.V. Limmon, W. Widjatmoro and M.J. Risk. 1998. Reef degradation and coral biodiversity in Indonesia: effects of land-based pollution, destructive fishing practices and changes over time. Marine Pollution Bulletin, 36:617-630.

Ehler, C. and D. Basta. 1993. Integrated Management of Coastal Areas and Marine Sanctuaries. Oceanus, 36:379-387.

El-Asmar, H.M. and K. White. 2002. Changes in coastal sediment transport processes due to construction of New Damietta Harbour, Nile Delta, Egypt. Coastal Engineering, 46:127-138.

Erkkila, A. and R. Kalliola. 2004. Patterns and dynamics of coastal waters in multi-temporal satellite images: support to water quality monitoring in the Archipelago Sea, Finland. Estuarine, Coastal and Shelf Science, 60:165-177.

Evans, T.P. and E.F. Moran. 2002. Spatial integration of social and biophysical factors related to landcover change. In: W. Lutz, A. Prskawetz, and W.C. Sanderson (Eds.), Population and Development Review, Supplement to Vol. 28.

Frazier, P. and K.J. Page. 2000. Water Body Detection and Delineation with Landsat TM Data. Photogrammetric Engineering and Remote Sensing, 66:1461-1469.

Geoghegan, J., L. Pritchard, Y. Ogneva-Himmelberger, R.R. Chowdhury, S. Sanderson and B. Turner. 1998. Socializing the Pixel and Pixelizing the Social in Land-Use and Land-Cover-Change. p. 51-69. In: D.Liverman, E. Moran, R. Rindfuss, and P. Stern (Eds.), People and Pixels, Washington, DC: National Academy Press, 256 pp.

Gottfried K. 2003. Geoinformation, Remote Sensing, Photogrammetry and Geographic Information Systems. Taylor and Francis, 272 pp.

Hochberg, E. and M. Atkinson M. 2003. Capabilities of remote sensors to classify coral, algae, and sand as pure and mixed spectra. Remote Sensing of Environment, 85:174-189.

Igbokwe, J. 1999. Geometrical processing of multi-sensoral multi-temporal satellite images for change detection studies. International Journal of Remote Sensing, 20:1141-1148.

Justice, C., Starr, J., Wickland, D., Privette, J. and T. Suttles. 1997. Science Working Group meeting for the AM Platform Land Validation Coordination meeting.

Kearney, M.S., A. S. Rogers, J. Townshend, W. Lawrence, K. Dorn, K. Eldred, F. Lindsay, E. Rizzo and D. Stutzer. 1995. Developing a model for determining coastal marsh "health." Proceedings of the Third Thematic Conference on Remote Sensing for Marine and Coastal Environments. 1:263-272.

Keiner, L. and X-H, Yan. 1998. A neural network model for estimating sea surface chlorophyll and sediments from Thematic Mapper imagery. Remote Sensing of Environment, 66:153-165.

Kepner, W. and T. Edmonds, Remote Sensing and GIS for Decision Analysis in Public Resource Administration: A Case Study of 25 years of Landscape Change in a Southwestern Watershed, Office of Research and Development, National Exposure Research Laboratory, P.O. Box 93478 Las Vegas NV 89193 Ti: EPA/600/R-02/039 June 2002.

Konecny, G. 2003. Geoinformation, Remote Sensing, Photogrammetry and Geographic Information Systems, Taylor and Francis, New York, 280 pp.

Lavery, P.S., C.B. Pattiaratchi, A. Wyllie, and P. Hick. 1993. Water quality monitoring in estuarine waters using the Landsat Thematic Mapper. Remote Sensing Environment, 46(3): 268-280.

Liand, S. and J. Townshend. 1997. Angular signatures of NOAA/NASA Pathfinder AHVRR Land data and applications to land cover identification, IGARSS, Singapore.

Lillesand, T.M., R.W. Kiefer and J.W. Chipman. 2003. Remote Sensing and Image Interpretation 5th edition. Wiley Publishing, 784 pp.

Longley P. A. and M. Batty. 1996. Spatial Analysis: Modelling in a GIS Environment, John Wiley and Sons Inc., 400 pp.

Lunetta, R. 1998. Applications, Project Formulation and Analytical Approach In: and C. Elvidge and R. Lunetta [eds.], Remote Sensing Change Detection – Environmental Monitoring Methods and Applications, Ann Arbor Press, Michigan, 350 pp.

Lunetta, R., R. Congalton, L. Fenstermaker, J. Jensen, K. McGwire, and L. Tinney. 1991. Remote Sensing and Geographic Information System Data Integration: Error Sources and Research Issues. Photogrammetric Engineering and Remote Sensing, 57:677-687.

Maynard, N.G. 2003. Satellites, Settlements, and Human Health In Ridd, M.[ed.], Remote Sensing of Human Settlements. American Society of Photogrammetry and Remote Sensing, 3rd Edition - 2003 Manual of Remote Sensing, 869 pp.

Miller, R. L. and B.A. McKee. 2004. Using MODIS Terra 250m imagery to map concentrations of total suspended matter in coastal waters. Remote Sensing of the Environment, 93:259-266.

Minnett, P.J. 2003. Radiometric measurements of the sea-surface skin temperature for the validation of measurements from satellites – the competing roles of the diurnal thermocline and the cool skin. International Journal of Remote Sensing, 24(24): 5033-5047.

Mumby, P., E. Green, A. Edwards and C. Clark. 1999. The cost-effectiveness of remote sensing for tropical coastal resources assessment and management. Journal of Environmental Management, 55:157-166.

Mumby, P. and A.J. Edwards. 2002. Mapping marine environments with IKONOS imagery: enhanced spatial resolution does deliver greater thematic accuracy. Remote Sensing of Environment, 82:248-257.

Osborne, L.L. and D.A. Kovacic. 1993. Riparian vegetated buffer strips in water-quality restoration and stream management. Freshwater Biology, 29: 243-258.

Revenga, C., Brunner, J., Henninger, N., Payne, R. and K. Kassem. Pilot Analysis of Global Ecosystems: Freshwater, 2000, ISBN: 1-56973-460-7.

Rizzo, E., A. Rogers, M. Kearney, J. Townshend and W. Lawrence. 1996. Changes in Blackwater Marsh, Maryland 1938-1993 as determined by aerial photography and Thematic Mapper data. In: 1996 ASPRS/ACSM Annual Convention and Exposition Technical Papers, Remote Sensing and Photogrammetry, 220-229.

Rindfuss, R. and P. Stern. 1998. Linking Remote Sensing and Social Science: The Need and the Challenges. In: D. Liverman, E. Moran, R. Rindfuss, and P. Stern (Eds.), People and Pixels. Washington, DC: National Academy Press, 256 pp.

Sasser, C.E., and J.G. Gosselink. 1984. Vegetation and primary production in a floating freshwater marsh in Louisiana. Aquatic Botany, 20: 245-255.

Space Science and Engineering Data Center (SSEC), University of Wisconsin-Madison Space Science and Engineering Center (SSEC), 2001, online publication: http://www.ssec.wisc.edu/datacenter/

Stehman, S. 1999. Comparing thematic maps based on map value. International Journal of Remote Sensing, 20:2347-2366.

Sulong, I., H. Mohd-Lokman, K. Mohd-Tarmizi and A. Ismail. 2002. Mangrove Mapping Using Landsat Imagery and Aerial Photographs: Kemaman District, Terengganu, Malaysia. Environment, Development and Sustainability, 4:135-152.

Thomas A.C., D. Byrne and R. Weatherbee. 2002. Coastal sea surface temperature variability from Landsat infrared data. Remote Sensing of Environment, 81:262-272.

Townshend, J. 1999. NASA Landsat Pathfinder Humid Tropical Deforestation Project, Geography Department, University of Maryland, College Park MD.

Townshend et al., 2000. A Landcover Earth Science Information Partnership, Global Land Cover Facility, Proposal Type 2-ESIP, University of Maryland, College Park, Md.

Van der Weide, J. 1993. A systems view of integrated coastal management. Ocean and Coastal Management, 21:129-148.

Walsh, S. J. and K. Crews-Meyer (Eds.) 2002. Linking People, Place, and Policy: A GIScience Approach, Kluwer Academic Publishers, Boston, 262 pp.

Wang Y., Y. Tian, Y. Zhang, N. El-Saleous, Y. Knyazikhin, E. Vermote and R. Myeni. 2001. Investigation of product accuracy as a function of input and model uncertainties: Case study with SeaWiFS and MODIS LAI/FPAR Algorithm. Remote Sensing of Environment, 78:296-311.

White, K. and H. M. El Asmar. 1999. Monitoring changing position of coastlines using Thematic Mapper imagery, an example from the Nile Delta. Geomorphology, 29:93-105.

Yamano, H. and M. Tamura. 2004. Detection limits of coral reef bleaching by satellite remote sensing: simulation and data analysis. Remote Sensing of Environment, 90:86-103.

Zhou, X. and S. Li. 2003. Comparison between in situ and MODIS-derived spectral reflectances of snow and sea ice in the Amundsen sea, Antarctica . International Journal of Remote Sensing, 24:5011-5032.

Zhou, J., D.L. Civco and J.A. Silander. 1998. A wavelet transform method to merge Landsat TM and SPOT panchromatic data. International Journal of Remote Sensing, 19:743-757.

Section III

Management Applications

Chapter 9

OBSERVING COASTAL WATERS WITH SPACEBORNE SENSORS

A Practical Guide for Management and Science

BRIAN G. WHITEHOUSE[1] AND DANIEL HUTT[2]
[1]*OEA Technologies Inc, 14 - 4 Westwood Blvd, Suite 393, Upper Tantallon, NS, B3Z 1H3 Canada*
[2]*Defence R&D Canada – Atlantic, P.O. Box 1012, Dartmouth, NS, B2Y 3Z7 Canada*

1. Introduction

Aquatic remote sensing technologies are used routinely to facilitate environmental research and monitoring, emergency response, national security, military operations, search and rescue, and meteorological forecasting. Emerging applications include ocean forecasting and climate modeling, among others. These technologies are also used to support certain industrial activities, such as oil and gas exploration.

This chapter addresses, from a management perspective, sensors mounted on Earth-observation satellites, focusing on practical aspects of their applications in coastal waters. Successful application requires appreciation of possible differences between a satellite sensor's ability to detect a coastal feature and its ability to fulfill a management or science requirement. Such differences can result in a satellite sensor, which is known to detect the feature of interest, being deemed impractical or of minimal incremental benefit from a management perspective. We address this subject through presentation of certain operational issues, identification of relevant Earth-observation satellite systems and their aquatic applications, and subsequent discussion of practical aspects of their performance in aquatic environments.

2. Platforms vis-à-vis Sensors

If one is responsible for managing many aspects of the aquatic environment, it is likely that use of an *in situ* aquatic sensor, whether it be free-floating, fixed, surficial, or submerged, is part of the management program. Such sensors are discussed throughout this book as they are the most widely used types of environmental sensors. All are capable of providing accurate point source data. The most commonly used are those that measure water temperature, salinity, optical properties, sea state, and currents. And as the atmosphere and oceans are linked like Siamese twins, it is also common to find meteorological sensors on surficial *in situ* platforms, especially wind speed and direction sensors. These environmental features can also be measured with environmental sensors mounted on aircraft and satellites. Indeed, there are five primary types of platforms for aquatic environmental sensors: satellites, aircraft, vessels, *in situ* platforms, and shore-based installations (Whitehouse and Hutt, 2004).

Once an aquatic feature of interest has been identified, and therefore the required sensor, managers and researchers soon discover that the choice of sensor platform influences their ability to monitor the feature. In practice, the question of which platform to employ for a given aquatic application comes down to a matter of available resources. This reality has resulted in the field of marine monitoring and surveillance being platform limited, and this will continue to be the case for the foreseeable future.

L.L. Richardson and E.F. LeDrew (eds.), Remote Sensing of Aquatic Coastal Ecosystem Processes: Science and Management Applications, 201-215.

However, the definition of available resources has changed dramatically over the past eight to ten years. Specifically, during this period various Earth-observation satellite sensors have been operationalized and a number of practical aquatic applications have been developed for these sensors.

Utilization of satellite sensors is becoming less of an option and more of an essential element of a comprehensive coastal sampling program, one reason being that satellites allow us to view coastal processes on a repeatable basis at spatial scales not sampled by any other platform. In so doing, spaceborne (i.e. satellite-based) sensors have helped us to realize that the aquatic environment is not as homogeneous as once believed. Additionally they have reminded us yet again that our understanding of environmental processes is often limited by the techniques available to observe them.

Once deployed, satellite sensors are designed to function for several years. Certain satellite programs launch replacements once the life of a given satellite has expired. Aquatic Earth-observation satellites have been launched since the 1970s, with the result that we are starting to realize time series on the order of a quarter of century. We caution, however, that some of these series are incomplete, and the inaugural sensors and calibration programs were not as advanced as those in use today.

In addition to satellites, airborne and shore-based platforms are capable of providing a synoptic view of coastal waters, albeit not the same view. Although airborne and shore-based sensors are beyond the scope of this chapter, they should not be viewed as redundant means of obtaining synoptic aquatic information. Certain coastal synoptic requirements, which can be fulfilled with existing airborne and shore-based sensors, cannot be providced with existing satellite sensors. Examples of this include aquatic surveys that require airborne hyperspectral or LIDAR sensors, and synoptic mapping of surface currents with coastal HF radar technologies.

We recognize two inherent limitations of sensors mounted on satellite, airborne, and shore-based platforms. First, sensors mounted on these platforms only measure surficial properties. The sensor's depth of penetration into the water column varies from a few microns to a few tens of meters, depending on the particular sensor, platform and body of water. Second, satellite, airborne and shore-based sensors are separated from coastal waters by the atmosphere, and can be influenced by or otherwise rendered unusable by atmospheric effects. The atmosphere contributes about 90% of the signal recorded by spaceborne ocean color sensors, and a correction must be applied to account for this contribution (Antoine et al., 2003). Indeed, the presence of clouds, which represents one aspect of this issue, is arguably the greatest practical limitation of this type of sensor in coastal waters as clouds typically cover 60% of the tropical ocean and 75% of the ocean at mid latitudes (Chelton et al., 2001).

3. Elements of Time

3.1 SATELLITE/SENSOR REVISIT TIME

Almost all aquatic environmental satellites are polar orbiting. An inherent feature of such a satellite is that it will pass over different areas of the planet as it progresses through its orbit cycle. In addition, the frequency with which it revisits a specific geographic area varies with latitude. A polar-orbiting satellite can only sample the aquatic area of interest when the satellite is overhead or thereabout. The length of time a satellite sensor requires to revisit a specific geographic area is referred to herein as its *revisit time* or *temporal resolution*.

For certain satellite programs, the satellite sensor's revisit time is the same as the satellite's orbit repeat cycle. In other cases, the satellite sensor is steerable, or has a wide swath (covers a wide area), or there are several satellites flying the same type of sensor at a given time (i.e. a constellation), or some combination thereof. In such cases, we refer to the satellite sensor's *effective revisit time*. In general, the effective revisit time of such a sensor will be shorter than the satellite's orbit repeat cycle. The Canadian Radarsat satellite, for example, orbits the planet every 100.7 minutes, completes 14 orbits per day, and repeats its orbit cycle every 24 days. However, Radarsat's sensor, which has various scanning and wide-swath modes, is capable of imaging a specific geographic area every three to five days at mid latitudes, and approximately daily at extremely high latitudes (e.g. polar regions). Thus, although Radarsat has an orbit revisit cycle of 24 days, its effective revisit time is much faster, ranging between one to several days.

The satellite sensor's effective revisit time is the critical time element for operational coastal management and science applications, as this is the element which tells a manager or researcher how often the sensor will image a specific area of interest. Engineering details, such as how long the satellite requires to orbit the planet or how many orbits it completes per day, are not required for purposes of practical application.

One of the best demonstrations of the satellite revisit time issue is courtesy of the private company QinetiQ (see Figure 1). They demonstrate that in the Gulf of Oman, available thermal IR satellite sensors provide several images of the region per day, almost every day. Similarly, existing multispectral sensors provide approximately daily

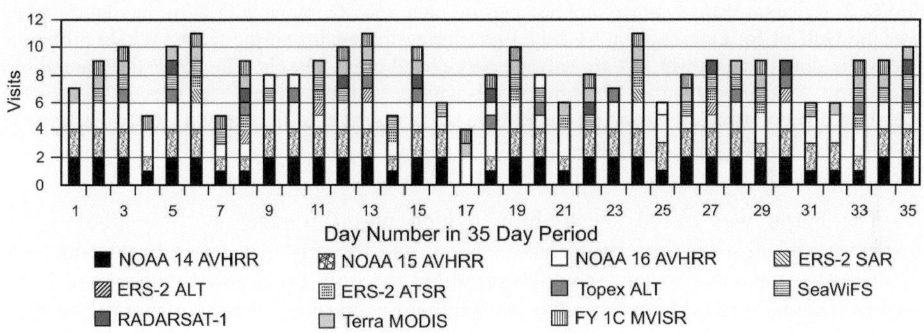

Figure 1. Gulf of Oman (20° N 65° E) 35-day revisit schedule for aquatic, polar-orbiting, Earth-observing satellites. Graph provided by and reprinted with the permission of N. Stapleton, QinetiQ, UK.

coverage. Existing synthetic aperture radar sensors, on the other hand, provide a total of 11 images during the 35 day demonstration period, about one every three days. All but two of these radar images are provided by one satellite (Radarsat), which has an effective revisit time of about four days at the latitude used in this example.

We conclude our discussion of the temporal resolution issue with an example, which took place in Canada in 1996. Figure 2 is a Radarsat image. It demonstrates the unique information content and synoptic perspective of satellite sensors, while also highlighting the impact of temporal resolution on operations. This satellite image was taken over the Gulf of St. Lawrence during the raising of a sunken oil barge, as a potential aid to oil spill response. Although Canadian government agencies applied various *in situ* and airborne monitoring techniques to this project, none of them detected the oil slicks to the extent or with the perspective of the satellite sensor.

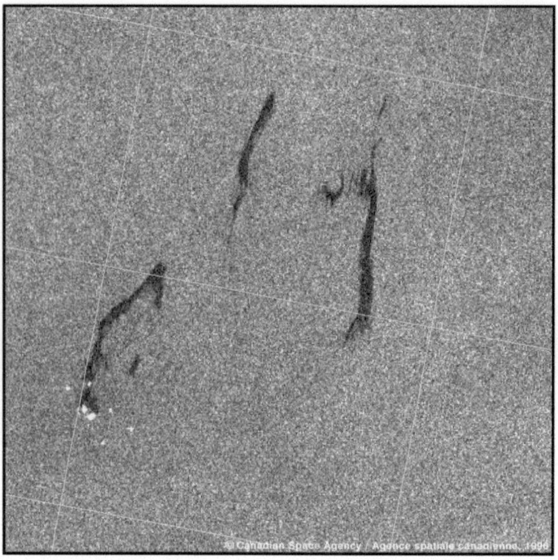

Figure 2. Radarsat wide-swath synthetic aperture radar (SAR) image with 27 m resolution taken over the Gulf of St. Lawrence on 31 July 1996 during the raising of the Irving Whale oil barge. The white dots in the lower left are salvage barges and other vessels. The black lines are slicks comprising a relatively small amount of oil which escaped during this successful salvage operation. Image provided by and reprinted with the permission of the Canada Centre for Remote Sensing. Original data © CSA 1996.

The barge was raised successfully, which was fortunate not only due to the environmental disaster that would have ensued had the oil-laden barge broken-up, but also because the satellite sensor which provided the only operational data would have been of little benefit had the Canadian government been required to launch a major spill response operation. This is because of the temporal resolution of satellite. The next available wide-swath Radarsat image of the site was four days after the barge was (successfully) raised. By then, the need for imagery had passed, thus averting a situation in which the spatial extent of the slick could not be assessed using alternate techniques. The project highlighted this sensor's ability to detect surface slicks while indicating that a constellation of such sensors is needed for low to mid-latitude operations that require an effective revisit time on the order of daily to several times per day. Such constellations of civilian radar sensors have since been proposed.

Geostationary satellites, which orbit the equator rather than near the poles, are capable of fulfilling the temporal resolution requirements of dynamic environmental processes and coastal operations. However, satellite sensors in geostationary orbit tend to have relatively coarse spatial resolution (i.e. on the order of tens of kilometres), which limits their application to coastal waters, cannot view very high latitudes, and are limited in terms of types of available sensors. As a result, although they fulfill meteorological requirements, geostationary Earth-observation satellite sensors are not common components of coastal aquatic management programs.

3.2 DATA RECEPTION AND PROCESSING TIME

Another element of time we wish to address is the amount of time required to receive and process satellite data and to subsequently deliver the derived environmental information to the user. The act of processing satellite data into useful environmental information is a critical function, the subject of which is of encyclopedic proportion. We make no attempt to address this subject, but simply make the point that ultimately the user requires environmental information, not unprocessed satellite data, and therefore there is a time delay between the satellite passing over the area of interest and the user receiving the resulting environmental information. Depending upon the system employed and given application, this may vary from minutes to days or weeks.

If satellite data can be obtained by the user within seconds or a few minutes of the satellite passing overhead, we say that the data can be obtained in *real time*. If it can be obtained by the user within two hours of overpass, we define it as being obtained in *near-real time*. This is subjective terminology and there appears to be no convention within the environmental satellite community on this matter.

This time delay limits certain applications but not others. If one is using multi-spectral satellite sensors to map mangroves for inventory purposes, for example, this is not likely to be an issue as the satellite imagery could arrive days to weeks after the satellite overpass without losing its value. If, on the other hand, you are mapping the same area in support of a developing toxic algal bloom, or national security or emergency operations, it could be an issue. Although not all research applications require real or near-real time satellite data, surveillance and emergency response applications almost always have this requirement.

3.3 DATA ORDERING TIME

The final element of time we wish to address is how far in advance one needs to request satellite data in order to have the satellite sensor image the user's area of interest. For most aquatic satellite systems, this is not a relevant issue as the sensor images Earth's surface continuously and therefore the area of interest is imaged without user intervention. NOAA's AVHRR sensor falls within this category as does NASA's multispectral Modis sensor, the private-sector SeaWiFS sensor, the various altimeters, scatterometers, etc. For certain systems, however, such as synthetic aperture radar (SAR) sensors, the satellite/sensor must be programmed to image a given area at a given time, and perhaps in a given mode. As a result, advance notice must be given by the user. This tends to be a critical factor for applications where the user has little warning of a need for the data, such as operational applications pertaining to emergency response, or science applications involving episodic events, such as heavy rains.

4. Sensors and Their Applications

This section identifies polar-orbiting Earth-observation satellite sensors of relevance to the aquatic environment. It also cross references these sensors with their aquatic applications, lists their revisit or effective revisit time, their spatial resolution and whether or not resulting environmental information is obtainable in near-real time. Spatial resolution refers to the smallest physical unit discernable by the sensor (Kramer, 2002).

Certain environmental satellite systems may have restrictions on data distribution. Just because the satellite exists does not necessarily mean you have access to its data for your particular area of interest. Usually, such restrictions involve systems that have security or military applications. Some systems may have been designed for restricted research purposes or simply does not produce data over certain areas. Often, simply knowing the country, agency or enterprise of origin (i.e. who owns the satellite), and what the satellite system was designed to detect provides insight into (a) the likelihood of you obtaining data and (b) whether or not you will need to pay for it. As a result, we also identify the sensor's country or agency of origin, and type (commercial, research and development, military, etc).

As this chapter is intended to be a practical guide for aquatic coastal managers, this section is limited to polar-orbiting satellites that are: (i) in orbit and functioning as of 2005, (ii) able to sense the Earth at mid-latitudes, and (iii) have a spatial resolution that is less than 70 km. There is no attempt to list satellites that have been launched but are no longer being used, or have been proposed or are being built for future launch. Nor do we review academic literature pertaining to these sensors and their applications, or provide engineering details of the satellites and sensors. Comprehensive presentations of this type of information are available elsewhere (e.g. Kramer, 2002 and other chapters in this volume).

Other than Figure 2, we provide no sample images derived from satellite data in this generalized chapter. For managers, the most immediately fulfilling means to appreciate this aspect of spaceborne remote sensing is to search the internet. Thus, we provide a primary Web address for each sensor. Note that many of these Web sites are hosted by the space agency which owns the satellite, and therefore the site may have a space engineering focus. Similarly, the sites for satellites owned by private companies tend to have a commercial focus. Most of these sites, however, have linkages to the academic and applied aquatic communities. We encourage managers to explore the linkages to these sites as a means of viewing the now countless examples of how these Earth-observation sensors are being employed in aquatic environments.

It should be evident that these satellite sensors have natural groupings, with each grouping having near-identical aquatic applications. As a result, although there is a buffet of satellites and at least thirty relevant sensors, they can be viewed as comprising seven distinct groups of satellite sensors: land-oriented multispectral sensors, marine-oriented multispectral sensors, thermal IR sensors, altimeters, synthetic aperture radars, passive microwave sensors and scatterometers.

4.1 MULTISPECTRAL SENSORS

Multispectral sensors are listed in Tables 1A and 1B. They fall into two categories: those that are designed primarily for terrestrial applications but are employed in coastal environments (Table 1A), and those whose spectral bands and sensitivities have been optimized for aquatic applications (i.e. to observe aquatic constituents – Table 1B). The sensors in both groups are restricted by the presence of clouds. As defined in the table caption, applications of a given sensor are identified by a grey or black bar. If the application is reasonably well founded, it is identified with a black bar. In practice, these are the only applications that are likely to be of immediate operational value to coastal managers. One relevant consideration is that most of these sensors are flown on commercial satellites and therefore their data must be purchased.

Table 1. Land-oriented (A) and marine oriented (B) multispectral polar-orbiting satellite sensors used to observe coastal features. A black bar indicates the application is reasonably well founded. A grey bar indicates the sensor is not optimal for this application but merits consideration. See the References section of this chapter for explanation of symbols, superscripts, Web sites and acronyms. Table reprinted with permission (Whitehouse, 2003). [Not included in B: three Chinese research sensors (CMODIS, COCTS, CZI), the data from which are only available within China, and one Taiwanese research sensor (OCI). See www.ioccg.org for technical details on these and other ocean color sensors.]

A **Satellite**	**ORBVIEW 3**	**QUICKBIRD**	**IKONOS**	**SPOT**	**LANDSAT & EO**	**SAC-C**
SENSOR	MS	MS	MS	HRG	ETM & ALI	MMRS
type[1]	C	C	C	C	C/R	R
revisit time (days)[2]	~3	~3	~3	~3	16	9
near-real time[3]	no	no	no	no	no	no
spatial resolution (m)	4	2.5	4	10	30 & 30	175/350
Web site	0	1	2	3	4 & 4b	5
Country or agency	USA	USA	USA	France	USA	Argentina
Floods/Storm Surge						
Fronts/Eddies (biological)						
Fronts/Eddies (thermal)						
Ice						
Surface Temperature						
Turbidity (coastal)						
Vegetation (littoral)						
Water Coloring Constituents						

Table 1, cont'd.

B **Satellite**	**KOMPSAT**	**OCEANSAT**	**ORBVIEW 2**	**ENVISAT**	**TERRA & AQUA**
SENSOR	OSMI	OCM	SEAWIFS	MERIS	MODIS
type[1]	R	R	C/R	R	T
Revisit time (days)[2]	~2	~2	~1	~3	~1
near-real time[3]	No	no	yes	no	yes
Spatial resolution (m)	1000	360	1100/4000	300/1200	1000
Web site	6	7	8	9	10
country or agency	Korea	India	USA	ESA	USA
Floods/Storm Surge					
Fronts/Eddies (biological)					
Fronts/Eddies (thermal)					
Ice					
Surface Temperature					
Turbidity (coastal)					
Vegetation (littoral)					
Water Coloring Constituents					

[1] Sensor type (C)ommercial, (Met)eorology, (T)ransitional, (Mil)itary, (R)esearch/development. See appendix for definitions of categories.
[2] Revisit time refers to either the satellite's orbit recycle time or the sensor's effective revisit time (identified with a "~") at mid-latitudes. Revisit time varies with latitude and sensor configuration.
[3] Near-real time refers to satellite data that can be delivered to the user within two hours of satellite overpass.

All of the multispectral satellite sensors listed in Table 1a are well known for their terrestrial mapping applications. SPOT, for example, was used extensively during the Gulf War in 1991. While they are not optimal sensors for observing aquatic constituents, each of these land-oriented multispectral sensors have proven to be useful in aquatic coastal projects involving tropical and subtropical waters, where the water is relatively clear. With spatial resolutions ranging from submetre to a few tens of metres, they usually meet the spatial resolution requirements of nearshore mapping programs. It is almost certain that if not for the cost of purchasing large numbers of images produced by these sensors, they would be used to a much greater extent by coastal managers and researchers.

NASA's Coastal Zone Color Scanner (CZCS), which was launched in 1978 and produced data for the period 1979 to 1986, forever changed the way we view the world's oceans, and subsequently encouraged the development and eventual launch of the marine-oriented multispectral sensors listed in Table 1B. Unfortunately, although the follow-on to the CZCS mission, SeaWiFS, is still in operation, it was launched in 1997, thereby leaving a more than ten year temporal gap in ocean color data. The more recent sensors listed in Table 1B, such as Modis and Meris, have additional spectral bands which improve our ability to distinguish various coloring constituents which may be present in coastal waters, such as constituents of terrestrial or benthic origin. Certain multispectral programs, such as Modis, also include thermal IR sensors which permit concurrent measurement of sea-surface temperature.

4.2 THERMAL IR SENSORS

In addition to the combined multispectral/thermal IR sensors listed in Table 1, relevant thermal IR sensors are found on other satellites, as listed in Table 2. This

Table 2. Thermal IR polar-orbiting satellite sensors used to observe aquatic features. Black and grey bars are as described in Table 1. Annotations are as defined in Table 1. Table reprinted with permission (Whitehouse, 2003).

Satellite	**NOAA**	**ENVISAT**
SENSOR	AVHRR	ATSR
type[1]	Met	R
revisit time (days)[2]	~0.25	~3
near-real time[3]	Yes	No
spatial resolution (m)	1100/4000	1000
Web site	12	13
country or agency	USA	UK
Floods/Storm Surge		
Fronts/Eddies (biological)		
Fronts/Eddies (thermal)		
Ice		
Surface Temperature		
Turbidity (coastal)		
Vegetation (littoral)		
Water Coloring Constituents		

grouping includes NOAA's AVHRR sensor, which is flown on a constellation of satellites and is the most widely used marine spaceborne sensor. The extensive operational usage of the AVHRR satellite program reflects a number of factors. NOAA has a constellation of these sensors in service at any given time, resulting in an effective revisit time of about six hours at mid-latitudes; it readily shows thermal structure in surface waters, such as that associated with certain eddies, fronts and terrestrial runoff; it images Earth's surface constantly (i.e. does not require user intervention); its data can be downloaded directly by the user with a relatively inexpensive L-band receiver; it is broadcast free-of-charge; and there are readily available processing and cloud-masking algorithms. In short, it responds very well to user requirements. The substantive drawback to this and other sensors operating in the thermal IR band is the influence of atmospheric effects upon performance. For example, as is the case for sensors listed in Table 1, sensors listed in Table 2 are inhibited by the presence of clouds.

4.3 ALTIMETERS AND SYNTHETIC APERTURE RADARS

Sensors presented in Table 3 operate in the microwave band of the electromagnetic spectrum. They are not restricted by clouds, atmospheric particles or the absence of sunlight (i.e. operate day and night). Note that the applications listed in Table 3 differ somewhat from those listed in the previous tables.

With the exception of altimeters, all of the satellite sensors identified in this chapter are restricted to providing information pertaining to surface waters. Altimeters are capable of providing information pertaining to the entire water column and therefore provide insight into the three-dimensional structure of the ocean. This results in altimeters being of considerable interest to oceanographers. The spatial and temporal scales at which these sensors operate, however, result in them having much less direct benefit to coastal managers. However, altimeter data are being incorporated into operational ocean models, which can result in these data being of indirect benefit to coastal managers.

Altimeters provide information pertaining to ocean circulation, significant wave height and wind speed (Chelton et al., 2001). This information is extracted from the range, shape and power of the sensor's returning signal, respectively. Ocean circulation information from altimeters is restricted to meso and basin scales, where mesoscale variability is defined as being on the order of 50-100 km and 10-100 days at mid-latitudes.

Presently, altimeters are restricted to providing information related to the time variant component of ocean circulation. This means that they are not able to provide information on absolute currents, only those components that vary with time. This is due to the present level of imprecision in our ability to measure the marine geoid, which is used as a reference surface for altimetric measurements. This situation is changing. In March 2002, the German/US GRACE satellite was launched with the objective of measuring this geophysical property with greater precision than previously possible.

All of the sensors identified in this chapter are able to detect aspects of ice in marine waters. However, the synthetic aperture radar sensors listed in Table 3 have emerged as the sensors of choice for operational ice monitoring. SAR sensors have been demonstrated in coastal flooding operations, and as discussed previously, they can detect surface oil slicks. Unfortunately, the limited temporal resolution (revisit time) of single-satellite SAR sensors appears to be constraining the extent to which they are used for operational purposes (i.e. they are platform limited).

Table 3. Altimeter and Synthetic Aperture Radar (SAR) polar-orbiting satellite sensors used to observe aquatic features. Black and grey bars are as described in Table 1. Annotations are as defined in Table 1. Table reprinted with permission (Whitehouse, 2003). [SAR sensors are also used to detect ships in aquatic environments, however, as ships are not a natural aquatic feature or process they are not included in this table.]

Satellite	TOPEX / POSEIDON	JASON	GFO	ENVISAT	RADARSAT	ENVISAT
SENSOR	altimeter	altimeter	altimeter	altimeter	SAR	SAR
type[1]	T	T	Mil	R	T/C	R/T
revisit time (days)[2]	10	10	17	35	~3 to ~5	~3 to ~5
Near-real time[3]	no	no	no	No	yes	no
spatial resolution (m)					10 to 100	30 to 100
Web site	14	15	16	17	18	19
Country or agency	France/USA	France/USA	USA	ESA	Canada	ESA
Floods/Storm Surge						
Mesoscale Eddies						
Ice						
Geostrophic Currents						
Density (3D)						
Surface Slicks						
Surface Temperature						
Wave Height						
Waves (internal)						
Wind (speed & direction)						

4.4 PASSIVE MICROWAVE SENSORS AND SCATTEROMETERS

As is the case with the sensors listed in Table 3, sensors listed in Table 4 operate in the microwave band of the electromagnetic spectrum, thus are not restricted by clouds and can operate day and night. Note that sensors listed in Table 4 have spatial resolutions on the order of kilometers whereas Tables 1 through 3 list spatial resolution in meters.

With the possible exception of the AMSR sensor, which is in a development phase, passive microwave and scatterometer sensors are used operationally for meteorological purposes, providing such information products as wind speed, wind direction, surface brightness temperature and ice location. Scatterometers, which are also listed in Table 4, are the only sensors listed in Tables 1 through 4 that provide both wind speed and direction, with all others only providing one component, usually wind speed. As a result, scatterometers are the only group of sensors designated with a black bar in the wind features area. They are already slated to graduate from being sensors flown on space agency R&D missions to sensors flown on operational meteorological satellites, such as those operated by EUMETSAT. The nominal 25 km spatial resolution of scatterometers is appropriate for meteorological and ocean forecasting purposes but restricts their use in nearshore and certain coastal environments, even though such data can be processed to 12.5 km resolution (Liu, 2001). SAR sensors (Table 3) have been demonstrated to provide wind information in such restricted environments, however, the temporal resolution of existing spaceborne SAR systems impedes their application to operational meteorology.

Table 4. Passive microwave and scatterometer polar-orbiting satellite sensors used to observe aquatic features. Black and grey bars are as described in Table 1. Annotations are as defined in Table 1. Table reprinted with permission (Whitehouse, 2003).

Satellite	DMSP	AQUA	QUICKSCAT	ERS
SENSOR	SSM/I	AMSR	Seawinds	scatterometer
Type[1]	Mil	R/T	R/T	R/T
revisit time (days)[2]	~0.25	~2	~1-2	~4
Near-real time[3]	yes	eventually	Yes	no
spatial resolution (km)	25km	24km to 56km	25km	25km
Web site	20	21	22	23
country or agency	USA	Japan/USA	USA	ESA
Floods/Storm Surge				
Fronts/Eddies				
Ice				
Geostrophic Currents				
Density (3D)				
Surface Slicks				
Surface Temperature				
Wave Height (surface)				
Waves (internal)				
Wind (speed & direction)				

5. Financial Issues

The observation that satellites augment rather than replace other coastal observing platforms has budgetary significance. Non-spaceborne sampling programs, such as airborne and *in situ* programs, may be run more effectively with the addition of a spaceborne sampling program, but they are not likely to be eliminated. In order to use satellite data operationally, additional financial and human resources are required to employ satellite data specialists, to utilize specialized hardware and software, and in certain cases, to purchase satellite data and external data processing services.

Government agencies cover capital and operating costs associated with many of the relevant satellite programs, a result being that coastal managers and researchers can often obtain satellite data free of charge or for a nominal fee. On the other hand, researchers and managers usually cover all costs associated with their airborne or *in situ* sampling program. The data cost advantage only applies to certain satellite sensors as several environmental satellites, as identified in Tables 1- 4, are operated on a commercial basis. Where a fee is levied, the cost of purchasing satellite imagery is not usually onerous for a few images. However, for operational applications in which data requirements can run into tens to thousands of satellite images over the life of a project, such costs can run into the tens of thousands to millions of dollars for certain types of imagery.

The issue of satellite data costs has been debated widely within the aquatic observing community. In our opinion, the issue arose largely as a result of overly optimistic projections of the commercial potential for marine environmental satellite sensors. Fortunately, we now have a more realistic understanding of the commercial vs. public-mandate applications of such sensors, and as a result, space agencies have a better understanding of the users' ability to pay for satellite data.

In the late 1980s to early 1990s, the temporal resolution requirements of coastal management programs were poorly assessed or comprehended by the space sector. In hindsight, this is not surprising given that several of today's aquatic applications of spaceborne sensors did not exist at that time. However, as a result, subsequent spaceborne environmental sensing programs did not live up to expectations from an operational usage perspective. The satellite systems that emerged in the 1990s tended to perform at or beyond engineering and academic research expectations, but usually fell short of operational utilization expectations. Similarly, projections of the commercial potential for such satellite systems proved to be widely optimistic. The latter has had significant impact on the extent to which the private sector is now willing to invest in spaceborne environmental satellite programs. Additionally, as several space agency development programs are focused on industrial development it is likely that this has had an indirect effect on public-sector budgets for Earth-observation programs.

6. Summary

Although sensors mounted on different platforms may observe similar phenomenon, such as water optical properties, one should not view these various platforms as being redundant or competitive. All sensors and platforms have limitations and these limitations vary for each specific sensor and platform. Of fundamental significance are the temporal and spatial scales at which these technologies operate relative to the aquatic coastal process of interest. There is often a trade off, with higher spatial resolution spaceborne systems having longer revisit times and vice versa.

As a means of providing additional practical information for the operational user, we classified sensors (superscripts in Tables 1- 4) as being (C)ommercial, (Met)eorological, (Mil)itary or (R)esearch and Development sensors. However, we also listed sensors considered to be in a (T)ransitional phase, and we suggest the aquatic satellite community itself is in a phase of transition. One could argue, correctly, that certain sensors listed in Tables 1 through 4, such as SeaWiFS and the AVHRR, are existing examples of operational oceanographic satellite sensors, even though we did not identify them as such. There are few indications, however, that NASA and the private sector intend to continue to use the SeaWiFS program model for future oceanographic satellite sensors. NOAA's AVHRR is designated as a (Met)eorological sensor, and it is feasible that the meteorology community will expand its mandate and infrastructure to include the operation of other dual use satellite sensors. Scatterometers, for example, are already destined for this path and we foresee further synergies between meteorology and oceanography.

Finally, as a word of caution, a limitation to publishing the type of tables included n this chapter is that they become dated. Fortunately, the internet now solves this problem with updates provided periodically online (e.g. www.oeatech.com and www.ioccg.org).

7. Acknowledgements

This work was supported in part by DRDC-Atlantic through PWGS Contract W7707-021909.

8. References

Antoine, D., A. Morel, B. Gentili, H.R. Gordon, V.F. Banzon, R.H. Evans, J.W. Brown, S. Walsh, W. Baringer and A. Li, In Search of Long-term Trends in Ocean Color, EOS, Vol. 84(32), 2003, 301-309.

Chelton, D.B., J.C. Ries, B.J. Haines, L-L Fu, and P.S. Callahan, Satellite Altimetry, in Satellite Altimetry and Earth Sciences, L-L Fu and A. Cazenave (eds), Academic Press, 2001, pp 1-131.

Kramer, H.J., Observation of the Earth and its Environment – Survey of Missions and Sensors, 4th Edition, Springer Verlag, 2002.

Liu, W.T., Wind Over Troubled Waters, Backscatter, Vol. 12(2), 2001, 10-14.

Whitehouse, B.G., Analysis and Recommendations for a Canadian Forces Maritime Environmental Assessment Program, DRDC-Atlantic Report No. CR 2003-176, DND/DRDC, Halifax, Canada, 2003, 64 pp.

Whitehouse, B.G., and D. Hutt, Ocean Intelligence in the Maritime Battlespace: the Role of Spaceborne Sensors and HF Radar, Canadian Military Journal, Vol 5(1), 2004, 35-42.

9. Appendix

The following reference information is provided in support of Tables 1 through 4:

Superscripts

[1.] Sensor type (C)ommercial, (Met)eorology, (T)ransitional, (Mil)itary, (R)esearch/development. Commercial satellites are owned or otherwise operated by private companies. R & D satellites are usually owned by space agencies. All listed Meteorology and Military satellites are owned by public agencies and meet a prescribed operational requirement. Transitional sensors are considered by the authors to be in a transition phase – progressing from research and development tools to satellites of another category. See text for additional discussion. Where a satellite program involves a series of satellites, the tables only list relevant sensors mounted on the latest satellite in the series.

[2.] The sensor's revisit time refers to either the satellite's orbit recycle time or the sensor's effective revisit time (identified with a "~") at mid-latitudes. Revisit time varies with latitude and sensor configuration.

[3.] Near-real time refers to satellite data that can be delivered to the user within two hours of satellite overpass.

Web Sites (for updates see www.oeatech.com or www.ioccg.org)

0	www.orbimage.com	1	www.digitalglobe.com	2	www.spaceimaging.com
3	www.spotimage.fr	4	landsat7.usgs.gov	4b	eol.gsfc.nasa.gov
5	www.invap.com.ar	6	kompsat.kari.re.kr	7	www.isro.org/irsp4.htm
8	seawifs.gsfc.nasa.gov	9	envisat.esa.int	10	modis.gsfc.nasa.gov
12	www.oso.noaa.gov/poes	13	envisat.esa.int	14-16	www.jason.oceanobs.com
17	www.esrin.esa.int	18	www.space.gc.ca	19	www.esrin..esa.int
20	dmsp.ngdc.noaa.gov/dmsp.html	21	aqua.gsfc.nasa.gov	22	winds.jpl.nasa.gov
23	earth.esa.int/ers/instruments/index.html				

Acronyms

ALI	Advanced Land Imager
AMSR	Advanced Microwave Scanning Radiometer
ATSR	Along Track Scanning Radiometer
AVHRR	Advanced Very High Resolution Radiometer
DMSP	Defence Meteorological Satellite Program
ERS	European Remote Sensing
ETM	Enhanced Thematic Mapper
GFO	Geosat Follow On
HRG	High Resolution Geometric
MMRS	Multispectral Medium Resolution Scanner
OCM	Ocean Color Monitor
OSMI	Ocean Scanning Multispectral Imager
SAR	Synthetic Aperture Radar
SSM/I	Special Sensor Microwave Imager

Chapter 10

THE ROLE OF INTEGRATED INFORMATION ACQUISITION AND MANAGEMENT IN THE ANALYSIS OF COASTAL ECOSYSTEM CHANGE

STUART PHINN, KAREN JOYCE, PETER SCARTH AND CHRIS ROELFSEMA
Centre for Remote Sensing & Spatial Information Science, School of Geography, Planning and Architecture, The University of Queensland, 4072, Australia
s.phinn@uq.edu.au

1. Introduction

Coastal aquatic ecosystems are often perceived as a complex and "difficult" area from a management perspective due to their dynamic nature and joint management by multiple local, state, and national level government agencies. Additionally, collecting information to understand, monitor, and manage these ecosystems necessitates the use of spatial information suited to the management agency and/or agencies (Belfiore, 2003; Treitz, 2003). The objective of this chapter is to demonstrate how environmental indicators can be used to provide a basis for designing and implementing mapping and monitoring programs using remotely-sensed and field data for coastal and coral reef environments. This chapter complements Whitehouse and Hutt (2005 – this book), by providing a practical approach for matching remotely sensed data sets within common requirements for monitoring and managing coastal environments.

2. Information Requirements for Understanding, Monitoring and Managing Coastal and Coral Reef Environments

2.1 NATURAL RESOURCE MANAGEMENT IN COASTAL AQUATIC ECO-SYSTEMS

The material covered in this chapter concerns the application of remote sensing technologies to coastal aquatic systems. Coastal aquatic ecosystems are defined here as substrate, benthos, water column, and water surface features extending from the mean-high water level to the edge of the continental shelf. This definition includes inter-tidal mangroves and saltmarsh, tidal flats, rocky shores, seagrass beds, coral reefs, organic and inorganic water column contents (seston), and water surface characteristics. As an interface between terrestrial, atmospheric, and aquatic environments this is a highly dynamic area characterised by processes and structures that change on hourly to daily time-scales due to tidal, wind, wave, and river processes.

Due to the number of human activities conducted in coastal environments, they are monitored and managed at local to regional, national, and in some cases, international levels. Common activities range from extractive resource use (fishing, tourism), recreation (boating, diving, swimming), and urban development (housing, port, industrial, commercial) to natural functions (habitat, aesthetics, shoreline stabilisation, flood reduction, nutrient sinks/sources). As a result, there is a critical requirement for information to support ecosystem management.

Numerous types of remote sensing approaches have been used for monitoring coastal environments around the world, mainly through a combination of field survey

L.L. Richardson and E.F. LeDrew (eds.), Remote Sensing of Aquatic Coastal Ecosystem Processes: Science and Management Applications, 217-249. © 2006 *Springer. Printed in the Netherlands.*

with aerial photography, and more recently in combination with satellite remotely sensed data (Green et al., 1996; Edwards, 1999a; Dadouh-Guebas, 2002; Joyce et al., 2002). Recent reviews outline the capabilities of remote sensing for environmental monitoring in terms of the data sets and technical approaches applicable for change analysis for terrestrial environments (Coppin, 2003; Treitz, 2003); water quality parameters; and for mapping substrate types, such as coral, seagrass and algae (Green et al., 1996, 2000; Edwards, 1999b; Dekker et al., 2001b). This chapter will place the information contained in these reviews within a context of typical requirements for managing coastal environments. With the notable exception of Edwards (1999a), and Green et al. (1996, 2000), there is little guidance provided to coastal resource managers as to how to practically integrate remotely sensed data within existing field programs for use in monitoring and management activities.

A number of useful surveys covering practical applications or evaluations of remotely sensed data have been published recently (Green et al., 1996, 2000; Wallace and Campbell, 1998; Edwards, 1999a; Phinn et al., 2001b, 2002a,b; Dadouh-Guebas, 2002; Joyce et al., 2002; Malthus, 2003; Belfiore, 2003; Trinder, 2003;). These applications were often in cooperation with field programs and provide a worthwhile overview of the capabilities of currently available remote sensing technologies. In the surveys of natural resource managers conducted by a number of these reviews, consistent responses were:

- a need for closer integration between existing monitoring programs and remotely sensed data.
- an onus on demonstrating the effectiveness, accuracy and cost efficiency of remote sensing approaches.

2.2 THE THREE "Ms" FOR MANAGEMENT: MAPPING, MONITORING AND MODELING

A central concept presented in this chapter is that environmental management in coastal zones is part of a continuum of applications. The continuum represents a progression of knowledge necessary for environmental management, and is termed the "three-M" approach. It starts with baseline Mapping and inventory, then progresses to Monitoring, and finally to Modeling a coastal environments' processes and structures (McCloy, 1994; Viles, 1995; Green et al., 1996; Smith, 2001; Phinn et al., 2003). The continuum of spatial data collection as it relates to management of coastal aquatic environments can be described as follows:

- ***Mapping*** – Baseline surveys or inventories are conducted to determine the presence and location of features. This most basic application level provides information, at one snapshot in time.
- ***Monitoring*** – A comparison of base-line maps of an environmental feature (e.g. substrate type or water depth) is carried out over a series of different points in time, enabling changes to be mapped and measured.
- ***Modeling*** – The highest level of spatial and non-spatial data integration is based on understanding and then replicating how an environmental system, or one of its components, operates. A model of a coastal environment (e.g. a hydrodynamic circulation model) enables parameters to be modified to determine how the system will change under certain environmental conditions.

Mapping and monitoring programs provide information that is combined with scientific knowledge of environments to increase understanding of the environmental function. In some cases this can provide a measure of how different the environment is to a known "healthy" condition. Models provide critical heuristic and planning tools for resource managers, by enabling "what if ?" questions to be posed, e.g. how will key coastal environment structures and processes be altered by certain management activities?

2.3 INTEGRATING REMOTE SENSING AND MANAGEMENT

The "three-M" concept for the integration of spatial data in coastal environmental management was derived from the authors' reviews of relevant literature in the field, and their experiences evaluating remote sensing solutions to natural resource management problems in Australian coastal and forest environments. In a number of the application projects discussed in this chapter, Ecologically Sustainable Development (ESD) and the United Nations Environment Program's (UNEP's) pressure-state-response model have been the basis for the adoption of State of Environment reporting frameworks using environmental indicators (Wallace and Campbell, 1998; Phinn et al., 1998a, 2002 a,b; Belfiore, 2003; Trinder, 2003). The use of environmental indicators as a central tool for environmental management enables the explicit linkage of the "three-M" approach to remote sensing for mapping, monitoring, and modeling coastal aquatic environments. The final goal is sustainable ecosystem management and Integrated Coastal Management (ICM).

The remainder of this chapter presents a progression of concepts and application examples to demonstrate how remotely sensed data can be integrated directly into coastal ecosystem management activities through the use of environmental indicators. The first two sections explain the concept of environmental indicators and their critical roles within coastal monitoring and management programs. The third section provides a worked example of matching suitable remotely sensed data to an environmental indicator for a coastal zone mapping application. This section uses environmental indicators as the key to specifying a component of the environment that can be mapped using remotely sensed data. Change and trend detection techniques are then reviewed, providing a logical expansion of some of the key techniques used for monitoring coastal environments from remotely sensed data. Two case studies are then used to present examples where specific coastal environmental management problems have been addressed by integrating field data, remote sensing techniques, and community involvement. The applications covered in the case studies progress in scale from the entire Great Barrier Reef to a local-scale harmful algal bloom in southeast Queensland. A concluding section draws attention to the need for integrating remotely sensed coastal environmental indicators within monitoring and management activities from local to national and international scales.

3. The Role of Environmental Indicators in Monitoring and Managing Coastal Environments

3.1 ENVIRONMENTAL INDICATORS

Environmental, ecosystem, or ecological indicators are variables considered to be representative of the biophysical or socio-economic status of a specific environment. Management agencies and governments from local to international levels commonly adopt the concept of ecological indicators. These are based on the Organization for Economic Cooperation and Development's (OECD) "pressure-state-response" model, where the indicators are selected for key environmental, economic, and social areas of concern (Vandermeulen, 1998). In an environmental context (Bromberg, 1990; McKenzie et al., 1992; and Australian and Queensland State of the Environment Reports), the indicators can be grouped into:

- ***Response Indicators***: quantify the condition of organisms, populations, communities, or ecosystem processes.
- ***Exposure Indicators***: physical, chemical or biological measurements that reflect pollutant exposure, habitat degradation, or other causes of poor ecosystem condition.
- ***Stressor Indicators:*** data documenting human activities and natural processes that can cause changes in exposure indices.

The utility of basing monitoring and management programs around ecological indicators is that agencies can then agree on a set of variables to measure, how to measure them, and then commit to doing this over time. The net result is agreement on variables that can be monitored and used to understand the current state, and short- to long-term changes in an environment. In turn, the definition of environmental variables to measure for each indicator at specific spatial and temporal scales provides a direct link to remotely sensed data (Wallace and Campbell, 1997; Phinn, 1998; Foody, 2003).

Management agencies with common environmental requirements (e.g. forest conservation, maintenance of water storage facilities), normally have met at local, state, national and international levels to agree on common indicators. They then set measurement protocols and develop coordinated mapping and monitoring programs. Examples of this approach include the Montreal Protocol for Sustainable Forests, and, within Australia, the State of Environment Reporting framework, which is used by local to national government agencies.

3.2 ENVIRONMENTAL INDICATORS FOR COASTAL AQUATIC ECOSYSTEMS

Indicators for coastal ecosystem monitoring have received significant attention in the last ten years, mainly through the global promotion and development of Integrated Coastal Management (ICM) activities (Table 1). A comprehensive summary of coastal ecosystem health indicator development for ICM and their application for monitoring environmental condition, and human impacts is provided in a special issue of Ocean and Coastal Management (Belfiore, 2003). This journal covers key presentations given at "The Role of Indicators in Integrated Coastal Management Conference" held in 2002. Coastal ecosystem health indicators are considered essential for tracking the implementation of ICM activities. A number of key papers in this special issue define relevant indicators for local, national, and global scale coastal monitoring and management programs (Rice, 2003) and successful processes for developing and implementing indicator-based monitoring programs (Kabuta, 2003). Recent reviews have identified several hundred types of environmental indicators with relevance to coastal environments.

Selecting an appropriate set of indicators for the environmental monitoring/ management issue and decision maker in question is considered a key task (Bromberg, 1990; McKenzie, 1992; Rice, 2003). The selection of indicators to use for monitoring and management applications should be based on scientific validity, clear links to management goals, incorporation into the management process, understand-ability, time required for implementation and maintenance, and cost efficiency. The majority of indicators presented in the literature are based on data collected through field surveys, with limited use of remote sensing techniques, except in the case of mapping the extent of surface cover features (e.g. vegetation communities, weeds or algal blooms).

3.3 LINKS BETWEEN ENVIRONMENTAL INDICATORS AND REMOTE SENSING

Environmental indicators provide one of most useful links from environmental monitoring and management programs to remotely sensed data. The utility of the indicators is that they define set environmental parameters to be mapped, at a specific spatial scale, and usually over a set time period. This information provides a basis for selecting suitable remotely sensed data sets and processing techniques to deliver map(s) of the requested environmental indicator (Phinn, 1998a; Foody, 2003). The framework proposed (Phinn, 1998a) and refined (Phinn et al., 200b, 2003) by Phinn provides a basis for making the link between environmental indicators and suitable remotely sensed data, and techniques for processing remotely sensed data. The framework is outlined in detail in the next section and defines the process used in other studies linking environmental indicators to remotely sensed data (Wallace and Campbell, 1998; Green et al., 2000; Foody, 2003; Trinder, 2003).

4. Linking remotely sensed data sets to environmental indicators, the community, and policy-makers

4.1 A FRAMEWORK FOR LINKING ENVIRONMENTAL INDICATORS TO REMOTELY SENSED DATA

The framework outlined in Figure 1 and described in Parts 1 – 2 below provides a guide to evaluating commercially available remotely sensed data sets for mapping or monitoring selected environmental indicators. Originally this approach was developed for use in coastal wetlands (Phinn, 1998a), then modified for tropical wetlands and tropical rainforest environments (Phinn et al., 2000b, 2002a, b). The key to this approach is linking the spatial and temporal scale(s) of data and information required to remotely sensed data with corresponding dimensions. Explicit consideration is also given to the full costs of processing image data to a map product in terms of necessary hardware, software, ancillary data, and skilled personnel. A worked example is provided in Tables 2-5 for monitoring one selected indicator of coastal ecosystem health, the extent of seagrass beds in Moreton Bay, Queensland, Australia (Phinn et al., 2001b). Implementation of the framework requires initial specifications of: 1) the indicator to map/monitor, the area to cover, and the timeframe; 2) available financial support; and 3) image processing capability. The framework is then used to select a suitable image data set and to evaluate the cost of various image-processing strategies.

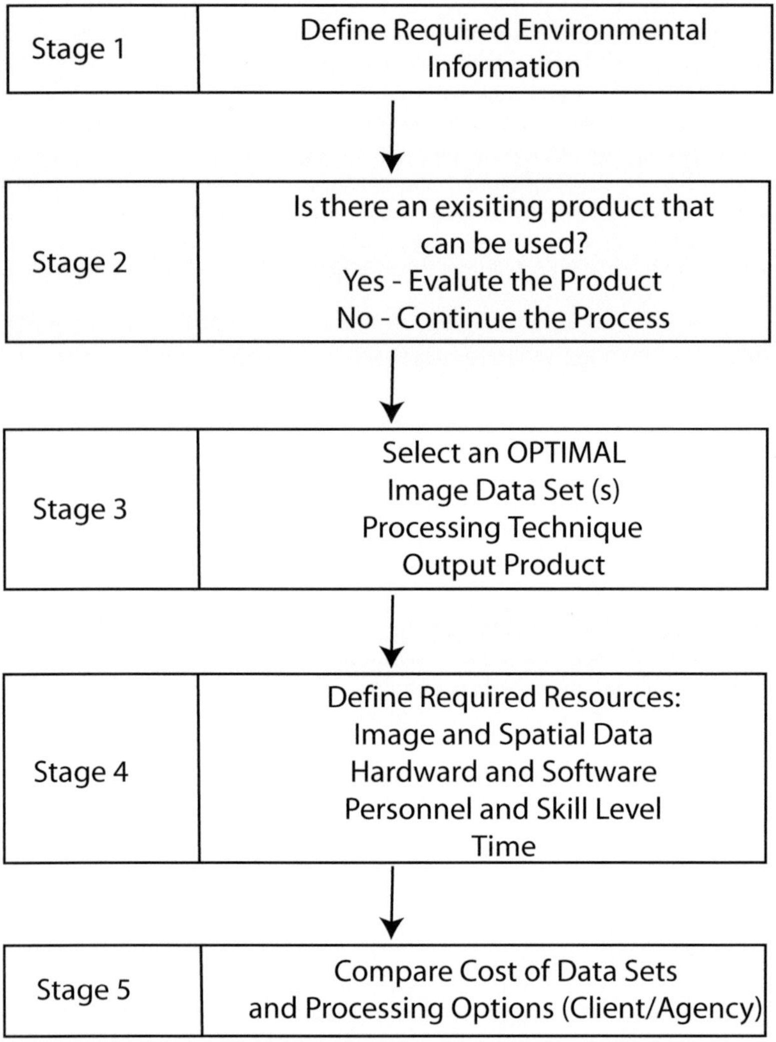

Figure 1. Conceptual framework for integrating remote sensing with environmental monitoring programs.

Table 1. Summary of the operational status (column 2) of coastal ecosystem indicators for use with remotely sensed data. Additional columns detail the remotely sensed data (column 3) required and a suitable image processing technique (column 4).

Indicator	Status of RS Estimates	Spectral Resolution	Spatial Resolution	Type of Analysis	Reference
Water Quality - Concentrations TSM/Tripton Chl *a* CDOM	 Feasible Feasible Feasible (clear/turbid)	 Multispectral Hyperspectral (high, med,low)	High, medium, low	 Analytic/radiative transfer models	(Dekker et al., 2001b)
Algal blooms	Operational (clear water)	Multispectral Hyperspectral (high, med,low)	High, medium, low	Image classification Analytic/radiative transfer models	
Toxic chemical spills	Feasible	Multispectral Hyperspectral (high, med,low)			
Depth	Feasible (clear water) Operational (clear water)	Multispectral Hyperspectral (high, med) Airborne LADS	High, medium	Analytic/radiative transfer models Ratio of Ln transformed data	(Green et al., 2000; Stumpf, 2003)
Substrate Type Estuary	Operational (clear water)	Multispectral Hyperspectral (high, med)	High, medium	Image classification Analytic/radiative transfer models	(Stumpf, 2003)
Substrate Type Coral Reefs	Operational (clear water)	Multispectral Hyperspectral (high, med)	High, medium	Image classification Analytic/ radiative transfer models	(Green et al., 2000; Joyce et al., in press; Palandro et al., 2003b; Palandro et al., 2003c)
Substrate type Rock platforms	Feasible (clear water)	Multispectral Hyperspectral (high)	High	Image classification Analytic/ radiative transfer models	
SAV Density	Feasible (clear water)	Multispectral Hyperspectral (high, med)	High, medium	Image classification Analytic/radiative transfer models	(Green et al., 2000)

Table 1, cont'd.

Indicator	Status of RS Estimates	Spectral Resolution	Spatial Resolution	Type of Analysis	Reference
SAV Biomass	Feasible (clear water)	Multispectral Hyperspectral (high, med)	High, medium	Analytic/ radiative transfer models	
SAV Live/Dead	Operational (clear water)	Multispectral Hyperspectral (high)	High, medium	Image classification Analytic/ radiative transfer models	(Green et al., 2000)
Coral Live/Dead	Feasible (clear water)	Multispectral Hyperspectral (high)	High	Image classification Analytic/ radiative transfer models	(Green et al., 2000)

Image Data:
Spectral Characteristics
Multi-spectral = less than 10 broad bands
Hyperspectral = greater than 10 narrow bands
Spatial Characteristics
Low = pixel size > 250m
Medium: pixel size 20m – 250m
High: pixel size < 20m

TSM: Total (organic + inorganic) Suspended Matter concentration in the water column
CDOM: Coloured Dissolved Organic Matter in the water column
Chl *a*: Chlorophyll a concentration in the water column
SAV: Submerged Aquatic Vegetation (seagrass, micro/macro-algae, coral)
RS: remote sensing
A number of local, national and international monitoring and management programs have built successful monitoring and management programs for coastal environments around sets of select indicators. The following list represents recognised coastal ecosystem status indicators and an established monitoring and management program using that indicator:
- Water quality parameters – Moreton Bay Ecological Health and Monitoring Program (Dennison and Abal, 1999) ; (- Algal bloom characteristics – Moreton Bay Lyngbya Task force (Roelfsema et al., 2001);
- Seagrass and benthic substrate community attributes – NOAA-Coastwatch; and
- Coral reef attributes – Great Barrier Reef Marine Park Authority, Global Coral Reef Monitoring Network (Wilkinson, 2000).

Table 2. Example evaluation matrix for the Moreton Bay indicator – sea grass extent and links to environmental variables that can be measured using remote sensing data and spatial-image analysis techniques.

Indicator ***Surrogate***	**Spatial Scale** **Extent** **Min.Map Unit**	**Temporal Scale** **Frequency** **Time of Year**	**Remotely Sensed Variable**
Extent of segrass beds	Moreton Bay – 30 x 60km (1000's km^2) < 1ha	Annual e.g. by June for August delivery or event driven	Land/benthic-cover

Table 3. Listing of remotely sensed variables and the indicators they can be used to measure for coastal aquatic ecosystems

Remotely Sensed Variable	**Indicator**
Inherent Optical Properties	Water Quality - Concentrations TSM/Tripton Chl *a* CDOM
Water Surface Characteristics	Algal blooms
Depth	Depth
Substrate Cover Type (benthos)	Substrate Type Estuary Coral Reefs Rock platforms
Image based indices	SAV Density Biomass Live/Dead Coral Live/Dead

Part 1. Identification of Remotely Sensed Data Sources and Image Processing Operations This is an inventory stage in the framework, relying on past published work. A comprehensive summary is provided elsewhere for currently available airborne and satellite image data sets suitable for use in coastal aquatic environments (Phinn et al., 2000b, 2002a,b, 2003). These references provide details in a table for each type of commercially available passive and active image data set in terms of:

- the area covered in one image
- the size of the smallest ground feature able to be mapped
- the type of measurement used to produce the image, e.g. active or passive, and the waveband measured
- how often the images are collected over the wet tropics
- how to obtain the data and its cost

The processing methods used to convert airborne and satellite images to maps of relevant environmental indicators (e.g. seagrass extent) are reviewed in a separate table (Phinn et al., 2001b). The results of the review explain type of input data required, their processing assumptions and the forms/reliability of output maps. For reasons of brevity, examples of the tables listing all the processing methods were not included in the text and the reader is referred to (Phinn et al., 2000a; Phinn et al., 2001b; Phinn, 1998b; Phinn et al., 2000b).

Part 2. Evaluation of Remotely Sensed Data and Processing Approaches for Indicator Monitoring Each indicator (e.g. seagrass extent) is directly compared to relevant remotely sensed data sets and processing approaches listed in Part 1 to determine the suitability of remotely sensed solutions for monitoring an indicator (i.e. Operational, Feasible, Likely/Possible or Unlikely/Impossible).

To arrive at a direct link between the specified indicator(s) and suitable remote sensing data and processing approaches, a three-stage procedure is implemented. At the completion of this procedure, *a clear link is established between each indicator and the remotely sensed data set that could be used for its measurement* (e.g. Table 1). This linkage includes specifications of the most appropriate remotely sensed data, image processing techniques, required personnel, hardware, and software necessary to complete the task. An estimated cost of mapping, verification, and monitoring for the indicator can be provided for each potentially suitable data type. A final assessment can then be made for each data type and processing operation in terms of its "feasibility" for operational monitoring of select indicators.

The first stage of this process involves determining a direct link between environmental variables that could be mapped, measured, and monitored from remotely sensed data, and relevant environmental indicators (Tables 2 and 3). If an indicator can not be matched with a remotely sensed variable or surrogate it is removed from the evaluation process and considered to be in the "Impossible" category. An extensive review of past and current remote sensing applications in coastal environments should be used as a basis for this evaluation (e.g Edwards, 1999a; Dekker et al., 2001b; Dadouh-Guebas, 2002; Malthus, 2003). This information is then condensed into Table 3, where the level of match between indicators and remotely sensed variables is identified. For example, processing of airborne or satellite image data sets to produce benthic cover maps provides the information required to assess several indicators.

The next stage is to link "appropriate" remotely sensed data sets to each remotely sensed variable. This is achieved in Table 4 by taking all of the commercially available remotely sensed data types and identifying the remotely sensed variable(s) they have been used to derive. Next, the most "appropriate" remotely sensed data set(s) for deriving remotely sensed variables linked to an associated indicator are identified (Table 5).

The final stage specifies the resources required to map and monitor indicators from the most appropriate form of remotely sensed data and image-derived variables. A direct assessment of the feasibility and costs of selected indicators derived through remote sensing is provided. Table 5 contains an example of the results of the assessment. The format of each table first specifies the relevant remotely sensed variable and its spatial and temporal dimensions. The most appropriate data sets selected for each remotely sensed variable are then added, along with their dimensions and a listing of:

- Processing technique(s) required to convert remotely sensed data to the relevant environmental variable and indicator
- Specifications (and costs estimates) for the necessary data, hardware and software systems required to complete the processing of remotely sensed data to map or monitor the relevant environmental variable and indicator

- Required type and level of skills (along with time to complete the task) for staff completing the processing of remotely sensed data to map or monitor the relevant environmental variable and indicator

The final table (e.g. Table 5) provides a complete assessment of the types of remotely sensed data suited to monitoring a coastal environmental indicator - seagrass extent in this context - and its accompanying resource requirements. As several remotely sensed data types are often considered at this stage, the summary table provides an effective comparison between the cost:benefit of each data set/approach. In theory, this provides the basis for selecting a suitable type of remotely senses data.

4.2 PRESENTING REMOTELY SENSED DATA AND DERIVED INFORMATION FOR USE BY POLICY MAKERS AND STAKEHOLDERS

A final consideration is that of how to present the derived spatial information on the state of selected coastal environmental indicators to decision makers and interested stakeholders. This is a critical consideration and will determine if, and how well, the data from a monitoring program are used.

In some cases, the management agency conducting the monitoring may have established reporting mechanisms through formal publications, (e.g. a newsletter series, and monthly or annual reports), or through on-line static websites or interactive internet map servers for delivering various themes of spatial data for an area. The main types of communication products to consider include:

- publications in hardcopy books, reports, newsletters and handouts
- softcopy static information (e.g. PDF files of reports, newsletters and handouts)
- softcopy interactive information (e.g. internet map servers for delivery of spatial data on digital base maps) for spatial and tabular data
- communications with print (newspaper) and electronic (television and radio) media through press release to draw attention to hard or softcopy publications
- posters for public and educational use

An example communication strategy is provided below to indicate how the results from an on-going coastal waterway monitoring program were presented to decision-makers and the public. The Moreton Bay Healthy Waterways Catchment Partnership (MBHWCP) is responsible for monitoring water quality in Moreton Bay, Queensland Australia. MBHWCP is funded by the 17 local councils with catchments draining into Moreton Bay, and coordinates the Ecological Health and Monitoring Program (EHMP). The EHMP monitors water quality parameters in Moreton Bay and its tributaries on a monthly-annual basis. The communication products used by this program include:

- Quarterly text and graphic reports with maps of key water quality parameters and explanatory text
- Annual reports presented in a "report card format" where each river and section of the Bay is given a rating from A to D (good to bad), and an indication of improving, stable or decreasing water quality
- Web-based access to all reports as PDF files at www.healthywaterways.org

Table 4. Assessment of remotely sensed data sets suitability against the spatial, spectral and temporal scales that are linked to selected environmental indicators. The spatial scale section of column has two scales, 1) regional – the extents of Moreton Bay (100 km^2-3000km^2); and 2) local (<100 km^2).

Data Type Sensor (platform)	**Spatial Scale Extent**	**Spatial Scale Min.Map Unit**	**Spectral Scale**	**Temporal Scale Frequency**	**Remotely sensed variable**
Field spectrometers	Site specific	Site specific	Very High	User defined	- Veg. Type - Structure/Biomass Index
Aerial photographs	Local - Regional	Local - Regional	Low	User defined Cloud restricted	- Land-cover - Land-cover change - Veg. type - Structure - Stanton and Stanton veg. maps
Airborne multi-spectral	Local - Regional	Local - Regional	Moderate – High	User defined Cloud restricted	- Land-cover - Land-cover change - Veg. Type - Veg.Index - Soil Index -Structure/ Biomass Index
Airborne Hyperspectral	Local - Regional	Local - Regional	Very High	User defined Cloud restricted	- Land-cover - Land-cover change - Veg. Type - Veg. Index - Soil Index -Structure/Biomass Index

Table 4, cont'd.

Data Type Sensor (platform)	**Spatial Scale Extent**	**Spatial Scale Min.Map Unit**	**Spectral Scale**	**Temporal Scale Frequency**	**Remotely sensed variable**
Satellite multispectral					
Ikonos (Space-Imaging) Quickbird (Earthwatch)	Local - Regional	Local - Regional	Low	At least 5 days Cloud restricted	- Land-cover - Land-cover change - Veg. Type - Veg. Index - Soil Index -Structure/ Biomass Index
Landsat ETM Landsat TM SPOT XS IRS	Regional	Regional	Moderate	At least 5 days Cloud restricted	As above
SPOT VMI NOAA AVHRR	Regional	Regional	Low	Daily Cloud Restricted	- Land-cover - Land-cover change - Veg. Index - Soil Index - Biomass Index
Satellite hyperspectral					
MODIS (EOS-AM)	Regional	Regional	High	Daily Cloud Restricted	- Land-cover - Land-cover change - Veg. Index - Soil Index - Biomass Index

Table 4, cont'd.

Data Type Sensor (platform)	Spatial Scale Extent	Spatial Scale Min.Map Unit	Spectral Scale	Temporal Scale Frequency	Remotely sensed variable
Field laser ranging	Site specific	Site specific	N/A	User defined	-Biomass/Structure Index
Airborne laser altimeters	Local - Regional	Local - Regional	N/A	User defined	-Biomass/Structure Index
Satellite SAR	Regional	Regional-	Low	Minimum of 5 days No cloud or smoke restrictions	- Land-cover - Land-cover change - Veg. Type -Structure/Biomass Index

Table 5. The remotely sensed variable vegetation type applies to coastal indicators: Substrate Type - Estuary, Coral Reefs, and Rock platforms) and the listing of data types, processing requirements and costs for mapping and monitoring this variable using several suitable types of remotely sensed data. MMU: Minimum mapping unit; GRE: Ground resolution element.

Vegetation type	**Indicator attributes**	**Data type #1** Landsat ETM	**Data type #2** Airborne Hyperspectral	**Data type #3** Aerial Photographs
Spatial Scale	Moreton Bay			
Extent	– 30 x 60km (1000's km^2)	185km x 185km per scene	Up to 100km^2	1.3 – 33km^2
MMU/GRE	< 1ha	15m panchromatic 30m multi-spectral 60m thermal	0.5 – 10m	5m – 250m
Temporal	Annual e.g. by June for August delivery or event driven	Approx 9.45am every 16 days	User controlled (subject to weather and aircraft availability)	User controlled (subject to weather and aircraft availability)
Variable	Land/benthic-cover	Reflectance in up to 7 spectral bands	Reflectance in up to 126 spectral bands	Contact prints (23cm x 23cm) requiring scanning and ortho-correction to produce a digital mosaic
Processing technique **(Output)**		Image classification or feature detection (Vegetation type map and target features) Note: The ability to map specific targets will depend on their growth form and extent.	Image classification or (hyperspectral) feature detection (Vegetation type map and target features) Note: The ability to map specific targets will depend on their growth form and extent.	Manual delineation of SAV types either on hard-copy photographs or on-screen digitizing. (Vegetation type map)
Resource – Equipment		PC Image processing software GIS with image classification module (e.g. Arc-View Image Analyst)	PC Image processing software capable of hyperspectral data processing.	PC A3 size or larger Scanner Softcopy photogrammetry software Image processing software GIS with image classification module (e.g. Arc-View Image Analyst)

Resource – Personnel		Trained in image classification Experience with Landsat data Knowledge of area to be mapped	Trained in image classification and spectral unmixing or matching. Experience with Hyperspectral data Knowledge of area to be mapped	Training in softcopy photogrammetry and image processing. Extensive knowledge of area to be mapped
Estimated task and times		Image pre-processing (1 day) Image classification to SAV Types (15 days per scene) Field/Photo verification for a select number of sample sites: (8 days) Map output production: (2 days) Total = 26 days per scene	Note: This estimate is for a 10km x 10km area Image pre-processing (2 days) Image analysis using classification, un-mixing or matching to define SAV Types: (8 days per area) Field/Photo verification for a select number of sample sites: (3 days) Map output production: (1 days) Total = 14 days per 10km x 10km scene	Note: This estimate is for a 20km x 20km area (10 x 10 photos) Aerial Photograph Scanning (1 day) Digital photographs ortho-correction (5 days) Photograph interpretation and digitizing boundaries (25 days) Build and clean up vegetation type layer Map output production: (2 days) Total = 33 days per 20km x 20km scene
Estimated Cost Note that these are estimates are flexible		Data acquisition: Image data = $1950 Aerial Photos (10) = $90/frame to acquire or less to hire from Dept. of Natural Resources Ancillary data (topo sheets) = $200 Processing = 28 days of technical officer @ $150/day = $4200 Total = $7250 Note: This assumes software have been purchased	Data acquisition: Image data = $15000 Aerial Photos (10) = $90/frame to acquire or less to hire from Dept. of Natural Resources Ancillary data (topo sheets) = $200 Processing = 14 days of technical officer @ $150/day = $1700 Total = $16900 Note: This assumes software have been purchased	Data acquisition: Aerial Photos (10) = $90/frame to acquire or less to hire from Dept. of Natural Resources = $9000 Ancillary data (topographic map sheets) = $200 Processing = 33 days of technical officer @ $150/day = $4950 Total = $14150 Note: This assumes software have been purchased
Evaluation Result		**Operational**	**Feasible**	**Operational**

5. Multi-temporal analysis techniques for mapping and monitoring changes in coastal and coral reef environments

Effective management and monitoring of coastal environments requires an integrative approach for selecting remotely sensed data to monitor changes, as demonstrated in earlier sections of this chapter (and see Phinn et al., 2001a). The ability of agencies to effect their monitoring requirements is dependent on the availability of timely, accurate, and comprehensive information on the type, distribution, and rate of change (Phinn et al., 2000a). Remotely sensed data are particularly suitable in change detection applications, as this approach is relatively cost effective (Mumby et al., 1999) and can provide repeated, non-intrusive sampling over large coastal areas (Green et al., 1996). Remotely sensed data have been used in coastal and aquatic environments to study reef geography and reef form, (Kuchler et al., 1986), assess water quality and benthic and inter-tidal flora (Phinn et al., 2001a), and to map littoral and shallow marine habitats, bathymetry, suspended sediment plumes, and coastal currents (Dekker and Seyhan, 1988; Green et al., 1996; Dekker et al., 2001 a,b). However, the successful monitoring of change and environmental processes requires significant additional analysis of remotely sensed data products (Jensen, 1996b; Coppin, 2003; Treitz, 2003). This section outlines the types of change that can be detected from remotely sensed data, the image pre-processing requirements and change and trend detection techniques required for operational change detection, and the presentation of change detection results for managers, agencies, and stakeholders.

5.1 TYPES OF ENVIRONMENTAL CHANGE AND PROCESSES THAT CAN BE DETECTED FROM REMOTELY SENSED DATA FOR COASTAL ECOSYSTEMS

The changes and processes that can be distinguished using remotely sensed data can be loosely classified into coastal landcover, water quality and substrate/benthos composition. The following sections outline selected previous work in these ecosystems.

5.1.1 *Coastal Landcover*

Studies of coastal landcover change have primarily used the Landsat series of sensors. Landsat data provides a synoptic view of landscape processes at a regional scale, however more detailed mapping can be achieved with high spatial resolution airborne and satellite sensors (Phinn et al., 2000a). Multi-temporal post-classification studies using both the Landsat Multispectral Scanner (MSS) and Thematic Mapper (TM) sensors to detect coastal landscape changes in the Majahual system, along the Mexican Pacific were conducted by Ruiz Luna and Berlanga Robles (1999). They classified change in six land-use classes (mangrove, lagoon, saltmarsh, dry forest, secondary succession, and agriculture) initially using four (Ruiz Luna and Berlanga Robles, 1999) and later six scenes (Berlanga Robles and Ruiz Luna, 2002) to evaluate trends of changes between the classes. Other examples of mangrove mapping include work by Hill et al. (1994), who used SPOT to classify mangrove change in the Ba River Delta, Fiji. Jinnahtul Islam et al. (1997) used ancillary data to enhance their change detection of mangrove forest of the Sunderbans region of Bangladesh over a 54 year period using interpreted aerial photography. More specialised work by Trepanier et al. (2002) used SPOT images to determining the accumulation-erosion budget for a 14 km portion of coastline in Vietnam.

5.1.2 *Water Quality*

Mapping change in water quality relies on the development of a predictive model relating the water quality variable of interest to the radiance received by the sensor. Early work in this area was completed in the Loosdrecht Lakes in The Netherlands by Dekker and Seyhan (1988), who used qualitative and quantitative assessment of satellite (Landsat TM, SPOT) and airborne (CAESAR-MSS and low-altitude aerial colour photography) remote sensing data to detect and study the temporal and spatial variations in water quality. Further work by Lathrop et al. (1991) investigated multi-date water-quality calibration algorithms for turbid inland water conditions using Landsat TM in Green Bay, Lake Michigan to estimate absolute values and change in total suspended solids and Secchi depth. Similar work in Egyptian lagoons using Landsat TM and locally calibrated regression models by Dewidar and Khedr (2001) have been similarly successful. Multitemporal classification approaches have also been found successful in detecting water quality change. Work by Pal and Mohanty (2002), using IRS-1B data from the Chilka Lagoon, East Coast of India, was used to predict selected water quality parameters and lagoon modification over an inter-annual cycle. However, the importance of accurate image calibration was demonstrated in work by Islam et al. (2003) in Moreton Bay, Brisbane, Australia. Their estimates of total suspended sediment and Secchi depth, based on empirical models derived from a Landsat TM reference image, were found to differ by 35-152% when applied to different images. They concluded that image calibration to like-values could be used to reliably map certain water quality parameters from multitemporal TM images, as long as the water type under study remains unchanged. To avoid the problem of multiple calibrations, more recent semi-analytical and analytic models that account for bottom depth have been developed (Brando and Dekker, 2003). Some, like the model of Lee et al. (2001) have been used to derive accurate estimates of chlorophyll, dissolved organic matter, and suspended sediment concentrations, but they rely on the availability of calibrated hyperspectral imagery. To date, this has been difficult to obtain for multi-temporal studies at regional scales. More recent work (Dekker et al., 2001b; Phinn, 2003) has addressed imitations of empirical approaches (Islam et al., 2003), and used atmospheric and air-water interface corrected multi-date Landsat ETM data with field measured optical properties to estimate organic and suspended matter concentrations.

5.1.3 *Substrate Composition*

The measurement of substrate composition and benthic cover is complicated by the spatial variations of water depth and water quality. These variations prevent the normalization of image data required for accurate mapping, estimation of biophysical properties, and change detection (Phinn et al., 2000c). In this sense, the classification of substrate composition, water depth, and the measurement of water quality parameters are explicably linked (Lee et al., 2001). In a desktop study using a radiative transfer code to simulate the effect of water column effects, Holden and LeDrew (2002), noted that the classification accuracy of benthic habitat type increased significantly when the effects of the water column were removed. Such effects can also be seen using the interactive WASI program available at http://extras.springer.com in association with Chapter 4 (Gege and Albert). Image based studies confirmed the importance of including depth effects (Mumby et al., 1998). The availability of radiative transfer equations has allowed the development of classification approaches based on simulated spectra derived at different depths (Louchard et al., 2003). Use of a radiative transfer approach allows the retrieval of the seafloor reflectance, which can then be used to classify the benthos or derive biophysical indicators of ecosystem health, such as the leaf area index of seagrasses (Dierssen et al., 2003). Supervised and unsupervised classification approaches have also been successful in mapping benthos at various spatial and temporal scales. Landsat TM and ETM+ have proved valuable for mapping reef characteristics, including morphological and ecological zonation and cover types (Neil et al., 2000) and for mapping change in coral, sand and algae cover (Palandro et al., 2001).

Historically, interpreted aerial photography has been used for fine scale mapping of change in coral reef communities, with this being supplemented in recent years by the advent of higher spatial resolution satellite sensors such as IKONOS and QuickBird (Palandro et al., 2003a). Mapping using these sensors can be quite accurate, with accuracies of 89% reported in a study to map sand, coral reef and seagrass features (Maeder et al., 2002). However, substrate mapping, particularly in coral reef ecosystems, is complicated by geometry and scale of reef feature variation, especially in relation to the vertical orientation and location of photosynthetic and productive components (Phinn et al., 2000c), and thus affects the spectral resolution requirements of sensor systems that can be used (Hochberg and Atkinson, 2003).

5.2 IMAGE PRE-PROCESSING REQUIREMENTS FOR CHANGE AND TREND DETECTION IN COASTAL ECOSYSTEMS

Change and trend detection in coastal ecosystems requires rigorous image pre-processing to ensure that the variable of interest is detected with sufficient signal to noise ratio. At a minimum, multi-temporal analysis requires sub-pixel precision georeferencing, atmospheric correction, multi-date normalization, and ground-truthing for accuracy assessment (Palandro et al., 2001). These processes need to be explicitly considered within the framework used to select the remotely sensed data for the specific monitoring requirement (Phinn, 1998b).

There has been little research on the effects of image mis-registration in aquatic environments on change detection accuracy, although research in terrestrial regions has found that significant and serious classification errors can be induced by a mis-registration of only one pixel (Townshend et al., 1992; Phinn and Rowland, 2001). These registration errors will become increasingly significant in the move towards higher spatial resolution imagery.

Correction of atmospheric effects is dependant on the analytical methods used in the change analysis. In many cases involving classification and change detection, atmospheric correction is unnecessary, as long as the training data and the data to be classified are in the same relative scale. Atmospheric correction is often unnecessary when using atmospherically resistant indices developed for the application of interest (Jensen, 1996b). Often atmospheric correction alone will not be adequate in images of aquatic environments due to whitecaps and/or sun glint, with the corrected images requiring additional empirical adjustment. Therefore, there is often no substantial benefit in performing an atmospheric correction compared to an empirical correction alone (Collins and Woodcock, 1996; Andréfouët et al., 2001).

When the purpose of processing is to derive change using semi-analytical modelling, corrections to a common radiometric scale are essential. (Song et al., 2001). Multi-date normalisation is used to minimise radiometric differences among images caused by changes in acquisition conditions, and require the use of reference and subject image pairs along with selected sample points. Normalisation methods include image regression, pseudo-invariant features, histogram matching, and radiometric control set and no-change set determined from scattergrams (Tokola et al., 1999). Yang and Lo (2000) found that normalisation methods that used a large number of samples exhibited a better overall performance, but reduced the dynamic range and coefficient of variation of the images and therefore reduced the accuracy of image classification.

5.3 CHANGE AND TREND DETECTION TECHNIQUES

The information requirements of the project and the environment of interest guide the choice of change or trend detection technique. No single change detection technique is suitable for the myriad of monitoring applications, with the various methods often giving differing map accuracy (Rogan et al., 2002). Change detection methods include direct image differencing, spectral index differencing, linear change enhancement techniques (e.g. selective principal components analysis), direct multi-date unsupervised classification, post-classification change differencing, and decision tree analysis (Coppin and Bauer, 1996; Mas, 1999). Typically, these techniques are applied to imagery collected at two dates, with the differencing and linear change enhancement techniques resulting in a continuous map product that is subsequently thresholded to provide change classes. The classification approaches are either applied individually to each image, where the change can then be classed as change from one cover type one to another, or to the entire image stack. In this case, the output classification will need careful interpretation to develop reliable change classes (Jensen, 1996b). Trend detection methods typically involve the analysis of absolute values of some variable such as a vegetation index or chlorophyll or TSS concentration, and rely on a form of per-pixel time-series analysis through fitting of polynomial functions such as Fourier or wavelet analysis (Ruiz Luna and Berlanga Robles, 1999; Li and Kafatos, 2000; Coppin, 2003). These deterministic trend detection models are advantageous, since they can be applied in the same way to a variety of similar trend detection situations, resulting in standardised reporting of the trend in the indicator of interest in different regions.

5.4 PRESENTATION OF CHANGE AND TREND DETECTION RESULTS

Accuracy assessment is an important feature of mapping, not only as a guide to map quality and reliability, but also in understanding thematic uncertainty and its likely implications to the end user (Czaplewski, 2003). Prior to image classification, calibration data must be sampled from appropriate areas, at an appropriate support size (Stehman and Czaplewski, 1998). However, sampling for change detection is more challenging than that found in single-date approaches (Biging et al., 1998). Typically, a first step in this process is to highlight areas of change vs. no-change. This can be accomplished using an optimal threshold value based on similar spectral band comparisons between dates, vegetation indices, or texture measures (Lunetta et al., 1998). To ensure appropriate sampling of no-change areas, the stratified adaptive cluster sampling (SACS) approach has been recommended (Brown and Manly, 1998). SACS has particular utility for sampling disturbed locations (changed land-cover and land-use) because they usually represent a minor portion of the target population (most of the land area has not changed) and are often clustered (Rogan et al., 2002).

Following classification, the accuracy of the change maps must be assessed. The total error in a thematic map is the sum of the following: 1) reference data errors; 2) sensitivity of the classification scheme to observer variability; 3) inappropriateness of the mapping process or the technological interpolation method; and 4) general mapping error. General (total) map error conveys map quality, or 'fitness for use' by end users (Chrisman, 1991). The conventional method of communicating 'fitness for use' for map users is the confusion or error matrix (Richards, 1996). The error matrix summarizes results by comparing a primary reference class label to the map land-cover or land-use class for the sampling unit and presents errors of inclusion (commission errors) and errors of exclusion (omission errors) in a classification.

6. Applications of Remote Sensing in Monitoring Programs

In the following two sections, information is presented from two perspectives to illustrate practical applications of the concepts discussed in the preceding sections for linking remote sensing and multi-temporal analysis techniques to coastal and coral reef environmental indicators. In the first section, the status of remote sensing for mapping and monitoring coral reefs is reviewed. A local scale example is then provided, demonstrating the development and transfer to a government agency of a combined field and remotely sensed system for mapping toxic algal blooms.

6.1 CORAL REEF MONITORING PROGRAMS USING REMOTELY SENSED DATA

6.1.1 *Potential Coral Reef Monitoring Capabilities Using Remote Sensing*

Given the large and often inaccessible areas of reef ecosystems on a global basis, remote sensing remains the only way to obtain synoptic data about coral reef ecosystem composition and dynamics. Remotely sensed data provide a mapping capability that would be impossible to replicate using traditional field survey techniques. Remote sensing permits construction of baseline maps depicting reef location, extent, structure and composition, and can also be used to provide information about water quality, temperature and hydrodynamics, all of which may affect reef processes and health.

Landsat image data have been used for reef mapping applications since the mid 1980s (Jupp, Mayo et al., 1985; Kuchler, Jupp et al., 1986; Bour, 1988). It is commonly accepted that these data are well suited to geomorphic and reef zonation studies, but finer description of reef habitats (e.g. coral and algal definition) requires higher spatial and/or spectral resolution imagery (Mumby and Edwards, 2002). However, to analyse changes in reef substrate composition over time, the opportunities available with Landsat data are yet to be fully exploited (but see Palandro et al., 2003a, c). Together, the Landsat time series and frequency of image acquisition provide an information source incomparable to other data types. However, while this is a more cost-effective option than high-resolution data, Landsata data cannot provide the spatial or spectral information required to map small-scale dynamics relevant to individual coral patches.

More recently, the increased availability of high spatial resolution satellite data (e.g. Ikonos, Quickbird) has presented the opportunity to map reef habitats in greater detail (see Robinson et al., Chapter 12, this book). Where analysis of Landsat imagery may be able to generate a benthic habitat map with up to eight classes, Ikonos can increase the definition to around thirteen classes with a similar accuracy level (Mumby and Edwards, 2002; Andréfouët et al., 2003). However, this high spatial resolution still may not be sufficient to provide information about many reef processes operating on a finer scale. For example, Andréfouët et al. (2003) suggest a pixel size of as little as 15 cm is needed to detect coral bleaching. In mass bleaching events, however, such as those occurring in early 1998 and 2002, timely Ikonos data should be sensitive enough to detect a benthic change (Andréfouët et al., 2003).

6.1.2 *Existing Coral Reef Monitoring Applications with Remote Sensing*

In a survey of 64 organizations involved in coral reef management and research conducted in 2001, 62% of respondents reported using remote sensing in some capacity for research, mapping, monitoring, or management activities (Joyce et al., 2002). The number of years of use of these data were relatively equally distributed between new users (less than 2 years) and longer term users. Remote sensing is used for a variety of purposes in coral reef environments, though the most common applications are benthic habitat mapping, coastal zone management, and change detection. The least common use is rehabilitation monitoring. Given the inherent difficulties in mapping submerged ecosystems, for example water column attenuation, spectral similarity between features, and high levels of heterogeneity, it is not surprising that the more complex tasks of rehabilitation monitoring are yet to be employed in reef environments.

Of note were the responses received from representatives of research and development organizations, including educational institutions, which commonly stated that remote sensing cannot as yet be used for effective monitoring of reef systems due to the current state of knowledge and lack of understanding of light interactions in these environments. The difficulties are seen as primarily related to the lack of algorithms that can be used to measure reef properties related to health, especially due to the spectral similarities of corals and algae in the limited portion of the spectrum able to penetrate the water column. Furthermore, questions were raised as to the definition of "reef health". It was noted that there is a need for a greater understanding of reef processes and a quantification of reef condition other than presence and absence of algae, and detection of coral bleaching. Once an effective indicator of reef health is established, then the process outlined in section 3 can be used to identify the most appropriate form of remotely sensed data to address monitoring and management of the ecosystem.

Remote sensing has been used most effectively in indirect coral reef monitoring, such as the sea surface temperature data used for developing the degree-heating week hot spot maps or coral bleaching index maps (see Chapter 2, this book). These data are readily available to the public via the National Oceanic and Atmospheric Administration's (NOAA) website. These types of information and data are used extensively by reef management agencies, e.g. the Great Barrier Reef Marine Park Authority (GBRMPA). At present GBRMPA conducts extensive annual field surveys along set transects covering the entire reef. However, the samples are limited to dive-based video along set transects and do not provide spatially extensive coverage. GBRMPA is currently mapping simple substrate types from Landsat ETM+ mosaic over the entire Great Barrier Reef and is investigating the use of multi-temporal data for mapping disturbance impacts.

6.1.3 *Developing Remote Sensing for Increased Use in Coral Reef Monitoring*

Effective monitoring of any environment requires reliability, repetition, and cost-effectiveness. In tropical environments prone to cloud cover, repeatable remote sensing image acquisition becomes a challenge. In addition, many reef locations are inaccessible for extensive and repeated field validation, so the validity and consistency of image-derived maps is a particularly pertinent question. The remoteness of some reefs also means that airborne data are either prohibitively expensive or logistically impossible, thus satellite imagery remains the only option. However, some satellite systems do not systematically acquire data over oceanic regions, thus specific tasking is required (e.g. Quickbird).

Effective field validation methods for image classifications remain a challenge in reef environments due to scales of heterogeneity from individual coral patches to entire reef ecosystems. Neither field campaigns nor image data can capture all scales, and integrating the two is difficult. Although global standards for reef substrate monitoring have been developed (e.g. Reef Check), using this classification scheme or field method for calibrating and validating image data presents problems with scaling and accuracy (Joyce et al., 2004) . The Reef Check methods are simple and can be implemented as a rapid assessment scheme by volunteers with little training (Mumby et al., 1995), thus are ideal for assisting with image data classifications. However the scaling challenges related to the differences between image and field data resolutions need to be overcome.

According to one survey, (Joyce et al., 2002), the main limitations to the use of remotely sensed data in coral reef environments were perceived to be cost of image acquisition and inadequate spectral resolution. Increased utilization of these data requires better integration with GIS and a greater capacity (human and computer) to effectively process and extract the information. Improved remote sensing technologies (e.g. data set development, higher spatial and spectral resolution) will be welcomed by the majority of coral reef monitoring and management agencies, however cost was noted as a potential constraint, with only 14% believing they would have both technical and financial capabilities to fully utilize new remotely sensed data sets. The majority believed they would have neither the technical nor financial capacity, while other organizations believed that finance would prove to be the only constraint. Although the majority of respondents identified the cost of remotely sensed data sets as a major limitation, most were unsure of the cost of their data, due to government provisions, special research allowances, or infrequency of purchase. This would suggest a limited knowledge of the full costs of acquiring and processing images for their purpose, or a limited knowledge of actual prices of image data sets. In either case this is an indication of the lack of detailed knowledge of the cost, time, and possible accuracy involved in applying remote sensing for coral reef monitoring activities. An example of this is aerial photography, where the cost of acquisition is not indicative of the total cost to integrate fully into a GIS database. As a large portion of the world's coral reefs occur in the waters of developing countries, financial constraints are a significant factor in the methods employed for coral reef management.

Another commonly identified limitation of remotely sensed data is the high degree of user expertise required to understand and extract the required information. The survey also indicated the strong need for further research into, and development of, techniques to best use remote sensing as a monitoring tool.

Based on the results of this survey, it appears that the combination of: 1) the greater range of image data sets now available, and 2) an inability to consistently identify indicators of reef condition that can be used to produce reliable change detection approaches, have placed coral reef remote sensing in a developmental stage. Recommendations for furthering the utility of remote sensing in coral reef environments should focus on: 1) the identification and development of algorithms (and related spectral resolution) to relate reef bio-optical properties with relevant biophysical controls; 2) further development of techniques to remove the attenuating effects of the overlying water column; 3) greater incorporation of biogeochemical cycles (e.g. climatic and oceanographic data) with remote sensing data to understand the processes that influence the biology of reefs and their subsequent bio-optical properties; and 4) evaluation and increased utilization of a greater range of image data sources (e.g. MODIS, IKONOS, and SeaWiFS).

6.2 A COMBINED FIELD AND REMOTELY SENSED PROGRAM FOR MAPPING HARMFUL ALGAL BLOOMS

6.2.1 *Characteristics of Lyngbya majuscula as a Harmful Algal Bloom*

Blooms of the toxic cyanobacterium *Lyngbya majuscula* have become a significant problem in Moreton Bay, southeast Queensland, Australia due to the large size of blooms that have been occurring since 1997 (Figure 2) (Dennison, 1999). As the algae produces skin and respiratory reactions, it has forced the closure of several net-fisheries in the Bay, and also forced the closure and clean up of beaches used for recreational purposes. From an ecological perspective, the smothering effect of the algae on seagrass may be impacting the health of turtles and dugongs in the area where they feed from the seagrass (Dennison, 1999; Watkinson, 2000). *L. majuscula* is a toxic, filamentous, non-heterocystous marine cyanobacterium that fixes nitrogen. It is found attached to seagrass, algae, and coral, and may rise to the water's surface by internal accumulation of gas bubbles (Watkinson, 2000). The *L. majuscula* blooms in Moreton Bay occur over areas from 8 - 80 km^2, with varying amounts of projected cover and vertical thickness (i.e. up to one meter thick). The blooms may last over periods of days to months (Watkinson, 2000; Roelfsema, 2001). *L. majuscula* blooms in the Moreton Bay region have most recently been observed in large areas (>1ha) on the seagrass beds of Deception Bay and the clear waters of the Eastern Banks, though they have also been observed in most locations throughout the Bay (Figure 2). Blooms have been observed in other locations along the Australian coastline and other tropical coastal regions throughout the world (e.g. Hawaiian islands, Cook Islands, Fiji, and Eastern Africa).

6.2.2 *Scientific and Community Monitoring Requirements*

The negative impacts of *L. majuscula* blooms on fisheries, beach conditions, and native marine fauna in Moreton Bay Marine Park were recognised quickly by the Local Councils and State Government Agencies responsible for managing Moreton Bay. These groups collaborated to form a "Lyngbya Taskforce" to collect baseline information, monitor, understand, and manage the blooms. Baseline mapping of the bloom extent and density for use in monitoring its changes was critical information. This spatial information was used by scientists attempting to understand the causes and dynamics of the bloom, and for the Marine Park Managers to restrict use of affected sites. For example, the baseline mapping information was used to:

- provide Marine Park Managers with a basis for initiating mitigation (e.g. selection of which beaches or areas to close and clean
- produce "Lyngbya Alert" maps on the Environmental Protection Agency's website to enable local residents to decide where to safely swim, fish or boat
- provide scientists with information to understand the bloom characteristics (e.g. dynamics and origin)

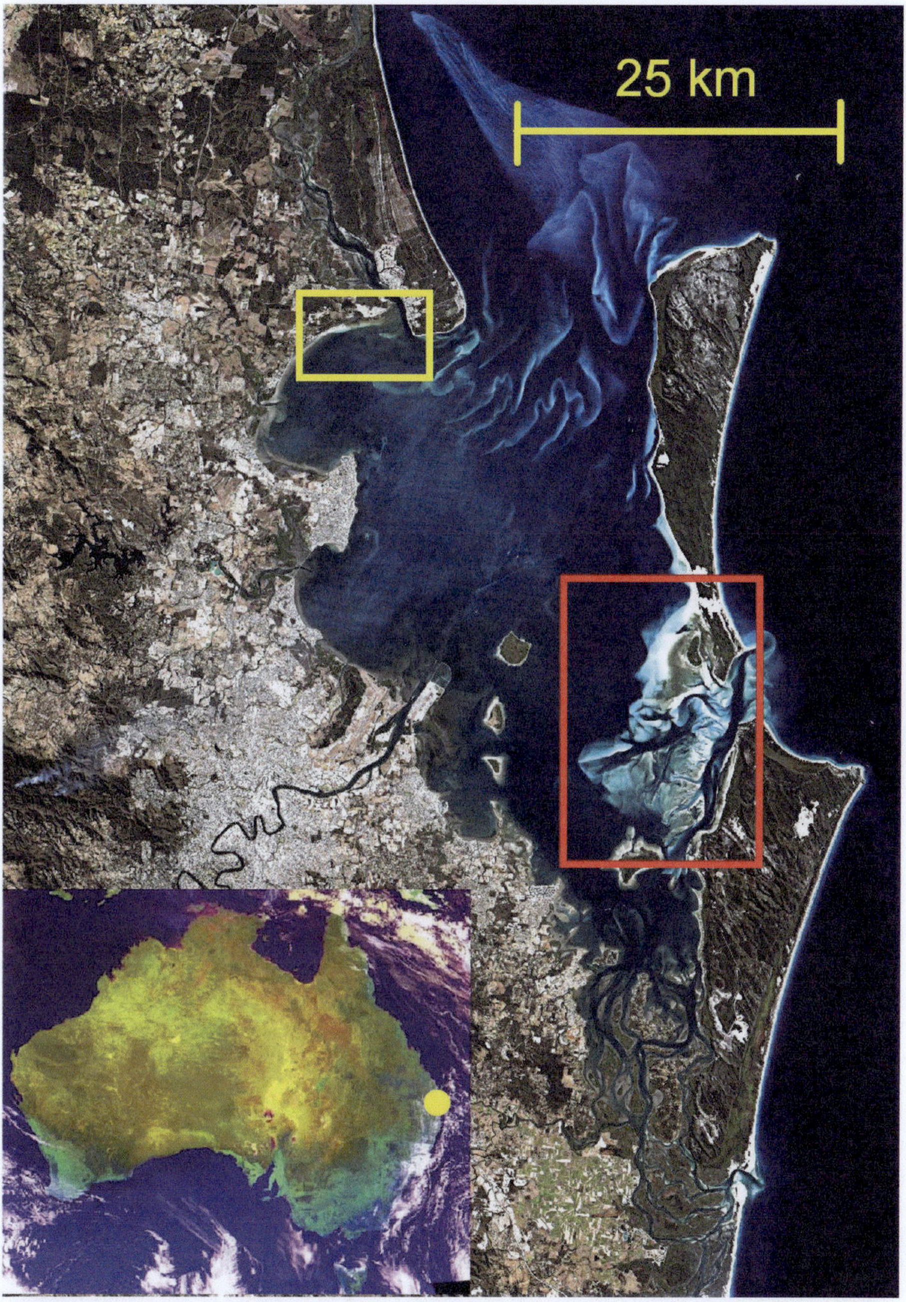

Figure 2. Landsat 7 ETM image of Moreton Bay collected on March 21, 2002. The main L. majuscula bloom sites at Deception Bay and Eastern Banks are shown. Yellow box = Deception Bay, Red Box = Eastern Banks. Image source: Geoimage/Geosciences Australia.

Hence, the monitoring program needed to meet multiple requirements, in addition to being an accurate, cost effective, and repeatable approach, so that management and monitoring agencies could implement it on a regular basis using their staff and resources.

6.2.3 *Remote Sensing for Mapping L. majuscula*

Due to the extensive area covered by Moreton Bay, mapping the extent and density of *L. majuscula* on a daily or weekly basis could not be done using standard field survey techniques (e.g. video transects, diver transects, or diver quadrats). Remotely sensed data from airborne or satellite imaging systems provided an alternative for synoptic coverage and could be used to map the location and density of *L. majuscula* to a depth of 3m in clear water (Roelfsema, 2001, 2002). Maps of *L. majuscula* collected over time could then be examined to assess bloom dynamics and potential controlling factors. A procedure to map *L. majuscula* in the Eastern region of Moreton Bay was developed from field-spectrometry of key substrate types, radiative transfer modelling to assess water depth effects on the ability to discriminate *Lyngbya*, and field data to verify the results of image based mapping (Roelfsema, 2001, 2002). The result of this work was a mapping program for use with Landsat TM/ETM+ data that could be used to map *L. majuscula* in areas of clear water < 3m deep. The mapping approach relied on coincident field survey data to train an image classification. A geometrically registered and dark-pixel corrected Landsat TM/ETM+ image was required, with the aid of field survey to map the location of *L. majuscula.* This work was implemented as a joint monitoring program, combining local field expertise of Marine Park authorities (Queensland Parks and Wildlife Service), with remote sensing capabilities of a research group at the University of Queensland. The aim of the program was to provide regular baseline maps, integrating field and image based approaches to present *L. majuscula* density and distribution in one section of Moreton Bay, Eastern Banks.

6.2.4 *Inclusion of Community Information with the Field and Image Based Mapping Program*

Community groups (e.g. oyster lease owners, seagrass watch groups) in Moreton Bay were informed about the characteristics of *L. majuscula* via websites, handouts, and public presentations. In each form of communication explicit instructions were provided on how information on bloom sightings (e.g. percent cover, colour, location) was to be recorded and then submitted to Queensland Parks and Wildlife Service (QPWS) for analysis.

A field-monitoring program was developed to provide a repeatable GPS based field survey to improve on an earlier QPWS subjective and non-geolocated survey technique. Boat driven survey tracks were first located over the sections of the Bay to be monitored, taking into account the most recent bloom locations and expected changes. Data were then collected at regular spatial and temporal intervals along these tracks. Data collected included: GPS coordinates for each sample point; substrate type; visual estimate of substrate percent cover; and digital images captured for those sites with *L. majuscula* present. The digital images were used to confirm the field estimation of % *L. majuscula* cover, and to provide archival information. The collected information was then processed in a GIS to produce quantitative field maps of *L. majuscula* distribution.

Field data collection activities were scheduled to coincide with overpasses of the Landsat TM/ETM+ sensors. This enabled the use of field data for both calibration and validation purposes. Depending on the severity of the bloom, the field and image acquisitions were scheduled on a monthly or fortnight basis. The satellite images used for the classification of *L. majuscula* blooms were subsets of the map-oriented, dark pixel corrected Landsat 7 ETM+ scenes (path 89, row 79) recorded at 9:45 am (AEST) on dates when field data collection and cloud free imagery coincided.

Mapping of *L. majuscula* patches followed a multistage process which ensured that only those areas in which *L. majuscula* could be reliably mapped were extracted from the image and used for mapping. The blue, green and red bands were selected for use in image classification because of the comparatively limited depth-related light attenuation effects and maximum signal from submerged features. Variations in light attenuation are particularly acute in this environment, due to the mixture of oceanic and coastal/estuarine water bodies (Morel, 1977). ERDAS Imagine software was used to process the imagery. Once geometrically corrected, the subset was corrected for additive path radiance by applying dark pixel subtraction (Jensen, 1996a).

Image classification to map L. majuscula. The field data collected by Marine Park authorities together with substrate coverage maps produced from previous studies (e.g. (Dennison, 1998)) were used to apply a supervised classification to map *L. majuscula* coverage. The substrate information used to select training pixels for classifying the Landsat TM or ETM+ image data were collected as close to the field survey date as possible. Statistics on substrate reflectance values were extracted from the image for field sites known to correspond to different *L. majuscula* density levels. The masked image of Eastern Banks was then subject to a "minimum distance to means" clustering routine to group pixels with similar reflectance values into three classes of varying *L. majuscula* cover and all other substrate. The final map for presentation to QPWS and for inclusion in their GIS, was made in Arcview 3.2, presenting the classification image results overlaid on the original image.

A pseudo "error matrix" was used to assess the accuracy of the classification and quantify the level of agreement between the classes identified from the image classification and the field data. The "pseudo" label, if applied as a true reference set, would have consisted of independently selected sites where *L. majuscula* cover had been measured and not used to train the image classification process. Hence, the error matrix is only a measure of how well the classification correctly identified the training data, not the whole study area.

6.2.5 *Maintaining the Mapping Program*

Currently the field component of this program is implemented on a regular basis, coinciding with Landsat 7 ETM+ over flights. The remote sensing component consists of an *L. majuscula* bloom contingency plan, which is initiated when the results of the field monitoring show medium to high levels of *Lyngbya.* At this time, a cloud free image of the study area will be purchased if available, and a classification using QWPS and community field data will be conducted. The results (in map and report format) of field and/or remote sensing monitoring are present on a website to be accessible for the community (Figure 3). The data itself are still analysed on a yearly basis and will be used as one of the parameters for a report card presenting the health of the local coastal areas. The presence, size and duration of a bloom are regarded as key indicators of coastal ecosystem health in Moreton Bay.

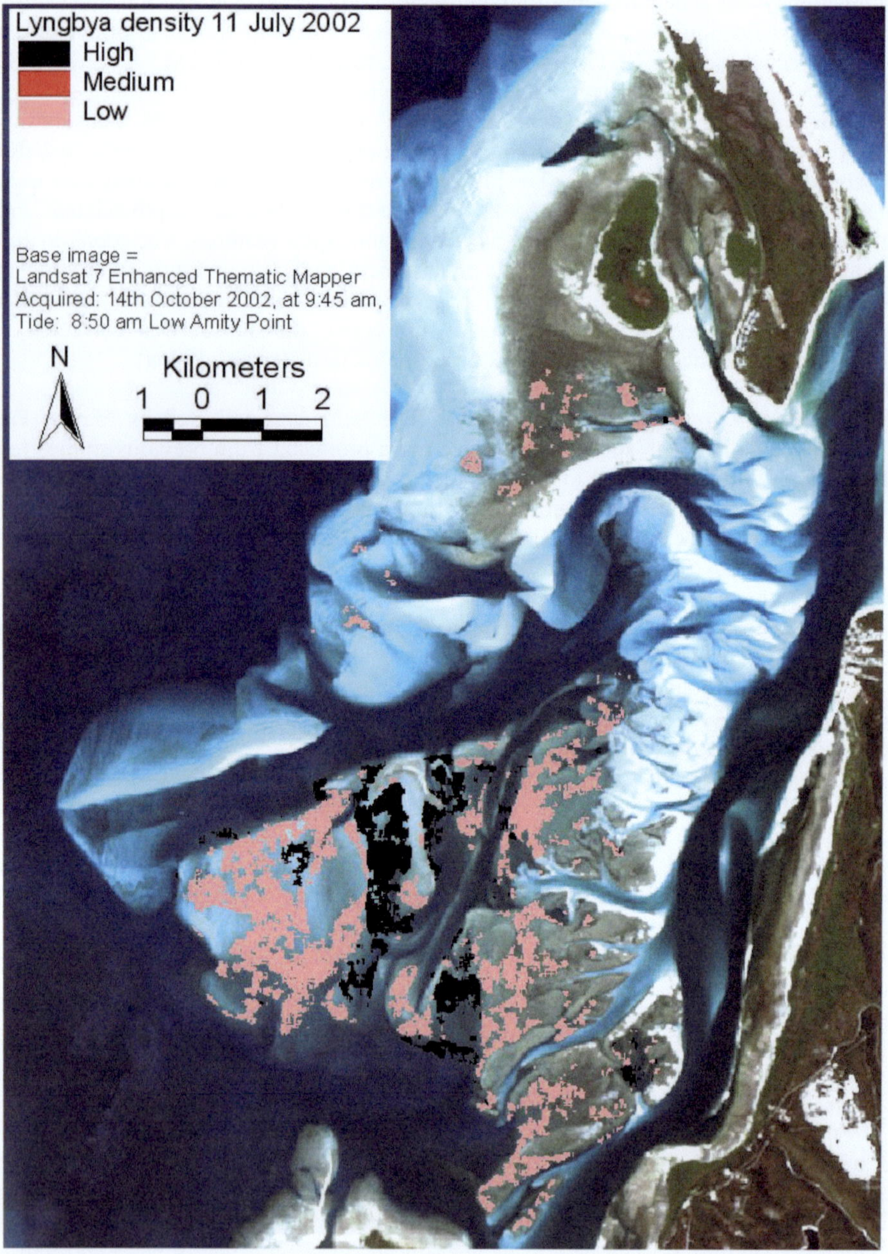

Figure 3. *Lyngbya majuscula* monitoring results: b) classification of satellite imagery resulting from field data and cloud free Landsat 7 ETM image.

7. Future Developments for Monitoring Coastal and Coral Reef Environments Using Remotely Sensed Data

7.1 CURRENT STATUS OF REMOTELY SENSED DATA/PROCESSING TECHNIQUES FOR MONITORING CHANGE IN COASTAL AND CORAL REEF ECOSYSTEMS

Other Chapters in this volume (e.g. Hendee et al., Newman et al., Skirving et al., Webster and Forbes) and recent review papers (Green et al., 2000; Dekker et al., 2001b; Joyce et al., 2002; Andréfouët et al., 2003; Coppin, 2003; Malthus, 2003) demonstrate that remote sensing techniques are operational for mapping and monitoring selected components and processes of coastal aquatic environments. In this context, the following applications from commercially available image data and image processing software are operational:

- mapping and monitoring changes of substrate type in relatively clear waters < 10m deep
- mapping depth in shallow clear waters < 10m deep
- mapping selected water quality parameters related to optical properties of water (e.g. total suspended matter, suspended organic material (e.g. chlorophyll) and dissolved organic material (e.g. CDOM) and
- mapping sea-surface skin temperature.

Substrate and water quality mapping applications of remote sensing perform accurately in clear, oceanic Case 1 waters. The accuracy and reliability of these mapping techniques is reduced significantly in coastal and estuarine waters, which are often a mix of Case 1 and Case 2 waters, unless hyperspectral data are acquired (see Chapter 3).

The majority of coastal management and monitoring programs are centered around measurement of ecosystem health indicators. Hence, it makes sense to focus on such indicators as a basis for selecting suitable remote sensing approaches towards the monitoring and management procedures of a region. Ecosystem health or status indicators often include environmental parameters that can be mapped directly or indirectly from passive and active image data sets. In this chapter we have presented a framework for developing remote sensing applications to map and monitor coastal ecosystem health or status indicators. The framework ensures explicit consideration is given to selection of an image data set suited to the indicator and its use in management. In addition, all of the considerations for using remotely sensed data are included in the evaluation process (data cost, software, hardware, personnel, etc.).

A key component of the use of remote sensing data for monitoring is the implementation of change and trend detection techniques. Our chapter provided an overview of the key pre-processing requirements and large range of processing options that are now available. The framework and change/trend detection sequence is an "ideal" approach and coastal managers and remote sensing practitioners will often be faced with a gap that persists between the expectations of both groups pertaining to the use of data. Our experience in this area, as demonstrated through the *L. majuscula* project, is to select one indicator and run through a trial project. It is critical that the trial project involves management and remote sensing scientists working together on image and field data collection, data analysis, error assessment, and presentation of results. This approach enables remote sensing to interface with existing activities and to be actively understood by those who will use the results.

Remotely sensed data continue to become more easily available in a range of different scales and data types, some of which are inherently suited to coastal environments. Continued incorporation of these technologies within integrated coastal management programs will not occur passively and requires demonstration that the data provide accurate, useful, timely, and cost effective information that meets the needs of management agencies. Careful application of the framework suggested in this chapter, along with cultivation of cooperative relationships with management agencies, should enable this to occur.

8. References

Andréfouët, S., F.E. Muller Karger, E.J. Hochberg, C. Hu, and K.L. Carder. 2001. Change detection in shallow coral reef environments using Landsat 7 ETM+ data. Remote Sensing of Environment, 78: 150-162.

Andréfouët, S., P. Kramer, D. Torres-Pulliza, K.E. Joyce, E.J. Hochberg, R. Garza-Perez, P.J. Mumby, B. Riegl, H. Yamano, W.H. White, M. Zubia, J. Brock, S.R. Phinn, A. Naseer, B.G. Hatcher, and F.E. Muller-Karger. 2003. Multi-sites evaluation of IKONOS data for classification of tropical coral reef environments. Remote Sensing of Environment, 88:128-143.

Belfiore, S., 2003. The growth of integrated coastal management and the role of indicators in integrated coastal management: introduction to the special issue. Ocean and Coastal Management, 46: 225-234.

Berlanga Robles, C.A. and A. Ruiz Luna. 2002. Land use mapping and change detection in the coastal zone of northwest Mexico using remote sensing techniques. Journal of Coastal Research, 18:514-522.

Brando, V.E. and A.G. Dekker. 2003. Satellite hyperspectral remote sensing for estimating estuarine and coastal water quality. IEEE Transactions on Geoscience and Remote Sensing, 41:1378-1387.

Bromberg, S.M. 1990. Identifying ecological indicators: An environmental monitoring and assessment program. Journal of Air and Waste Management Association, 40: 976-978.

Brown, J.A. and B.J.F. Manly. 1998. Restricted adaptive cluster sampling. Environmental and Ecological Statistics, 5:49-63.

Chrisman, N.R. 1991. The error component in spatial data. In: D.J. Maguire and et al. (Editors), Geographical information systems. Longman-Wiley, pp. 165-174.

Collins, J.B. and C.E. Woodcock. 1996. An assessment of several linear change detection techniques for mapping forest mortality using multitemporal Landsat TM data. Remote Sensing of Environment, 56: 66-77.

Coppin, P.R. and M.E. Bauer. 1996. Digital change detection in forest ecosystems with remote sensing imagery. Remote Sensing Reviews, 13:207-234.

Coppin, P., Jonckheere, I., Nackaerts, K., Muys, B., and E. Lambin. 2004. Digital change detection methods in ecosystem monitoring: a review. International Journal of Remote Sensing, 25(9):1565-1596.

Czaplewski, R.L. 2003. Can a sample of Landsat sensor scenes reliably estimate the global extent of tropical deforestation? International Journal of Remote Sensing, 24:1409-1412.

Dadouh-Guebas, F. 2002. The use of remote sensing and GIS in the sustainable management of tropical coastal ecosystems. Environment, Development and Sustainability, 4:93-112.

Dekker, A.G. and E. Seyhan. 1988. The Remote Sensing Loosdrecht Lakes project, International Journal of Remote Sensing, 9:1761-73.

Dekker, A.G., V.E. Brando, J. Anstee, N. Pinnel, and A. Held. 2001a. Preliminary assessment of the performance of Hyperion in coastal waters. Cal/Val activities in Moreton Bay, Queensland, Australia, IGARSS 2001. Scanning the Present and Resolving the Future. Proceedings. IEEE 2001 International Geoscience and Remote Sensing Symposium Cat. No.01CH37217. 2001. IEEE, Piscataway, NJ, USA, pp. 2665-7 vol.6.

Dekker, A.G., V.E. Brando, J.M. Anstee, N. Pinnel, T. Kutser, E.J. Hoogenboom, S. Peters, R. Pasterkamp, R. Vos, C. Olbert, and T.J.M. Malthius. 2001b. Imaging spectrometry of water. In F. Van der meer and S. M. de Jong, Kluwer (Eds.), Remote Sensing and Digital Image Processing, Kluwer, pp. 307-359.

Dennison, W.C. and E.G. Abal. 1999. Moreton Bay Study: A Scientific Basis for the Healthy Waterways Program, 1. South East Queensland Water Quality Management Strategy/Brisbane City Council, Brisbane, 246 pp.

Dennison, W.C., J.W. Udy, J. Rogers, C. Collier, J. Prange. 1998. Benthic Flora Nutrient Dynamics Final Report., Brisbane River & Moreton Bay Wastewater Management Study.

Dennison, W.C., E. Duffy, P. Oliver, and G. Shaw. 1999. Blooms of the cyanobacterium *Lyngbya majuscula* in coastal waters of Queensland, International Symposium on Marine Cyanobacteria. Bulletin de l'Institut Oceanographique, Monaco, pp. 632.

Dewidar, K. and A. Khedr. 2001. Water quality assessment with simultaneous Landsat-5 TM at Manzala Lagoon, Egypt. Hydrobiologia, 457:49-58.

Dierssen, H.M., R.C. Zimmerman, R.A. Leathers, T.V. Downes, and C.O. Davis. 2003. Ocean color remote sensing of seagrass and bathymetry in the Bahamas Banks by high-resolution airborne imagery. Limnology and Oceanography, 48:444-455.

Edwards, A.J. (Editor), 1999a. Applications of satellie and airborne image data to coastal management. Coastal Regions and Small Islands Papers, 4. UNESCO, Paris, 185 pp.

Edwards, A.J. (Editor), 1999b. Applications of satellite and airborne image data to coastal management. Coastal Regions and Small Islands Papers, 4. UNESCO, Paris, 185 pp.

Foody, G. 2003. Remote sensing of tropical forest environments: towards the monitoring of environmental resources for sustainable development. International Journal of Remote Sensing, 24:4035-4046.

Green, E.P., P.J. Mumby, A.J. Edwards, and C.D. Clark. 1996. A review of remote sensing for the assessment and management of tropical coastal resources. Coastal management, 24:1- 40.

Green, E.P., P.J. Mumby, A.J. Edwards, and C.D. Clark, C.D. 2000. Remote sensing handbook for tropical coastal management. UNESCO, Paris, 316 pp.

Hill, G.J.E., G.D. Kelly, and S. Phinn, S. 1994. Mangrove mapping in the Ba River Delta, Fiji, using SPOT data. Asian-Pacific Remote Sensing Journal, 7:1-8.

Hochberg, E.J. and M.J. Atkinson. 2003. Capabilities of remote sensors to classify coral, algae, and sand as pure and mixed spectra. Remote Sensing of Environment, 85:174-189.

Holden, H. and E. LeDrew. 2002. Measuring and modeling water column effects on hyperspectral reflectance in a coral reef environment. Remote Sensing of Environment, 81:300-308.

Islam M.A., J. Gao, W. Ahmad, D. Neil, and P. Bell. 2003. Image calibration to like-values in mapping shallow water quality from multitemporal data. Photogrammetric Engineering and Remote Sensing, 69:567-575.

Jensen, J.R., 1996. Introductory Digital Image Processing: A Remote Sensing Perspective. Second edition. Prentice Hall; Series in Geographic Information Science, 316 pp.

Jinnahtul Islam, M., M. Shamsul Alam, and K. Maudood Elahi. 1997. Remote sensing for change detection in the Sunderbans, Bangladesh. Geocarto International, 12:91-100.

Joyce, K.E., M. Stanford, and S.R Phinn, S.R. 2002. A Survey of the Coral Reef Community: Assessing its Remote Sensing Needs. Backscatter, 13:20-24.

Joyce, K.E., S.R. Phinn, C. Roelfsema, D.T. Neil, and W.C. Dennison. 2004. Combining Landsat ETM+ and Reef Check classifications for mapping coral reefs: A critical assessment from the southern Great Barrier Reef, Australia. Coral Reefs, 23:60-73.

Kabuta, S.H. and R.W. Laane. 2003. Ecological performance indicators in the North Sea: development and application. Ocean and Coastal Management, 46:277-297.

Kuchler, D.A., D.L.B. Jupp, D.B.V.R. Claasen, and W. Bour. 1986. Coral reef remote sensing applications. Geocarto International, 4:3-15.

Lathrop, R.G., T.M. Lillesand, and B.S. Yandell. 1991. Testing the utility of simple multi-date Thematic Mapper calibration algorithms for monitoring turbid inland waters. International Journal of Remote Sensing, 12:2045-2063.

Lee, Z., K.L. Carder, R.F. Chen, and T.G. Peacock. 2001. Properties of the water column and bottom derived from airborne visible infrared imaging spectrometer (AVIRIS) data. Journal of Geophysical Research C: Oceans, 106:11639-11651.

Li, Z. and M. Kafatos 2000. Interannual variability of vegetation in the United States and its relation to El Nino/Southern Oscillation. Remote Sensing of Environment, 71:239-247.

Louchard, E.M., R.P. Reid, F.C. Stephens, C.O. Davis, R.A. Leathers, and T.V. Downes. 2003. Optical remote sensing of benthic habitats and bathymetry in coastal environments at Lee Stocking Island, Bahamas: A comparative spectral classification approach. Limnology and Oceanography, 48:511-521.

Lunetta, R.S., Lyon, J.G., Guindon, B. and Elvidge, C.D., 1998. North american landscape characterization dataset development and data fusion issues. Photogrammetric Engineering and Remote Sensing, 64(8): 821-829.

Maeder, J., S. Narumalani, D.C. Rundquist, R.L. Perk, J. Schalles, K. Hutchins, and J. Keck. 2002. Classifying and mapping general coral-reef structure using Ikonos data. Photogrammetric Engineering and Remote Sensing, 68:1297-1305.

Malthus, T., and P. Mumby. 2003. Remote sensing of the coastal zone: an overview and priorities for future research. International Journal of Remote Sensing, 24:2805-2815.

Mas, J.F. 1999. Monitoring land-cover changes: A comparison of change detection techniques. International Journal of Remote Sensing, 20:139-152.

McCloy, K. 1994. Resource Management Information Systems. Process and practice. Taylor and Francis, Sydney, 415 pp.

McKenzie, D.H., D.E. Hyatt, and V.J. McDonald (Eds.). 1992. Ecological Indicators. Volumes 1 and 2, Proceedings of an International Symposium, Elsevier Applied Science, Fort Lauderdale, FL, USA.

Morel, A. 1977. Analysis of variations in ocean color. Limnology and Oceanography, 22:709-722.

Mumby, P.J. and A.J. Edwards. 2002. Mapping marine environments with IKONOS imagery: enhanced spatial resolution can deliver greater thematic accuracy. Remote Sensing of Environment, 82:248-257.

Mumby, P.J., A.R. Harborne, P.S. Raines, and J.M. Ridley. 1995. A critical assessment of data derived from Coral Cay conservation volunteers. Bulletin of Marine Science, 56:737 - 751.

Mumby, P.J., C.D. Clark, E.P. Green, and A.J. Edwards. 1998. Benefits of water column correction and contextual editing for mapping coral reefs. International Journal of Remote Sensing, 19:203-210.

Mumby, P.J., E.P. Green, A.J. Edwards, and C.D. Clark. 1999. The cost-effectiveness of remote sensing for tropical coastal resources assessment and management. Journal of Environmental Management, 55: 157-166.

Neil, D.T., S.R. Phinn, and W. Ahmad. 2000. Reef zonation and cover mapping with Landsat Thematic Mapper data: intra and inter-reef patterns in the southern Great Barrier Reef region, IGARSS 2000. IEEE 2000 International Geoscience and Remote Sensing Symposium. Taking the Pulse of the Planet: The Role of Remote Sensing in Managing the Environment. Proceedings Cat. No.00CH37120. 2000. IEEE, Piscataway, NJ, USA, pp. 1886-8 vol.5.

Pal, S.R. and P.K. Mohanty. 2002. Use of IRS-1B data for change detection in water quality and vegetation of Chilka Lagoon, East Coast of India. International Journal of Remote Sensing, 23:1027-1042.

Palandro, D., S. Andrefouet, F.E. Muller Karger, and P.Dustan. 2001. Coral reef change detection using Landsats 5 and 7: A case study using Carysfort Reef in the Florida Keys. Scanning the Present and Resolving the Future. Proceedings, IEEE International Geoscience and Remote Sensing Symposium, 2:625-7.

Palandro, D., S. Andréfouët, P. Dustan, and F. Muller-Karger. 2003a. Change detection in coral reef communities using Ikonos satellite sensor imagery and historic aerial photographs. International Journal of Remote Sensing, 24:873-878.

Palandro, D., S. Andréfouët, P. Dustan, and F. Muller-Karger. 2003b. Detection of changes in coral reef communities using Landsat 5/TM and Landsat 7/ETM+ Data. Canadian Journal of Remote Sensing, 29:207 - 209.

Phinn, S.R. 1998a. A framework for selecting appropriate remotely sensed data dimensions for environmental monitoring and management. International Journal of Remote Sensing, 19:3457-3463.

Phinn, S.R. 1998b. A framework for selecting appropriate remotely sensed data dimensions for environmental monitoring and management. International Journal of Remote Sensing, 19:3457-3463.

Phinn, S. and T. Rowland. 2001. Geometric misregistration of Landsat TM image data and its effects on change detection accuracy. Asia-Pacific Remote Sensing Journal, 14:41-54.

Phinn, S., A. Dekker, Eds. 2003. An Integrated Remote Sensing Approach for Adaptive Management of Complex Coastal Waters: The Moreton Bay Case Study. Coastal CRC Technical Report No. 23, 142 pp.

Phinn, S., C. Menges, G.J.E. Hill, and M. Stanford. 2000a. Optimising remotely sensed solutions for monitoring, modelling and managing coastal environments. Remote Sensing of Environment, 73: 117-132.

Phinn, S.R., C. Menges, G.J.E. Hill, and M. Stanford. 2000b. Optimizing Remotely Sensed Solutions for Monitoring, Modeling and Managing Coastal Environments. Remote Sensing of Environment, 72: 117-132.

Phinn, S.R., D.T. Neil, K.E. Joyce, and W. Ahmad. 2000c. Coral reefs: a multi-scale approach to monitoring their composition and dynamics, IGARSS 2000. IEEE 2000 International Geoscience and Remote Sensing Symposium. Taking the Pulse of the Planet: The Role of Remote Sensing in Managing the Environment. Proceedings Cat. No.00CH37120. 2000. IEEE, Piscataway, NJ, USA, pp. 2672-4 vol.6.

Phinn, S., M. Stanford, A. Held, and C. Ticehurst. 2001a. Approaches for monitoring benthic and water column biophysical properties in Australian coastal environments, IGARSS 2001. Scanning the Present and Resolving the Future. Proceedings. IEEE 2001 International Geoscience and Remote Sensing Symposium Cat. No.01CH37217. 2001. IEEE, Piscataway, NJ, USA, pp. 616-18 vol.2.

Phinn, S., M. Stanford, A.Held, and C. Ticehurst. 2001b. Evaluating the feasibility of remote sensing for monitoring state of the wet tropics environmental indicators, Cooperative Research Centre for Tropical Rainforest Ecology and Managment, Cairns, 71 pp.

Phinn, S., A. Held, M. Stanford, C. Ticehurst, and C. Simpson. 2002a. Optimising State of Environment Monitoring at multiple scales using remotely sensed data, Proceedings of the 11th Australasian Remote Sensing and Photogrammetry Conference. Causal Publications, Brisbane (CD).

Phinn, S.R., J.M. Nightingale, and M. Stanford. 2002b. A national survey of remote sensing for environmental monitoring and management applications in Australia. GIS User, 51:26-27.

Phinn, S., D. Stow, J. Franklin, L. Mertes, and J. Michaelsen. 2003. Remotely sensed data for ecosystem analyses: Combining hierarchy and scene models. Environmental Management, 31:429-441.

Rice, J. 2003. Environmental health indicators. Ocean and Coastal Management, 46:235-239.

Richards, J.A. 1996. Classifier performance and map accuracy. Remote Sensing of Environment, 57:161-166.

Roelfsema, C., S. Phinn, W.C. Dennison, A. Dekker, and V. Brando. 2001. Mapping *Lyngbya majuscula* blooms in Moreton Bay, Proceedings of the International Geosciences and Remote Sensing Symposium. IEEE-Piscataway NY, USA, Sydney, Australia (CD).

Roelfsema, C., S. Phinn, W.C. Dennison, A. Dekker, and V. Brando. 2002. Monitoring cyanobacterial blooms of *Lyngbya Majuscula* in Moreton Bay, Australia by combining field techniques with remote sensing, Proceedings of the 11th Australasian Remote Sensing and Photogrammetry Conference. Causal Publications, Brisbane (CD).

Rogan, J., J. Franklin, and D.A. Roberts. 2002. A comparison of methods for monitoring multitemporal vegetation change using thematic mapper imagery. Remote Sensing of Environment, 80:143-156.

Ruiz Luna, A. and C.A. Berlanga Robles. 1999. Modifications in coverage patterns and land use around the Huizache-Caimanero lagoon system, Sinaloa, Mexico: A multi-temporal analysis using LANDSAT images. Estuarine Coastal and Shelf Science, 49:37-44.

Smith, T., M. Sant, and B. Thom. 2001. Australian Estuaries: A Framework for Management. Cooperative Research Centre for Coastal Zone, Estuary and Waterway Management, Brisbane, 64 pp.

Song, C., C.E. Woodcock, K.C. Seto, M.P. Lenney, and S.A. Macomber. 2001. Classification and change detection using Landsat TM data: When and how to correct atmospheric effects? Remote Sensing of Environment, 75:230-244.

Stehman, S.V. and R.L. Czaplewski. 1998. Design and analysis for thematic map accuracy assessment: Fundamental principles. Remote Sensing of Environment, 643:331-344.

Stumpf, R., K. Holderied, and Sinclair, M. 2003. Determination of water depth with high resolution satellite imagery over variable bottom types. Limnology and Oceanography, 48(1 part 2):547-556.

Tokola, T., S. Lofman, and A. Erkkila. 1999. Relative calibration of multitemporal landsat data for forest cover change detection. Remote Sensing of Environment, 68:1-11.

Townshend, J.R.G., C.O. Justice, C. Gurney, and J. McManus. 1992. The impact of misregistration on change detection. IEEE Transactions on Geoscience and Remote Sensing, 30:1054-60.

Treitz, P. (Ed.), 2003. Remote sensing for mapping and monitoring land-cover and land-use change. Progress in Planning, 61:267.

Trepanier, I., J.M.M. Dubois, and F. Bonn. 2002. Study of the features of coastal evolution using remote sensing HRV and SPOT images: Application to the Red River Delta, Viet Nam. International Journal of Remote Sensing, 23:917-937.

Trinder, J.C. and T.K. Milne. 2003. Determining sustainability indicators by remote sensing. ISPRS-Highlights, 8:23-25.

Vandermeulen, H. 1998. The development of marine indicators for coastal zone management. Ocean and Coastal Management, 39:63-71.

Viles, H. and T. Spencer. 1995. Coastal Problems: Geomorphology, Ecology and Society at the Coast. Edward Arnold, New York, 350 pp.

Wallace, J. and N. Campbell. 1998. Evaluation of the feasibility of remote sensing for monitoring national state of the environment indicators, Department of Environment, Canberra, 93 pp.

Watkinson, A. 2000. Ecophysiology of the marine cyanobacterium *Lyngbya Majuscula* (Oscillatoriacea). Honours Thesis, Queensland, Brisbane, 42 pp.

Wilkinson, C., 2000. Status of Coral Reefs of the World: 2000. Australian Institute of Marine Science, Townsville, 363 pp.

Yang, X. and C.P. Lo. 2000. Relative radiometric normalization performance for change detection from multi-date satellite images. Photogrammetric Engineering and Remote Sensing, 66:967-980.

Chapter 11

MAPPING OF CORAL REEFS FOR MANAGEMENT OF MARINE PROTECTED AREAS IN DEVELOPING NATIONS USING REMOTE SENSING

CANDACE M. NEWMAN, ELLSWORTH F. LEDREW, AND ALAN LIM
Department of Geography, University of Waterloo, Waterloo, ON, N2L 3G1, Canada

1. Introduction

In 1998, the Australian high commission joined other national efforts, such as the US Coral Reef Task Force, to meet the challenge of developing a remote sensing-integrated management plan to 'conserve and protect' coral reef ecosystems. Today, in Australia alone, over 408 Marine Protected Areas (MPAs) have been developed conserving and protecting the country's reef heritage (www.mpaglobal.org). These MPAs represent the combined efforts of hundreds of scientists and managers to integrate some of the most advanced satellite technology into efforts to preserve reef ecosystems.

Today this same challenge exists within developing countries. Such countries reside outside the 'digital environment,' yet application of remote sensing in these areas would be beneficial. In particular, new advancements in remote sensing technologies that allow for the mapping of underwater features in shallow coastal environments may enhance local management efforts that seek to sustain essential marine resources that provide sustenance and drive economies in countries throughout the world. For many developing countries, these efforts are focused on MPA development, an increasingly common strategy used to control human activity where, for example, coral species and the aquatic organisms that inhabit them are at risk of exploitation and subsequent elimination.

Many remote sensing projects now aspire to the dual mission of identifying underwater features at high spatial scales and integrating this information with local management strategies. On the ground, this is an enormous task, requiring extensive training, specialist knowledge, and financial and time investments. Regardless of the effort required, however, environmental representatives, activists, planners, and managers claim that the returns far outweigh the costs. On the other hand, some critics note a lack of success both in developing remote sensing technologies to separate underwater features accurately, and in interpreting mapped underwater features in a biologically and culturally significant manner. This criticism has led to questions concerning the usefulness of remote sensing for management. Others have argued that despite the sometimes-limited success of the integration and use of remote sensing technologies, the integration of remote sensing into management plans is conceptually possible and may yield considerable benefits if the ideal integration is found.

In this chapter, we attempt to illustrate that such integration, as a means to affect management, is in fact possible. Based on our experiences in Fiji, we suggest that effective and appropriate communication is required, in particular in regions where language and cultural barriers bedevil attempts at cooperation between distinct groups such as remote sensing scientists and local coral reef managers. We begin with an examination of the status of the World's coral reefs, then, the use of Marine Protected

L.L. Richardson and E.F. LeDrew (eds.), Remote Sensing of Aquatic Coastal Ecosystem Processes: Science and Management Applications, 251-278. © 2006 *Springer. Printed in the Netherlands.*

Areas as a management model. Next, we explore the technology of remote sensing, where we highlight the differences between aerial and digital imagery, the past and present coastal sensors, and the advancements in mapping coral reef environments. We wish to equip the reader with an understanding of the complexity of mapping coral reefs and to acquaint the reader with the limitations that hinder high levels of mapping accuracy and precision. Following these explanations, we examine the place for 'communication' when remote sensing is applied to coral reef management. From our experiences in Fiji, we identify 'lessons learned' and examine the importance of incorporating local knowledge, when the goal is to utilize remote sensing effectively and appropriately in local management strategies. We advocate 'designing with culture' to enhance the utilization of remote sensing in many developing countries that are inexperienced with the remote sensing process, unfamiliar with interpreting remotely sensed information from imagery, and uncertain about how remote sensing can contribute to local management strategies.

2. Effective Management as a Response to the Coral Reef 'Crisis'

There is general agreement amongst scientists that the majority of reefs around the world are, with increasing frequency and magnitude, being negatively impacted by human and natural events (Kleypas et al., 2001). As a result, reefs are degrading at a rate faster than their rebounding and recruitment potential (Souter and Linden, 2000; Porter and Tougas, 2001). Reefs are at risk; this is no longer in question (Table 1).

Table 1. Location and status of reefs at risk. (From Bryant et al., 1998.)

Location	Reef Status
Caribbean and Atlantic Ocean	$2/3^{rds}$ of reefs are at risk and $1/3^{rd}$ at high risk
Indian Ocean	> 1/2 of the region's reefs are at risk
Middle East	$2/3^{rds}$ of the reefs are at risk in the Gulf
Southeast Asia	> 1/2 are at high risk and $>2/3^{rds}$ are at risk
Pacific	$2/3^{rds}$ are at low risk

On June 26, 2004, over 1000 scientists, managers, and government officials from around the world gathered in Okinawa, Japan, for the 10th International Coral Reef Symposium to discuss some of the greatest problems facing reefs today and the technologies that may provide solutions. The symposium theme, *Stability and Degradation of Coral Reef Ecosystems*, echoed real concerns facing tropical and subtropical coastal communities, from the well-developed tourist infrastructures along the beaches of Hawaii to the small island fishing villages of Indonesia. Many speakers highlighted complex issues that, over the years, have prevented scientists from attaining a comprehensive understanding of coral reef ecosystems and, as a result, identifying a manner in which to effectively stabilize or improve existing conditions.

One consensus that circulated throughout the conference was that although coral reef ecosystems are recognized to be both locally and globally valuable, they continue to face threats from coastal development, overexploitation, destructive fishing practices, and marine-based pollution. Several scientists projected that if all degradative impacts on coral reef ecosystems were halted today, reefs would still require decades to return to a level of productive stability with balanced trophic levels. Furthermore, projections indicate that, in light of currently known recovery rates of reef ecosystems from stresses, if impact by human activities continues at the present rate there may not be enough time, knowledge, or resources available to implement appropriate management strategies to preserve what remains of coral reefs (Hoegh-Guldberg, 1999; Risk, 1999).

Within the last decade, there has been an obvious realization of the dangers human activity can cause to reef ecosystems, and a concerted international effort to manage coral reef ecosystems in the most effective ways possible has ensued. One management response to the reef 'crisis' has been to develop a completely comprehensive global view of reef health (Bryant et al., 1998) for the purposes of (1) illustrating areas where reefs are at risk; (2) devising protected areas where human activity is not permitted; and (3) increasing awareness of human impact on coral reefs and their subsequent response to impacts. Government agencies, coral reef organizations, researchers, and scientists worldwide have been involved in developing strategies that address these issues in the most effective and appropriate manner possible (Hatcher, 1999). However, arguments over effectiveness have slowed response times and in some cases have generated suspicion about appropriate protection and remediation strategies. Fortunately, in recent years there has been an explosion in the amount of coral reef information. Though this information is scattered in databases all over the world (e.g. Green et al., 1996), there is now renewed confidence in scientific abilities to devise strategies that can bring about management of human activity and changes in our attitudes towards coral reef ecosystems.

Today, a myriad of knowledge bases exists to assist scientists and managers in studying and describing the condition of reefs within spatially isolated and dotted seascapes. These knowledge bases span biological to socio-economic approaches to the study of reefs, have become more focused, detailed, and comprehensive over the years, and are increasingly available to international researchers, managers, and planners (Wilkinson, 2000). In addition, these knowledge bases are evolving to serve the information needs of coastal managers, whose decision-making is a function of timely, accurate, and relevant information. Ideally, managers would utilize scientifically gathered and analyzed information to appropriately guide the development of management strategies in unique social and biological settings. In light of reef degradation, demands for science-based information are immense and growing, as management faces the task of stopping and reversing reef decline.

Over the last decade, remote sensing technology has come to play an important part in the development of this knowledge base. The remote sensing data set continuously expands with the collection of more satellite and aircraft imagery and data from field surveys. To the benefit of both scientists and managers, the combined effort of scientists undertaking image collection and concurrent field validation has enhanced the precision and accuracy of remotely sensed information. In more recent remote sensing efforts, reef habitat types and reef inhabitants have been detected at finer spatial scales, while changes in reef processes have been identified using more frequent repeat cycles. Improvements to sensor capabilities have broadened the capabilities of remote sensing: it can now provide more than the traditional maps of habitat distribution, as important as they might be, and provide additional information that allows reef scientists and managers to establish baseline data sets. Such data can be used for comparisons to assess the status of reefs, as well as the efficacy of current management strategies. In service of scientists' increasing interest in devising monitoring strategies to evaluate management plans, remote sensing also provides information on a regular and timely basis. As a result, remote sensing may be used to provide information that can complement or improve management strategies that seek to conserve coral reef resources (Klemas, 2001).

2.1 CORAL REEF MARINE PROTECTED AREAS AS A MANAGEMENT MODEL

Many organizations have evolved to tackle various aspects of coral reef management. These include organizations construct data bases of coral reefs and their resources, monitor events at regional global levels, coordinate reef research and conservations strategies, and seek to raise public awareness of reefs. One common strategy that has been promoted is the creation of Coral Reef Marine Protected Areas (CRMPAs) for managing the impact of human activities on coral reefs and the services that reefs provide. CRMPAs are seen as a key component of a much larger integrated system that requires extensive monitoring to determine patterns and causality of change. Results can be used to initiate and confirm management interventions. Although CRMPAs are not all created with the same purpose in mind, each requires relevant physical and social information for design, maintenance, and long-term sustainability. In 2003, over 100 countries had some form of coral reef protected areas, each motivated by a different vision and each incorporating different biophysical elements (Gubbay, 1995; Agardy, 1997; Kelleher, 1999). These protected areas were designed for a number of different purposes: (1) to limit human activities in deteriorating areas; (2) to prohibit fishing in regions in which large predators spawn; and (3) to permit selective activities that would not impact sustained levels of existing biodiversity. There is increasing interest in monitoring CRMPASs and evaluating their effectiveness on ecosystem health to satisfy social, commercial, planning, and government interests.

2.2 MARINE PROTECTED AREA DEFINED

Explicit calls for MPAs were made during the 17th International Union for Conservation of Nature and Natural Resources (IUCN) General Assembly in 1988, the IVth World Parks Congress in 1992, the World Summit on Sustainable Development in 2002 and more recently at the Vth World Parks Congress in 2003 (IUCN 2003). At each assembly, the definition of a Marine Protected Area was drawn from the IUCN, which describes an MPA as "…an area of intertidal or subtidal terrain, together with its overlying waters and associated flora and fauna, historical and cultural features, which have been reserved by legislation or other effective means to protect part or all of the enclosed environment" (Kelleher 1999). Although this definition provides considerable room for interpretation, the IUCN focused the definition by establishing six categories of protected areas, which apply to both marine and terrestrial environments and reflect the diversity of practical uses and ecological settings that MPAs include (Table 2). In essence, MPAs are tools through which marine areas can be managed.

Table 2. Categories of MPAs established by the IUCN (Kelleher, 1999).

Category	Description
I	Protected area managed mainly for science or wilderness protection (Strict nature reserve/Wilderness area).
II	Protected area managed mainly for ecosystem protection and recreation (National Park).
III	Protected area managed mainly for conservation of specific natural feature (National Monument).
IV	Protected area managed mainly for conservation through management intervention (Habitat/Species Management Area).
V	Protected area managed mainly for landscape/seascape conservation and recreation (Protected landscape/seascape).
VI	Protected area managed mainly for the sustainable use of natural ecosystems (Managed Resource Protected Area).

2.3 A CORNERSTONE OF MARINE PROTECTED AREAS

In the early stages of developing a CRMPA, managers seek information from databases for the design of maps containing spatially explicit information. Often the intent is to construct a map that combines locally significant regions with meaningful and categorized seascape features. Participatory mapping is a normal feature of CRMPAs: stakeholders define and identify significant areas, which in turn generates a greater awareness of the condition of these areas, and elicits a stronger sense of resource ownership. Another increasingly common feature of CRMPAs is the integration of information taken from images captured by sensors onboard satellites or aircraft. With greater capabilities to extract biophysical information from images, including habitat identification, reef health assessment, and change, scientists view images as an ideal tool for illustrating and assessing large spatial areas, a process that would take weeks or months using traditional ground surveys.

A benefit of remote sensing that befits coral reef management strategies is that it can be used to describe multiple biophysical components within a coral reef system. Given the complexity of managing a multi-component system in which everything is connected to everything else, managers are faced with the challenge of bringing together those components that are appropriate and drive changes in the aquatic ecosystem. Based on the physical characteristics of these components, remote-sensing technologies can target and then illustrate them in a spatially and temporally explicit manner. A tool of this strength, that is, with the ability to identify and assess key drivers of change for CRMPAs, may enable scientists to take key steps in the quest to manage natural resources.

3. Mapping Coral Reef Environments

The fundamental tools for mapping coral reef environments include both aerial photography and digital imaging. Both techniques involve capturing reflected light energy. Processing, calibrating, and correcting the imagery produces a classified image that extracts features of interest, which are selected by the operator. For imagery of tropical islands, this often means identification of coral geomorphology, biota, and nearby ecosystem counterparts that reside within a depth to which light can penetrate. Certain types of imagery are more useful than others to particular projects. In this section, we explore the differences between aerial photography and satellite imagery and while focusing on the advancements in remote sensing technologies over the past 10 years. Challenges presented by the sensors are described at the end of this section.

3.1 AERIAL PHOTOGRAPHY VERSUS DIGITAL IMAGERY

Aerial photography lags behind digital imagery in several areas, especially in multi-variable image extraction capabilities, but its comparatively lower cost and simple technology are particularly attractive when mapping reef environments in developing countries (Table 3). Aerial photography has been successful in mapping structural loss (Lewis, 2002), reef extent (Thamrongnawasawat and Hopley, 1994), and benthic communities (Cuevas-Jimenez and Ardisson, 2002). It has also been used as a tool for proactive management (Riegl and Piller, 2000). Generally, an aerial photograph is considered successful at mapping when (1) the spatial resolution of the map or image conforms to the complexity of the environment so that details of most interest are revealed; (2) the film is sensitive to wavelengths that penetrate the overlying water and illuminate the architecturally complex bottom structure; and (3) the final product is reliable (Chauvaud et al., 1998). With the overwhelming volume of archived, available aerial photography, there is a trend towards integrating digitized aerial photographs with digital imagery to improve the delineation of features and, thus, image accuracy (Green et al., 2000).

Table 3. Comparison between aerial photography and digital imagery of shallow-water coral reef areas. (Compiled from Chauvaud et al., 1998, Mumby et al., 1997, and Sheppard et al., 1995)

Issues	Aerial Photography	Digital Imagery
Discrimination obstructed by overlaying water	Difficulty with discrimination due to wave action particularly from low level photographs	Difficulty with discrimination, but can apply radiative transfer correction
Tendency for confusion between bottom types	Improved bottom type discrimination with input from an informed operator	Improved bottom type discrimination with algorithms and *in situ* data
Geomorphological Separability	Provides more accuracy for fine descriptive resolution	Provides detailed and broad zonal information
Wavelength data	These data are lost following digitizing	Wavelength data are retained and can be used to extrapolate biological or physical information
3-D perspective	A 3-D portrait is formed following the overlay of stereo pairs	Can be achieved by overlaying imagery on a bathymetric image.
Familiarity with the aquatic system	Need for remote sensing operator to be familiar with the spatial patterns within the aquatic system	Need for the remote sensing operator to be familiar not only with spatial patterns but also with biological processes that may be revealed within wavelength data
Availability and costs	High availability and costs when coverage is required for large areas	Requires scheduled satellite Passover with minimum purchase order leading to high costs

3.2 PAST AND PRESENT SENSORS

Over the past 20 years, sensors have evolved in their capability to capture spatial and temporal characteristics of coastal resources inhabited by coral reef ecosystems (Table 4). The evolutionary route of mapping coastal environments began with measurements of ocean temperatures, chlorophyll, and sediment concentrations (Robinson, 1985). Features with a large spatial extent, consistent behavior over time, and strong and distinct spectral response, were readily observed using passive remote sensing techniques (Robinson, 1985). These techniques offered the advantage of quickly retrieving environmental information for a region that normally could be examined on the ground only within a timeframe of weeks or months. Areal coverage, timely data acquisition, digital representation of marine features, and sensitivity to temporal changes in environmental attributes are still prominent advantages of airborne or spaceborne imaging.

In recent years, there has been an explosion of interest in hyperspectral and high spatial resolution satellite imagery (Table 5) (Green et al., 2000). The appeal is in the enormous versatility of these sensors, as operators mix and match spatial and spectral resolutions (Klemas, 2001) to create an ideal suite of specifications for mapping reef features. The finer spatial and spectral detail is useful for improving discrimination between similar reef features, while the temporal flexibility allows the operator to avoid unfavorable environmental conditions (Green et al., 1996). In 1999, the Ikonos satellite was launched and has since collected radiometric detail in 1-m panchromatic and 4-m multi-spectral detail. This increased spatial resolution has proven to be beneficial in the classification of reef structure (Mumby et al., 2004).

A disadvantage of some of these newer satellites with commercial mandates is a minimum coverage order requirement for non-US regions that is typically large and costly; hence satellite use is primarily suited to well-funded organizations (Green et al., 2000). In addition, computer storage space and processing requirements increase exponentially with higher spatial and spectral resolution data (Aplin et al., 1997). A further disadvantage to some airborne systems is that they are often restricted to their country of origin (Green et al., 2000, Whitehouse and Hutt, this book). Transporting sensors is difficult, and even when a system is permitted to enter a country, necessary modifications to the host aircraft may not be permitted (Monaco, Personal Communication, February 1999).

3.3 ADVANCEMENTS IN MAPPING TECHNIQUES

The differences between remote sensing efforts for mapping reefs just a few years ago and today is revealed in widely available literature (Table 6) and can be categorized into four separate areas: (1) clearly defining a reef environment within a biological context, but modified to match resolution capabilities of a pre-defined sensor; (2) identifying the most appropriate techniques for discriminating the spectral wavelengths of coral reef classes; (3) designing the most appropriate field survey techniques to match spatial and spectral characteristics of selected sensors; and (4) capitalizing on the increasingly available multi-temporal image data sets to detect changes in reef features. The focus of most recent studies has been to improve image-feature delineation, thereby identifying feature type, extent, and change with heightened precision and accuracy.

Table 4 Sensors with spatial resolutions greater than 10m.

Sensor	Satellite	Operation	Spatial Resolution	Spectral Resolution	Temporal Resolution	Further Information
Hyperion	EO-1	2001 – present	30 m (MS)	220 bands 0.4 to 2.5 μm	A target can be imaged 5 times during 16 days	http://eo1.gsfc.nasa.gov/Technology/Hyperion.html
ALI	EO-1	2000 – present	30 m (MS) 10 m (pan)	10 Bands	16 days	http://eo1.usgs.gov/ali.php
ETM +	Landsat 7	1999 – present	30 m (MS) 15 m (pan)	7 Bands 3 visible	16 days	http://geo.arc.nasa.gov/sge/landsat/daccess.html
OPS	JERS-1	1992 – present	18.3 m	8 Bands 2 visible	44 days	http://www.nasada.go.jp/index _ e.html
LISS	IRS	1988 – present	23.5 – 73 m (MS)	LISS-I and II: 4 Bands 3 visible	22 – 24 days	http://www.spaceimage.com/
MESSR	MOS	1987 – present	50 m	4 Bands 2 visible	17 days	http://www.nasda.go.jp/index_e.html
TM	Landsat 5	1984 – present	30 m (MS)	7 Bands 3 Visible	16 days	http://landsat7.usgs.gov/index.php
MSS	Landsat 1 – 7	1972 – present	~ 80 m (MS)	7 Bands 2 visible	16 days	http://geo.arc.nasa.gov/sge/landsat/daccess.html

Table 5. Sensors with spatial resolution less than 10 m.

Sensor	Agency	# of Bands	Spatial Resolution	Spectral Range (nm)	Temporal Resolution	Further Information
AISA	Karelsilva Oy	286	.2 –5 m	450 – 900 nm	Sensor	http://www.specim.fi/products-aisa.html
AVIRIS	NASA/JPL	224	< 10 m	400 – 2450 nm	Sensor	http://makalu.jpl.nasa.gov/html/overview.html
CASI 2	Itres Research Ltd.	288	.2 – 5 m	400 – 1000 nm	Sensor	http://www.itres.com/docs/casi2.html
HRG-SPOT 5	Spot Image	1 (PAN) 4 (MS)	5 m (PAN) 10 m (Multi)	480 – 710 nm (PAN) 500 – 1750 nm (MS)	26 days	http://www.spotimage.fr/
HYDICE	Naval Research Lab	206	0.75 - 3.75 m	400 – 2500 nm	Sensor	http://ltpwww.gsfc.nasa.gov/ISSSR-95/hydiceop.htm
HyMap	Integrated Spectronics	100 – 200 bands	2 – 10 m	450 – 2480 nm	Sensor	http:// www.intspec.com/
IKONOS	Space Imaging	1 (PAN) 4 (MS)	1x1m (PAN) 4x4 m (MS)	450 – 900 nm (PAN, MS)	< 3 days	http:// www.transavia.co.id/tip/geomatics/ikonos.html

Table 5, cont'd.

Sensor	**Agency**	**# of Bands**	**Spatial Resolution**	**Spectral Range (nm)**	**Temporal Resolution**	**Further Information**
NEMO	Office of Naval Research	200	5x5m (PAN)	400 – 2500 nm	7 days	http://nemo.nrl.navy.mil/public/concept.html
Orb-View 3	Orbital Sciences Corporation	1 (PAN) 4 (MS)	1x1m (PAN) 4x4m (MS)	450 – 900 nm (PAN, MS)	3 – 10 days	http://geo.arc.nasa.gov/sge/health/sensor/sensors/orbview.html#spec
PROBE-1	Earth Search Sciences	128	1 – 10 m	400 – 2450 nm	Sensor	http://www.earthsearch.com/technology/about_probe1.html
Quickbird	Ball Aerospace Global Imaging System 2000	1 (PAN) 4 (MS)	0.61m (PAN) 2.44 (MS)	450 – 900 nm	1 – 3.5 days	http://www.ballaerospace.com/quickbird.html

Table 6. Remote sensing applications in a coral reef environment.

Author	Objective	Sensor/Satellite	Ancillary Data	Approach
Andréfouët et al. (2002)	Evaluation of the ability of various sensor to monitor bleaching events	Aerial photographs	Assessment of coral damage per quadrant	Minimum distance algorithm based on comparison of Euclidean distance means
Liceaga-Correa and Euan-Avila (2002)	Assessment of accuracy in bathymetry of a shallow coral reef environment	Landsat TM	Echosounded profiles	1. Single linear regression model 2. Multiple linear regression 3. Classification techniques
Mumby and Edwards (2002)	Identification of different geomorphological reef features	Ikonos	Field data of reef features based on spatial scales	Image filters
Palandro et al. (2002)	Detection of changes in coral cover	Landsat TM Landsat ETM+ Ikonos Aerial-Photo	Visual estimates of coral cover along transects.	Mahalanobis distance classifications
LeDrew et al. (2000)	Identification of bleached corals	SPOT	Radiometric profiles over different bottom types	Getis statistic
Neil et al. (2000)	Determination of reef zones based on visual patterns	Landsat TM	*In situ* identification of major reef patterns	Supervised and unsupervised classifications
Stumpf et al. (2000)	Estimation concentrations of chlorophyll, other pigments and suspended sediments	SeaWiFS	Collection of reflectance at the surface.	Semi-analytical solution using Polynomial regression of log-transformed ratio of reflectances
Allee and Johnson (1999)	Development of predictive models for chlorophyll *a*	Landsat TM	Secchi disk depth estimations	Regression analysis and single linear model
Gower et al. (1999)	Detection of phytoplankton using a fluorescence signal from chlorophyll	MERIS	*In situ* optical properties	Radiative transfer algorithm

3.3.1 *Defining Reef Environments*

Ambiguous terminology concerning features within a coral reef ecosystem has threatened to limit the effectiveness of remote sensing. The problem is that accurate identification of reef features is a function of the operator's knowledge of reef terminology as well as the spatial and spectral capabilities of the sensor (Mumby and Harborne, 1999). The differences in terminology pose difficulties when comparisons are drawn between different types of classified imagery, especially within images of higher levels of detail (Mumby and Harborne, 1999). In an effort to minimize these effects, several authors have suggested that scientists adopt a systematic, objective, and replicable classification of reef habitats (Done, 1999; Andréfouët and Claereboudt, 2000; Edinger and Risk, 2000; Phinn et al., 2000; Mumby, 2001).

Couched within biological terminology, the term habitat is defined as "An aggregation of different species of organisms living and interacting within the same area" (Wasserman, 1975). Although defining a habitat explicitly would improve classifications of remotely sensed data, the above definition implies the aggregation of variable reef structures and biological conditions at varying spatial scales that are non-decipherable with current imaging technologies (Andréfouët and Claereboudt, 2000). A solution to variable habitat terminology has been revealed in the development of a framework that combines remote sensing capabilities and concepts from landscape ecology (Phinn, 1998). Moreover, developments in the "scaling theory" have provided direction for scaling field to image data and matching remotely sensed data to relevant state variables (Curran et al., 1998). The procedure involves several steps that begin with a thorough examination of the spatially distinct natural groupings of features visible within a coral reef environment.

A study pioneering this research derived benthic classes based on an objective approach using Agglomerative Hierarchical Classification of field data and Similarity Percentage Analysis (Mumby and Harborne, 1999). Results indicated that the descriptive resolution of remote sensing techniques is still dependent upon the limited subjective decisions of the operator. In addition, the method is useful only for turbid free images, in areas in which Jerlov 1 waters are consistent throughout the scene. Although subjectivity plays a role in defining the habitat of interest, the limitations of sensors abilities control the range of subjective decisions and thus influence the selection of the more appropriate sensor (Andréfouët et al., 2002).

3.3.2 *Using the Spectral Signature for Class Discrimination*

Normally we assume that relatively large differences between the intrinsic spectral signatures of corals will provide a relatively high probability that coral reef features within an image will be successfully delineated. However, from a radiometric or sensor perspective, often the electromagnetic energy of the target of interest can overlap between substrate types. The problem is that the optical signal leaving the coral surface is a function of that feature's pigmentation, structure, branch orientation, and internal shadowing characteristics (Holden and LeDrew, 1998; Hochberg and Atkinson, 2000; Joyce and Phinn, 2002; Lubin et al., 2001; Hedley and Mumby, 2002), elements that are difficult to separate out.

Studies of coral spectral signatures measured *in situ* typically fall into two categories. First, signatures are often examined as a function of variations in pigment densities that characterize the visual color of coral species (Falkowski et al., 1998). Several studies have examined the contribution of color to measured radiance, particularly comparing structures that lack color, which results from bleaching, to coral structures saturated with zooxanthellae, a measure of health (Holden and LeDrew,

1998). Color has also been used as a comparison measure between three coral species, five algal species, and three benthic (on sand) communities (Hochberg and Atkinson, 2000) and as a means of differentiating between dead coral in various stages of algae colonization (Clark et al., 2000). Fluorescence, a color that results when light is absorbed in one wavelength and emitted at a longer wavelength, has also been found to contribute to the reflectance received at a sensor. Five pigments have been found to fluoresce in Caribbean reef corals (Mazel, 1997). Remote detection of these pigments, using activation of florescence, has been used for inter-species differentiation and detection of bleached colonies (Hardy et al., 1992). Second, signatures are examined as a function of morphological characteristics. Reflectance values measured over varying angles and azimuths were examined to determine the bi-directional reflectance distribution function of coral species and their inter-species variation between rounded and branching types (Joyce and Phinn, 2002).

To discriminate between coral signatures, derivative spectroscopy has been used successful to differentiate between coral, algae, and sand (Table 7) (Holden and LeDrew, 1998; Hochberg and Atkinson, 2000; Joyce and Phinn, 2002). First and second order derivatives have been effective for broad class distinction, such as to differentiate between live coral, dead coral, debris and algae (Holden and LeDrew, 1998; Clark et al., 2000). Taking advantage of higher order derivatives that are relatively insensitive to noise and spectral variations caused by sunlight and skylight variations, Tsai and Philpot (1998) and Hockberg and Atkinson (2000) have illustrated that 4th order derivatives can be used to discriminate between five spectrally similar species of coral.

Table 7. Derivative techniques used to discriminate between features within a reef environment.

Author	Spectral Discriminating Feature	Discriminating Technique	Discriminatory wavelength peaks
Clark, Mumby, Chisholm, Jaubert, Andréfouët (2000)	Corals at various stages of mortality and algal colonization	Derivative Analysis (1st and 2nd order) of reflectance	First order derivative: 550 nm (live from dead coral) Second order derivative: 596 nm (live, dead, and algae covered)
Hochberg and Atkinson (2000)	Reef components, primarily coral, algae, and sand	Derivative Analysis (4th order) of reflectance	Coral: 573, 604, 652, 675 nm Algae: 556, 601, 649 nm Sand: 416, 448, 585, 652, 696 nm
Holden and LeDrew (1998)	Healthy and bleached coral	Derivative analysis (1st and 2nd order) of reflectance	First order derivative: 500-590 nm Second order derivative: 434, 524, 652 nm

The water column overlying coral and the column's apparent and inherent optical properties affect a remotely sensed signal (Mobley, 1994; Kirk, 1996) and thereby complicate spectral discrimination. Radiative transfer algorithms are often applied to imagery to remove the effects of variable upwelling and downwelling light streams. This variability causes confusion between similar bottoms types (Ohde and Siegel, 2001) at various depths or in regions with dissimilar concentrations of suspended solids (Dekker et al., 1996). Thus, the signal is not exclusively associated with a single environmental variable; rather, it possesses attributes of a combination of parameters. Several studies have attempted to separate these parameters to improve the ability to discriminate between coral reef features.

In 1978, a simple image-based approach to compensate for the influence of variable depth on water leaving radiance was developed (Lyzenga, 1978). The technique involved removing scattering in the atmosphere and variation in the surface of the water, connecting depth to radiance using a linear algorithm, calculating the ratio of attenuation coefficients for different band pairs, and generating a depth-invariant index of bottom type (Lyzenga, 1978). More recent approaches have focused on expanding this technique (Philpot, 1989; Maritorena et al., 1994) (Table 8). Typically, however, they are founded on several assumptions drawn from statistical relationships between reflectance and selected attributes of the aquatic environment. Using these techniques requires that researchers make the following assumptions: (1) that water turbidity is consistent, and low concentrations of suspended solids are present throughout an image scene; (2) that light attenuates exponentially with depth regardless of depth or bottom type; (3) that downwelling and upwelling light streams can be characterized in an identical manner and that there is no contribution by fluorescence or backscatter; and (4) that the ratio of bottom reflectances in two bands is the same for all bottom types within the scene (Newman, 2001). The assumption of log-linear attenuation with depth, in particular, is problematic for bright substrates in cases in which there may be multiple reflections between the suspended material and the surface, with the result that the rate of change of attenuation with depth changes (Newman and LeDrew, 2002)

One way of avoiding such assumptions is to use a radiative transfer algorithm such as Hydrolight, which was designed for computing radiance distributions for ocean water bodies. The algorithm is defined at the following website: http://www.sequoiasci .com/pdf/H42Description.pdf. (The WASI program of Gege and Albert, Chapter 4 of this book, also can be used for modeling the spectral distribution throughout the water column.) This time-independent model computes water-leaving radiance as a function of depth, direction, and wavelength within the water. The model requires several inputs that are measured at the time of image capture, where possible, and include (1) absorption and scattering coefficients; (2) water surface conditions; (3) benthic feature characteristics such as structure and depth, and (4) sky radiance distribution.

3.3.3 *Designing the Field Survey*

Collecting ground confirmation data (often misleadingly referred to as 'ground truth') remains a significant component of assessing the accuracy of remotely sensed imagery. In most cases, it is unlikely that depictions of reef characteristics in imagery alone will be sufficiently accurate for marine park planning and management, regardless of the image analyst's skill in the identification of marine habitats. Although ground confirmation information is often costly and more time consuming than airborne and satellite imaging, adequate field surveys can eliminate confusion and avert inappropriate planning strategies based upon poor information.

A field survey of coral reef environments must be used to support image analysis for three primary reasons: (1) to define habitats of interest; (2) to identify the spatial

Table 8. Analytical approaches to identify bottom reflectance in shallow coastal waters.

Author	Technique	Assumptions
Seafloor Reflectance Estimations		
Tassan (1996)	Linear transformation algorithm. The correlation of two wavelength bands yields the ratio of the attenuation coefficients.	Water column optical properties of deep water are similar to the ones of the shallow water area.
Maritorena et al. (1994)	Algorithm derived from the two-flow equations and Monte Carlo simulations	Backscattering coefficient, and vertical diffuse attenuation coefficient are not depth dependent. Bottom is a lambertian reflector (completing absorbing). Bottom contrast is exponentially attenuation in a two-way light path.
Estep and Holloway (1992)	Inverted single-scatter irradiance	Inputs for radiative transfer algorithm are provided from Case 1 Jerlov tables.
Spitzer and Dirks (1987)	Two-flow radiative transfer	Waters are vertically well mixed Coefficients are depth independent.
Lyzenga (1978)	Linear transformation algorithm. The correlation of two wavelength bands yields the ratio of the attenuation coefficients.	Bottom reflected radiance is approximately a linear function of the bottom reflectance and an exponential function of the water depth.
Bathymetry Estimations		
Philpot (1989)	Single-scatter irradiance model. One scalar variable is designed to respond linearly with depth and a second to be sensitive only to variations in bottom type.	Water quality and atmospheric conditions are stable within an image scene.
Clark et al. (1987)	Application of a single-band reflectance model and then dual-band ratio method. Both are linear band methods.	Bottom reflectance is constant over the bottom type. Atmosphere, sea state, and other effects are uniform or constant.
Jain and Miller (1977)	Two-flow approximation model	Water of uniform optical properties and thickness Pre-defined seafloor albedo.

location of the habitats, where spectral signatures may be taken; and (3) to test the accuracy of a classified image (Green et al., 2000). A variety of field survey methods may be employed (Table 9).

3.3.4 *Capitalizing on Multi-Temporal Coverage*

Following the launch of Landsat 7 ETM, a series of images of coral reefs from various regions has been collected systematically as a part of the Long Term Acquisition Plan (Gash et al., 2000). The Plan is the first major attempt to target isolated coral reef ecosystems repetitively. Users are being afforded a unique opportunity to focus their energies on change detection methods without the complications associated with mixing different image types of varying spatial or spectral specifications (Palandro et al., 2001).

Andréfouët et al. (2001) made an early assessment of the potential of the Landsat 7 ETM+ to detect change within a reef environment. Images were acquired within a brief time frame during which there was no major disturbance to reefs. The lack of disturbance permitted researchers to assess the stability of the images through time and to estimate biases that may be associated with change detection strategies. Having assumed that areas are large enough to provide an unmixed signal, the authors suggested that there is difficulty in detecting changes from one object to another even when the acquired images have identical specifications. The reason for this is that any occurring change must be visible and cover a large enough areal extent to saturate pixels. The effects of a hurricane, for example, would be easily identified since such a storm would destroy living structures and transform a heterogeneous environment into a homogeneous platform. Therefore, depending on the physical or biological shift within a coral reef ecosystem–disturbance, phase, or strategy–spectral differences may or may not be dissimilar enough to be detected. Un-mixing techniques are cited as a potential solution because they would allow observers to detect intra-pixel changes, but may be realistically applicable to no more than three classes at a time–sand, background, and foreground, for instance.

Several change-detection techniques applied to images of coral reef environments are derivatives of land or ice-based techniques (Lunetta and Elvidge, 1999). The Getis statistic, for example, can be used to examine the change in reef homogeneity. This measure considers the value of the reflectance within a single pixel and the relationship between that pixel and the surrounding reference pixels. It has been hypothesized that a healthy coral reef ecosystem will be heterogeneous and display negative autocorrelation, while a disturbed reef similar in bottom type over a large area will be spatially homogeneous and display a positive autocorrelation (LeDrew et al., 2004).

Two other measures available to researchers are principal components analysis (Mas, 1999) and the Mahalanobis distance classification (Palandro et al., 2001), which are regularly used prior to comparing images to detect change. It is generally accepted that there are four aspects of change detection that are important when monitoring natural resources: (1) detecting that a change has occurred; (2) identifying the nature of the change; (3) measuring the areal extent of the change; and (4) assessing the spatial pattern of the change (Klemas, 2001).

Table 9. Current field based techniques used to monitor coral reef characteristics. (Compiled from English et al., 1997.)

Method	Description	Strengths	
Measurement of Ambient Environmental Parameters	Data reflect the health of a reef and includes temperature, salinity, turbidity, light penetration, cloud cover and wind.	Objective analysis that provides standardized measures.	Expensive and time-consuming. Requires access to instruments specific to the environment.
Manta Tow Survey	This semi-quantitative technique is used to assess broad changes in the benthic communities of coral reefs where the unit of interest is often an entire reef.	Visual assessment of large areas of reef within a short time. Good for determining the effects of large-scale disturbances.	Accuracy and precision are a function of the identification abilities of the data collector. The driver who views the reef from above the water controls two-path.
Line Intercept Transect	This semi-quantitative technique is used to assess the sessile benthic community of coral reefs.	Requires little equipment and is a reliable and efficient sampling method for obtaining quantitative percent cover data.	Potential bias estimate of the proportion of the total area covered. Difficult to standardize some of the life form categories.
Permanent Quadrant Method	Designed to monitor change in the biological condition, growth, mortality and recruitment of corals in a permanently marked area.	Sampling is non-destructive. Detailed and careful observation, photography, and mapping of the fixed area provides a detailed record.	Method is slow and can be difficult with strong currents. Requires a flat area and coverage is small.
Coral Reef Fish Visual Census	Assessments are conducted along transects that are censed during daylight hours using SCUBA. Reef fishes are assessed using abundance categories, and individual counts allow estimates of standing stock and population size.	Quantitative and qualitative nondestructive method utilizes a minimum number of personnel. The same area can be resurveyed through time.	Observers must be well trained and experienced. Fish may be alarmed by the presence of divers and swim away. Technique is restricted to shallow depths.

3.4 OBSTACLES TO SUCCESSUL MAPPING OF CORAL REEFS

In the past, some believed that remote sensing of coral reefs was being oversold (Green et al., 1996). It is evident today, however, that aerial and digital mapping tools may offer some of the most important data sources for support of management strategies. Prior to mapping a coastal shoreline inhabited by reefs, researchers must understand the limitations of remote sensing techniques. The limitations of remote sensing can be grouped into three categories: (1) fundamental limitations; (2) operational limitations; and (3) user limitations (Table 10). Fundamental limitations are those that exist simply because of the nature of the environment. For example, the presence of water over coral complicates a remotely sensed signal. Operational limitations include technological limitations or mismatches that exist between field and image data. User limitations include imagery costs and necessary imagery expertise, as well as from shortcomings in the communication of information to the user. In all cases, current research is focused on finding solutions that will allow researchers to overcome remote sensing's limitations.

Table 10. Limitations that prevent accurate identification of reefs within satellite imagery.

Limitation	Potential Solution
Fundamental Limitations	
Variable Water Attenuation	Radiative transfer theorem
Variable water composition through time and space	Non-linear multi-layered algorithms
Variations in depth	Non-linear algorithms
Limited wavelength depth of penetration	Hyperspectral sensors
Heterogeneous bottom	Spectral unmixing algorithms
Air-water interface	Analytical techniques
Variable atmospheric conditions	Radiometric correction
Cloud cover	Masking techniques
Isolated reefs	Airborne imagery
Operational Limitations	
Course spatial resolution	Spectral unmixing
Course spectral resolution	Hyperspectral sensors
Disparity between date of image acquisition and *in situ* data collection	Analytical formulae
Satellites do not collect data in targeted area	Airborne imagery
Lack of *in situ* data	Combined analytical techniques and historical imagery
User Limitations	
Inadequate remote sensing knowledge on the part of managers and planners	Communication between experts and users
Differences between scientist generated image accuracy and required manager image accuracy	Pre-image consultation between scientist and user
Disparity between information images provide and information needed by managers	Consultant agencies
Information needs of coastal users are site specific	Variations in image displays
Cost of imagery, hardware, and software	International cooperation

4. Management Applications of Remotely Sensed Information

It has been suggested that remotely sensed images are powerful visual tools for display and communication. Their striking visual qualities are demonstrated, for instance, in the publication of and demand for several arresting coffee table books depicting exotic satellite and airborne imagery. Such images may form the foundation of powerful advocacy for remedial action regarding coastal resources. However, the manner in which these visual tools can be incorporated into coral reef management remains in question. In the west, a digital environment where remote sensing technologies thrive, remotely sensed images have been used by agencies, corporations, and academic institutions to describe weather conditions, examine topographical variations, and highlight terrestrial or aquatic features of interest. Outside of the digital environment, airborne and satellite images are used for purposes other than those intended by remote sensing scientists. In most instances, satellite images are used primarily for their visual appeal, as opposed to functional or operational purposes. Thus, a considerable challenge remains – how to interpret meaning and extract value from satellite imagery and transfer that information to people who reside outside of a remote-sensing familiar environment.

4.1 COMMUNICATION OF REMOTE SENSING INFORMATION

When attempting to communicate remotely sensed information about coral reef features to managers in developing countries, it is important to understand that while the analysis of airborne and satellite images is typically coordinated through personnel and facilities within developed countries, the application of the product occurs within socially and culturally different regions. Coral reef remote sensing scientists, to date, have rarely undertaken research to determine how to effectively communicate with stakeholders and apply research findings in socio-culturally distinct regions. To be effective, the analysis should involve the stakeholder from the beginning, and this should take place at the management level. In reality, the onus is often placed upon the end user (the local community member, manager, or planner in the stakeholder's region) to interpret the visual representation of coastal features and to integrate the information into the local management strategy. However, in the absence of scientific experts to guide interpretation and application of the information and significant funding to purchase satellite imagery, managers must depend on images produced by remote sensing scientists who have distinctly different objectives for the imagery. As a result, managers may disregard the satellite imagery, or minimally use the imagery in management projects. Consequently, the imagery is not fully realized.

4.2 OBSTACLES TO SUCCESSFUL COMMUNICATION OF REMOTELY SENSED INFORMATION

'Communication' involves conveying information. Within the context of typical applications of remote sensing, communication may be viewed as conveying the relevant quantitative information interpreted from airborne or satellite imagery. The effective or successful communication of information, derived from imagery, presumes that the end-user understands the conveyed information (Populus et al., 1995). Fostering this understanding is a challenge in applying remotely sensed data to management plans in developing countries. The sufficiency of the effort expended on communication can make the difference between success or failure. Insufficient effort, otherwise known as 'parachute science', involves minimal, if any, contact between a science and a manager. Unfortunately, this method still regularly remains in use for assessing management strategies and deciphering the most appropriate application of the technology. Other factors that challenge effective communication of remotely sensed information are summarized in Table 11.

From a map-maker standpoint, the accurate interpretation of features extracted from satellite imagery and their interconnectedness often requires significant operator knowledge and skill. A study involving the transfer of technology found that the inclusion of experienced personnel is an essential factor if the technology is to be applied successfully (Specter and Gayle, 1990). In addition, extensive fieldwork experience and knowledge of the key drivers of ecosystem change and set-up are essential not only for designing field surveys, but also for accurately discussing biological and operational processes with people on the ground. From a map-user standpoint, traditional knowledge influences a person's ability to interpret a satellite image. User attitudes, knowledge, and perspectives drive the interpretation and understanding of visually depicted elements (Specter and Gayle, 1990). In several traditional societies, local people do not fully understand coral reef environmental issues or the scientific explanations of ecosystem interrelationships (Veitayaki, 1998). As a result, differences in understandings' of the environment and in familiarity with technologies, tools, and terms can deter and often halt the communication of remotely sensed information to the user and management practices to the scientist.

Table 11. Major obstacles to communicating remotely sensed information to the user. (Modified from Specter and Gayle, 1990.)

Obstacle	Possible Solution
Technological Dimension	
Image development: Specialized techniques and processes Data volume	Involvement of experienced personnel familiar with calibrating and correcting airborne or satellite imagery Appropriate hardware and software processing capabilities and personnel trained in data reduction techniques
Field data collection: Complex linkages between multiple environmental variables	Involvement of trained personnel familiar with the dynamic physical and biological processes within a coral reef environment Involvement of local people in identifying targets of interest or areas under stress
User Dimension	
Visual perceptions: User's unfamiliarity with visual depiction of elements	Construction of alternative visual displays of information Incorporation of traditional forms of communication
Interpreter's knowledge: Different socio-cultural benchmarks between user and communicator	Involvement and cooperation of local community members and image operators from identification of objectives to final stages of image construction

4.3 IMPORTANCE OF INCORPORATING LOCAL KNOWLEDGE

The technical process of accurately imaging coral reefs, referred to as the 'construction stage', begins with defining the objective of the study and deciding on the most appropriate and feasible image type to meet pre-determined goals (Green et al.,

2000). In projects with restricted budgets or limited access to the region, archived data may be selected as the primary data for analysis. Regardless of whether archived or ordered images are selected, operators target images that possess minimal cloud cover and best represent the feature of interest in both spatial, spectral, and temporal dimensions (Green et al., 2000). If images that include the desired information are unavailable and/or if data for a particular event or season is required, then a scheduled flyover is planned, taking into consideration costs and image processing turn-around time. *In situ* field data may be collected simultaneously with image capture to evaluate the accuracy of the image and to enhance the information interpretation (Green et al., 2000). Once both field and image data are gathered, corrections and integration of data sets begins, and a final image is produced with a defined specified level of accuracy and information content (Lillesand and Kiefer, 1994).

During these preliminary steps, local communities can be involved. However, such involvement may introduce complexity and slow the process of image construction (Veitayaki, 1998). Nevertheless, local community involvement in the 'construction stage' considerably improves the relevance and practicality of research projects. When the objectives of a remote sensing study are being defined, for example, local community members should identify their own research needs (Johannes, 1998). Often these needs are in conflict with those of the technically trained who often wish to (1) address purely scientific questions concerning variables that have remained largely unexamined and/or (2) tackle specific organizational directives from the funding agency. Although scientists cannot be blamed for the direction of academic research, managers would welcome almost any scientific information and would benefit from opportunities to define their own agendas (Hof, 2002).

Once local communities are involved in the beginning stages of the project, on-going dialogues during subsequent stages of image construction can continue and increase the appropriateness of the final product. Local communities will benefit from the process by increasing their knowledge of environmental variables and linkages, and by better understanding how substrate features are identified, all the while appreciating the amount of time and effort required to achieve various levels of image accuracy. Although involving local communities creates complexity, it is essential to the planning and implementation of appropriate resource management strategies (Cooke, 1994). There are considerable issues to face when bringing together remotely sensed information and local knowledge, but incorporating a series of guidelines (Newman and LeDrew, 2005) may enhance the success of the integration process and bring about greater utilization of remotely sensed information.

4.3.1 *Building an Image with Local Input*

Communicating environmental information using satellite imagery, or an image-based map, is not a universal strategy (Johannes, 1981). For example, on several islands within Indonesia, it is typical for the local managers to learn about spatial and temporal changes to coral reef features without the aid of visual devices. Avoiding a spatial context, stakeholders describe features in the form of lists emphasizing type and abundance (World, 1994; Cesar et al., 1997; Fearnside, 1997; Pet-Soede et al., 1999). When viewing or working with geographic maps or images, different people obtain different amounts and kinds of information. This difference is the result not of the subjectivity of the information but rather of the different degrees of the viewers' ability to extract information. There are differences between the interpreter's image of reality, parts of reality that have been mapped, and what actually exists (Salichtchev, 1977). Therefore, remote sensing operators are challenged to illustrate complex environmental linkages in a locally appropriate manner to users who, in some cases, are unfamiliar with visualizing spatial patterns.

Building upon cartographic principles, researchers can devise an appropriate display to effectively communicate remotely sensed information (Veitayaki, 1998). Cartography as a practice is almost second nature to image analysts. It is a form of communication that presents locations and environmental features with scaled spatial proximities in a two-dimensional structure (Bartels and Beurden, 1998). In essence, maps or images are conceived as elements in a process of communication that involves both the mapmaker and map user (Eckert, 1977). The role of cartography in communicating remotely sensed information is rooted in the mapmakers' ability to express features in a solid, generally accepted visual representation that promotes successful comprehension by the user. Basic elements that include symbology, classes and class boundaries, colors, scale and projection, and basic map features (Bartels and Beurden, 1998) are used to express information to the user in a social-culturally appropriate manner. For this discussion, the critical concepts are 'generally acceptable' and 'culturally appropriate'. This means that several culturally relevant issues should be explored prior to product design (Table 12) because of their ability to influence significantly the success of image communication.

Table 12. Issues to consider prior to undertaking map construction.

Issues	Questions
Attitudes, Perceptions, Beliefs	How do people perceive and value the marine environment and its use? Who supports and who opposes visual representations of coral reef data and why? What do people expect of images ecologically and socioeconomically, over the short and long terms?
Taboos	How do people perceive the organisms within their environment? Which reef organisms are considered sacred, and what is their story? What criteria are in place when a "supernatural" feature is present?
Use Patterns	Who uses the marine environment in and near the identified boundaries and how? What do resource users know about the resource and its use? How might resource users, groups, and communities adapt, thus changing their use patterns in the future? What are the socioeconomic implications, including distribution of costs and benefits, of the above changes?

4.4 REMOTE SENSING FOR MANAGEMENT: A CASE STUDY IN SAVUSAVU BAY, FIJI

An experience of the authors in communicating remote sensing image-based information in Fiji illustrates the necessity of understanding the cultural context and the potential difficulty of presenting spatial information to local stakeholders. In a recent paper, LeDrew et al. (2004) discussed the extensive damage of a coral reef in Savusavu Bay, Fiji, as a consequence of the overflowing of settlement lagoons from a gold mine during a major precipitation event. SPOT images before and after the event were obtained. The site was visited and the damage recorded on video. A basic form of change detection analysis was performed to highlight regions that could be identified as stressed. Color-coded maps were generated from the analysis. From our experience in Fiji during prior field experiments, we understood that the village chiefs were the ultimate authority in management of the reefs. We decided to hold a 'town-hall' type of discussion with the villagers and the chiefs in the region affected by the coral damage and another village on the other side of the Bay that was only peripherally affected by the event. This village was also close to the main town of Savusavu, and inhabitants had interacted with tourists for many years through their employment at the local seaside resorts. The gold mine was far from any tourism influence. The villagers in the Savusavu town region could understand the concept of reef damage, had a good grasp of 'western' approaches to reef conservation and could identify their location on the map. The villagers in the region of the gold mine, however, showed little interest in the map and associated discussion, even though they appeared to understand the spoken English language. Although they were fascinated by a video of the damaged reef, they expressed no concern, only amusement at the moving pictures. There was no discussion of the nature of the damage or the impact on their livelihood, possibly since their livelihood was heavily dependent on the operating gold mine.

We learned that the traditional management of reef resources is based upon taboos (Knight et al., 1997) and that taboos exert pressure on local communities that directly influences their impressions and use of the reefs. For example, the concept of sacred ground is extended to sacred fishing regions where particular rules are followed. These regions are though to be under the influence of supernatural forces, which ensures fishing grounds are respected and protected at all times, even in the absence of enforcement officers. If a certain predator species is caught, for example, fishing stops. According to the villagers, if they act appropriately, their gods will provide them with what they have asked for (Jennings and Polunin, 1996). On the ground, the predator is there because of a downturn in the "health" of the coral reef. Some time in the past, the predator was associated with lower fish catches and stress on the villagers. In Fiji, it is essential that such cultural traditions be understood to avoid offensive visual representation of features that are respected and valued.

It is also possible that with the reliance upon gold mine work, the villagers from this region may have lost cultural practices based on traditional reliance on the reef. Others more qualified than we would be able to verify this. Our lesson was that the colorful map we presented had no impact on the villagers near the gold mine, since maps were not part of their normal interaction with the local environment. If we understood their taboos and illustrated the problem and potential remedial action within a comparable oral tradition, we might have been able to make our messages resonate within their cultural environment.

This example illustrates the importance of understanding local culture and local knowledge in effective communication. It is generally understood that local people possess close ties to their surroundings, and therefore, possess an inherent understanding of their environment: indeed, they have tended to ensure the long-term sustainability of their resource base (Veitayaki, 1998). Therefore, it is important for researchers to consider and incorporate local knowledge and understanding into a remote sensing project to ensure effective communication.

Being attentive to local traditions, knowledge, and beliefs provides significant benefits, as the local people:

- are directly aware of the current status of the marine resource on which they depend
- possess an inherent temporal knowledge of patterns within and changes to the marine environment
- are valuable resources for developing monitoring and managing strategies.

It is also important to be aware of the challenges that accompany local knowledge and the limitations of local knowledge's usefulness, which may include:

- the local knowledge is contained in a verbal medium and is often unverified
- the attitudes of local people towards their resource may not include a conservation outlook (which can discourage proper management plans)
- the local knowledge is exclusive to the area, with minimal overlap between regions within small spatial proximities
- the variability of human and physical events from one shallow coastal water environment to another may prevent the incorporation of an approach that was deemed successful elsewhere

Regardless of the challenges traditional knowledge may present to a remote sensing project, considerations to include this information must be made if greater utilization and appropriate applications of satellite imagery to local management is to be achieved. As the expert, remote-sensing scientists must develop effective and appropriate communication strategies to collaborate with local communities whose livelihood depends upon effective management of their marine resource.

5. Conclusions

Remote sensing alone will not solve the problems faced by coastal communities struggling to secure a balance between marine resource exploitation and conservation. The value of remote sensing lies in its ability to improve decisions and negotiations by providing better information to various stakeholders so that more alternatives can be generated and evaluated. Remote sensing has a role to play, and there is an urgent need to strengthen the communication basis for applying remotely sensed information for marine resource management plans in developing nations. The potential for remote sensing to contribute to specialized plans, particularly in small island states, will not be realized by simply gathering more satellite imagery or conducting more field surveys for validation and algorithm development. Realization and utilization will come only with on-site research that delivers appropriate products that have evolved through numerous meetings and discussions with local community members, planners, managers, government officials, and members of a local marine resource management board. Successful communication of remotely sensed information, which has been interpreted by the researcher in a fitting manner, and continuously discussed with local management authorities, is possible. Research must become part of management–a melding that demands an understanding of how the remote sensing processes work, where they don't, and how they can be interpreted, adapted, and communicated–a task that challenges scientists to 'design with culture'.

6. References

Agardy, T. 1997. Marine protected areas and ocean conservation. Academic Press, 244 pp.

Allee, R.J. and J. E. Johnson. 1999. Use of satellite imagery to estimate surface chlorophyll-a and Secchi disc depth of Bull Shoals Reservoir, Arkansas, USA. International Journal of Remote Sensing, 20:1057-1072.

Andréfouët, S., R. Berkelmans, L. Odriozola, T. Done, J. Oliver, and F. Muller-Karger. 2002. Choosing the appropriate spatial resolution for monitoring coral bleaching events using remote sensing. Coral Reefs, 21:147-154.

Andréfouët, S. and M. Claereboudt. 2000. Objective class definitions using correclation of similarities between remotely sensed and environmental data. International Journal of Remote Sensing, 21(9): 1925-1930.

Aplin, P., P. Atkinson and P. Curran. 1997. Fine spatial resolution satellite sensors for the next decade. International Journal of Remote Sensing, 18(18):3873-3881.

Bartels, C. and A. Beurden. 1998. Using geographic and cartographic principles for environmental assessment and risk mapping. Journal of Hazardous Materials, 61:115-124.

Birkeland, C. 1997. Life and Death of Coral Reefs. Chapman and Hall, 536 pp.

Bryant, D., L. Burke, J. McManus, and M. Spalding. 1998. Reefs at Risk: A Map-Based Indicator of Threats to the World's Coral Reefs. World Resources Institute, 56 pp.

Cesar, H., C. Lundin, S. Bettencourt, and J. Dixon. 1997. Indonesia coral reefs-an economic analysis of a precious but threatened resource. Ambio, 26:345-350.

Chauvaud, S., C. Bouchon and R. Maniere. 1998. Remote sensing techniques adapted to high resolution mapping of tropical coastal marine ecosystems (coral reefs, seagrass beds and mangrove). International Journal of Remote Sensing, 19(18):3625-3639.

Clark, C., P. Mumby, J. Chisholm, J. Jaubert, and S. Andréfouët. 2000. Spectral discrimination of coral mortality states following severe bleaching event. International Journal of Remote Sensing. 21(11): 2321-2327.

Clark, R., T. Fay, and C. Walker. 1987. Bathymetry calculations with Landsat 4 TM imagery under generalized ratio assumption. Applied Optics, 26(19):4036-4038.

Cooke, A. 1994. The qoliqoli of Fiji - some preliminary research findings in realtion to mangement, p. 179-182. In G. South, D. Goulet, S. Tuqiri and M. Church [eds.], Traditional marine tenure and sustainable management of marine resources in Asian and the Pacific. International Ocean Institute, South Pacific, 312 pp.

Cuevas-Jimenez, A. and P. Ardisson. 2002. Mapping of shallow coral reefs by colour aerial photography. International Journal of Remote Sensing, 23(18):3697-3712.

Curran, P., E. Milton, P. Atkinson, G. Foody. 1998. Remote sensing from data to understanding, p. 33-59. In P. Longley, S. Brooks, R. McDonnell and B. MacMillan, Geocomputation: A Primer. John Wiley and Sons, 278 pp.

Dekker, A. G., Z. Zamurovic-Nenad, H. Hoogenboom, and S. Peters. 1996. Remote sensing, ecological water quality modeling and *in situ* measurements: a case study in shallow lakes. Hydrological Sciences. 41(4): 531-547.

Done, T. 1999. Coral Community adaptability to environmental change at the scales of regions, reefs, and reef zones. American Zoologist, 39:66-79.

Eckert, M. 1977. On the nature of maps and map logic. Cartographica: The Nature of Cartographic Communication, 19:1-7.

Edinger, E. and M. Risk. 2000. Reef classification by coral morphology predicts coral reef conservation value. Biological Conservation, 92:1-13.

English, S., C. Wilkinson, and V. Baker. 1997. Survey manual for tropical marine resources: 2nd Edition. Australian Institute of Marine Science, 390 pp.

Estep, L. and J. Holloway. 1992. Estimators of bottom reflectance spectra. International Journal of Remote Sensing. 13(2):393-397.

Falkowski, P., P. Jokiel, and R. Kinzie III. 1998. Irradiance and corals, p. 89-108. In Z. Dubinsky [eds.], Ecosystems of the World: Coral Reefs. Elsevier, 550 pp.

Fearnside, P. 1997. Transmigartion in Indonesia: Lesssons from its environmental and social impacts. Environmental Management, 21(4):553-570.

Gash, J., T. Arvidson, S. Goward, S. Andréfouët, C. Hu, and F. Muller-Karger. 2000. An assessment of Landsat 7/ETM+ coverage of coral reefs worldwide. International Geosciences and Remote Sensing Symposium, 24-28 July 2000, Hawaii, USA.

Gower, J., R. Doerffer, and G. Borstad. 1999. Interpretation of the 685 nm peak in water-leaving radiance spectra in terms of fluorescence, absorption, and scattering, and its observation by MERIS. International Journal of Remote Sensing, 20(9):1771-1786.

Green, E., P. Mumby, A. Edwards, C. Clark. 1996. A review of remote sensing for the assessment and management of tropical coastal resources. Coastal Management, 24:1-40.

Green, E., P. Mumby, A. Edwards, and C. Clark. 2000. Remote sensing handbook for tropical coastal management. Unesco Publishing, 316 pp.

Gubbay, S. 1995. Marine protected areas - past, present and future, p. 1-14. In S. Cubbay, Marine Protected Areas: Principals and Techniques for Management. Chapman and Hall.

Hardy, J.T., F.E. Hoge, J. Yungel, and R. Dodge. 1992. Remote detection of coral bleaching using pulsed-laser fluorescence spectroscopy. Marine Ecology-Progress Series, 88(2-3):247-255.

Hatcher, B. 1999. Varities of science for coral reef management. Coral Reefs, 18:305-306.

Hedley, J. and P. Mumby. 2002. Biological and remote sensing perspectives of pigmentation in coral reef organisms. Advances in Marine Biology, 43:279-317.

Hochberg, E. and M. Atkinson. 2000. Spectral discrimination of coral reef benthic communities. Coral Reefs, 19:164-171.

Hof, T. 2002. Recruiting research that is useful to your MPA: Advice from experts. MPA News: International News and Analysis on Marine Protected Areas, 3(10). http://depts.washington.edu/mpanews/

Holden, H. and E. LeDrew. 1998. The Scientific Issues Surrounding Remote Detection of Submerged Coral Ecosystems. Progress in Physical Geography, 22:190-221.

Holden, H. and E. LeDrew. 1998. Spectral Discrimination of Healthy and Non-Healthy Corals Based on Cluster Analysis, Principal Components Analysis, and Derivative Spectroscopy. Remote Sensing of Environment, 65:217-224.

Jain, S. and J. Miller. 1977. Algebraic expression for the diffuse irradiance reflectivity of water from the two-flow model. Applied Optics, 16(1): 202-204.

Jennings, S. and N. Polunin. 1996. Fishing strategies, fishery development and socio-economics in traditionally management Fijian fishing grounds. Fisheries Management and Ecology, 3:335-347.

Johannes, R. 1981. Words of the Lagoon: Fishing and Marine Lore in the Palau District of Micronesia. University of California Press.

Johannes, R. 1998. Government-supported, village-based management of marine resources in Vanuatu. Ocean and Coastal Management, 40(2-3):165-186.

Joyce, K. and S. Phinn. 2002. Bi-directional reflectance of corals. International Journal of Remote Sensing. 23(2):389-394.

Kelleher, G. 1999. Guidelines for marine protected areas. IUCN, 79 pp.

Kirk, J.T.O. 1996. Light and Photosynthesis in Aquatic Ecosystems. Cambridge University Press, 509 pp.

Klemas, V. 2001. Remote sensing of landscape-level coastal environmental indicators. Enviromental Management, 27(1):47-57.

Kleypas, J., R. Buddemeier,and J. Gattuso. 2001. The future of coral reefs in an age of global change. International Journal of Earth Sciences, 90(2):426-437.

Knight, D., E. LeDrew, and H. Holden. 1997. Mapping submerged corals in Fiji from remote sensing and *in situ* measurements: applications for integrated coastal management. Ocean and Coastal Management, 24(2): 153-170.

LeDrew, E., H. Holden, M. Wulder, C. Derksen, and C. Newman. 2004. A spatial statistical operator applied to multidate satellite imagery for identification of coral reef stress. Remote Sensing of Environment, 91:271-279

Lewis, J. B. 2002. Evidence from aerial photography of structural loss of coral reefs at Barbados, West Indies. Coral Reefs, 21(1):49-56.

Liceaga-Correa, M. and J. Euan-Avila. 2002. Assessment of coral reef bathymetric mapping using visible Landsat Thematic Mapper data. International Journal of Remote Sensing, 23(1):3-14.

Lillesand, T. and R. Kiefer. 1994. Remote Sensing and Image Interpretation. John Wiley, 750 pp.

Lubin, D., W. Ki, P. Dustan, C. Mazel, K. Stamnes. 2001. Spectral signatures of coral reefs: Features from space. Remote Sensing of Environment. 75(1): 127-137.

Lunetta, R. and C. Elvidge. 1999. Remote sensing change detection: environmental monitoring methods and applications. Taylor and Francis, 318pp.

Lyzenga, D.R. 1978. Passive remote sensing techniques for mapping water depth and bottom features. Applied Optics, 17(3):379-383.

Maritorena, S., A. Morel, and B. Gentili. 1994. Diffuse reflectance of oceanic shallow waters: influence of water depth and bottom albedo. Limnology and Oceanography, 39(7):1689-1703.

Mas, J.-F. 1999. Monitoring land-cover changes: a comparison of change detection techniques. International Journal of Remote Sensing, 20(1):139-152.

Mazel, C. 1997. Coral fluorescence characteristics: excitation - emission spectra, fluorescence efficiencies, and contribution to apparent reflectance. SPIE, 2963:240-245.

Mobley, C. D. 1994. Light and Water: Radiative Transfer in Natural Waters. Academic Press, 592 pp.

Mumby, P. 2001. Beta and habitat diversity in marine systems: a new approach to measurement scaling and interpretation. Oecologia, 128:274-280.

Mumby, P. and A. Edwards. 2002. Mapping marine environments with IKONOS imagery: enhanced spatial resolution can deliver greater thematic accuracy. Remote Sensing of Environment, 82:248-257.

Mumby, P. and A. Harborne. 1999. Development of a systematic classification scheme of marine habitats to facilitate regional management and mapping of Caribbean coral reefs. Biological Conservation, 88(2):155-163.

Mumby, P., E. Green, A. Edwards, and C. Clark. 1997. Coral reef habitat mapping: how much detail can remote sensing provide. Marine Biology, 130(2):193-202.

Mumby, P., W. Skirving, A. Strong, J. Hardy, E. LeDrew, E. Hochberg, R. Stumpf, and L. David. 2004. Remote sensing of coral reefs and their physical environment. Marine Pollution Bulletin, 48(3-4): 219-228.

Neil, D. T., S. R. Phinn, and W. Ahmad. 2000. Reef zonation and cover mapping with Landsat Thematic Mapper data: intra and inter-reef patterns in the southern Great Barrier Reef region, IGARSS 2000. IEEE 2000 International Geoscience and Remote Sensing Symposium. Taking the Pulse of the Planet: The Role of Remote Sensing in Managing the Environment. Proceedings Cat. No.00CH37120. 2000. IEEE, Piscataway, NJ, USA, pp. 1886-8 vol.5.

Newman, C. 2001. Testing the Assumptions Surrounding Information Extraction from Remotely Sensed Imagery of Coral Reef Environments. Masters of Environmental Studies, Thesis, University of Waterloo, 132 pp.

Newman, C. and E. LeDrew 2002. Assessing the Uncertainty of Radiometric Properties in Coral Reef Environments. International Geosciences and Remote Sensing Symposium, 24-28 June 2002, Toronto, Canada.

Newman, C. and E. LeDrew. 2005. Towards Community- and Scientific-Based Information Integration in Marine Resource Management in Indonesia: Bunaken National Park Case Study. Environments Journal (in press).

Ohde, T. and H. Siegel. 2001. Correction of bottom influence in ocean colour satellite images of shallow water areas of the Baltic Sea. International Journal of Remote Sensing, 22(2-3):297-313.

Palandro, D., S. Andréfouët, F.E. Muller Karger, and P.Dustan. 2001. Coral reef change detection using Landsats 5 and 7: A case study using Carysfort Reef in the Florida Keys. Scanning the Present and Resolving the Future. Proceedings, IEEE International Geoscience and Remote Sensing Symposium, 2:625-7.

Pet-Soede, C., J. Cesar, and J. Pet. 1999. An economic analysis of blast fishing on Indonesian coral reefs. Environmental Conservation, 26(2):83-93.

Philpot, W. D. 1989. Bathymetric mapping with passive multispectral imagery. Applied Optics. 28(8): 1569-1578.

Phinn, S. 1998. A framework for selecting appropriate remotely sensed data dimensions for environmental monitoring and management. International Journal of Remote Sensing, 19(17):3457-3463.

Phinn, S. R., D. T. Neil, K. Joyce, and W. Ahmad. 2000. Coral reefs: a multi-scale approach to monitoring their composition and dynamics, IGARSS 2000. IEEE 2000 International Geoscience and Remote Sensing Symposium. Taking the Pulse of the Planet: The Role of Remote Sensing in Managing the Environment. Proceedings Cat. No.00CH37120. 2000. IEEE, Piscataway, NJ, USA, pp. 2672-4 vol.6.

Populus, J., W. Hastuti, J. Martin, O. Guelorget, B. Sumartono, and A. Wibowo. 1995. Remote sensing as a tool for diagnosis of water quality in Indonesian seas. Ocean and Coastal Management, 27(3):197-215.

Porter, J. and J. Tougas. 2001. Reef ecosystems: threats to their biodiversity. Encyclopedia of Biodiversity, 5:73-95.

Riegl, B. and W. Piller. 2000. Mapping of bethic habitats in northern Safaga Bay (Red Sea, Egypt): A tool for proactive management. Aquatic Conservation: Marine and Freshwater Ecosystems, 10:127-140.

Robinson, I. S. 1985. Satellite Oceanography. John Wiley, 455 pp.

Salichtchev, K. 1977. Some reflections on the subject and method of cartography after the sixth international cartographic conference. Cartographica: The Nature of Cartographic Communication, 19:111-116.

Sheppard, C., K. Matheson, J. Bythell, P. Murphy, C. Myers, and B. Blake. 1995. Habitat mapping in the Caribbean for management and conservation: use and assessment of aerial photography. Aquatic Conservation: Marine and Freshwater Ecosystems, 5:277-298.

Siegal, H., M. Gerth, and T. Neumann. 1999. Case studies on phytoplankton blooms in coastal and open waters of the Baltic Sea using Coastal Zone Color Scanner data. International Journal of Remote Sensing, 20(7):1249-1264.

Spitzer, D. and R. W. J. Dirks. 1987. Bottom influence on the reflectance of the sea. International Journal of Remote Sensing, 8(3):279-290.

Stumpf, R., V. Ransibrahmanakul, R. Arnone, R. Gould, P. Martinolich, P. Tester, R. Steward, A. Subramaniam, M. Culver, and J. Pennock. 2000. SeaWiFS ocean color data for US Southeast coastal waters. International Conference on Remote Sensing for Marine and Coastal Environments, 1-3 May 2000, Charleston, USA.

Souter, D. and O. Linden. 2000. The health and future of coral reef systems. Ocean and Coastal Management, 43:657-688.

Specter, C. and D. Gayle. 1990. Managing technology transfer for coastal zone development: Caribbean experts identify major issues. International Journal of Remote Sensing, 11(10):1729-1740.

Tassan, S. 1996. Modified Lyzenga's method for macroalgae detection in water with non-uniform composition. International Journal of Remote Sensing, 17(8):1601-1607.

Thamrongnawasawat, T. and D. Hopley. 1994. Digitised aerial photography applied to small area reef management. Recent Advances in Marine Science and Technology, 94:389-394.

Tsai, F. and W. Philpot. 1998. Derivative Analysis of Hyperspectral Data. Remote Sensing of Environment, 66:41-51.

Veitayaki, J. 1998. Traditional and Community-Based Marine Resources Management System in Fiji: An Evolving Integrated Process. Coastal Management, 26:47-60.

Wasserman, A. 1975. Biology. Addison-Wesley, 1217 pp.

Wilkinson, C. 2000. Status of Coral Reefs of the World: 2000. Australian Institute of Marine Science, Townsville, 363 pp.

World Bank. 1994. Indonesia: Environment and Development. A World Bank Country Study.

Chapter 12

DATA SYNTHESIS FOR COASTAL AND CORAL REEF ECOSYSTEM MANAGEMENT AT REGIONAL AND GLOBAL SCALES

JULIE A. ROBINSON[1], SERGE ANDRÉFOUËT[2] AND LAURETTA BURKE[3]
[1]*Earth Sciences and Image Analysis Laboratory, Johnson Space Center, 2101 NASA Parkway, Mail Code SA15, Houston, TX 77058, USA*
[2]*UR Coreus, Institut de Recherche pour le Développement, BP A5, 98848 Nouméa cedex, New Caledonia*
[3]*Information Program, World Resources Institute, 10 G Street, NE, Washington, DC 20002, USA*

1. Introduction: The Need for Synthesis Information in Management

Coastal habitats have been subjected to dramatic changes as land is converted to agricultural, urban and industrial uses. Key coastal habitats, including mangroves, wetlands, seagrasses, and coral reefs, are rapidly disappearing. An estimated 50% of the global original mangrove area is believed to have been lost in the last 50 years (Kelleher et al., 1995) and some reports suggest losses as high as 85% in several regions (Burke et al., 2000). Other types of coastal wetlands are more difficult to quantify. Peat swamps in Southeast Asia have declined by 46 to 100% in various countries (MacKinnon, 1997), and the loss of coastal wetlands in the United States is estimated to have been almost 50% by the mid-1970s (Bookman et al., 1999). Seagrasses worldwide have similarly declined (Short, 1999). Approximately 10% of the world's coral reefs have been seriously degraded, with nearly 60% of reef areas potentially threatened by human activities (Bryant et al., 1998).

Most of the changes in coastal habitats are human-induced, and are a direct result of rapid human population growth concentrated along coasts (Hinrichsen, 1996). Although coastal land (within 100 km of a coast) accounts for only 20% of all land area in the world, 40% of the world's population lives in these areas (Burke et al., 2000). Uplands and coasts are ecologically linked so that changes in uplands affect the coastal habitats, and loss of coastal habitats endangers uplands (Kennish, 2000). Both retrospective and predictive studies of coastal change are required to anticipate habitat degradation and determine management strategies to mitigate the problems (Dahdouh-Guebas, 2002).

Modification of coastal habitats is varied in terms of ecosystem and impact. Commonly modification has economic ramifications, and can affect both fisheries and the human populations that depend on them (Lindén, 1990). Other economic impacts arise from the increased vulnerability to flooding, storm surges, and sea level rise (Nicholls et al., 1999). Key habitats such as coral reefs, seagrasses and mangroves are each uniquely impacted by coastal modification. Major human-induced threats to coral reef habitats include urban and industrial coastal development, destructive fishing practices, overexploitation of marine resources, marine pollution, and sedimentation resulting from deforestation, mining and agricultural activities (Hodgson, 1999). Seagrasses can be lost as a result of coastal development which thin increases shoreline erosion, water column turbidity and degraded water quality (Bookman et al., 1999). Mangrove habitats are important in the reproduction of many marine species and

L.L. Richardson and E.F. LeDrew (eds.), Remote Sensing of Aquatic Coastal Ecosystem Processes: Science and Management Applications, 279-305.

provide important economic value, but are threatened by pollution, defoliants, harvesting for wood products, urbanization, reclamation, water diversion, and conversion to aquaculture or salt ponds (Farnsworth and Ellison, 1997; Spalding et al., 1997). Degradation of coral reef and mangrove habitats often leads to a "lose-lose" situation in which short-term economic gains evaporate and economic losses follow habitat destruction (Dadjouh-Guebas, 2002). Assessing the impacts of coastal degradation on fisheries and prioritizing management actions is limited by inadequate knowledge of basic parameters such as habitat locations, extent and characteristics (McManus, 1997). Moreover, there is often a lack of quantitative knowledge on the more complex processes that link terrestrial processes to the coastal zone.

In this context of habitat modification and degradation, extensive information on previous condition, current status, and change of coastal habitats is critical in order to make appropriate management decisions and monitor future change. In most regions worldwide, Burke et al. (2000) noted that existing information was not adequate to assess conversion of coastal habitats (to urban, agricultural and industrial uses) and corresponding changes in sediment flows and shoreline erosion. Even countries that have extensive efforts in natural resource mapping and monitoring, such as the United States, are struggling to understand loss of coastal habitats and how to better manage them (e.g. Bookman et al., 1999). The U.S. National Academy of Sciences is currently completing a report on "National Needs for Coastal Mapping and Charting" to identify: 1) primary data sets that need to be integrated; 2) gaps and overlaps among agencies collecting the information needed; and 3) technologies to acquire, archive, and disseminate information to the user community (National Research Council, 2004). The struggle to map, manage, and monitor coastal ecosystems is even more challenging in developing countries where technological and scientific resources are more limited.

Obtaining the information needed to set management priorities for coastal habitats requires synthesis of various data sets. Among them, remotely sensed data from airborne and satellite platforms are a cost-effective way of collecting information on coastal habitats for regional management objectives so as to optimize coverage of different spatial scales, and to be able to integrate processes on land, at the coast, and in the sea (Mumby et al., 1999). The objective of this chapter is to summarize the synthesis of different data types for meeting the information needs for management and conservation. Our emphasis is on efforts to provide regional or global datasets that are of value to managers as baseline information, the systems that are being deployed to make that information available (particularly across international boundaries), and how combining information from different data sources will contribute to better resource management.

2. Global Datasets Derived from Primarily Cartographic Origins

Natural resource managers, conservation organizations, and international planning organizations rely on map data and commonly use geographic information system (GIS) tools in supporting their planning. A number of major global datasets are commonly employed in these endeavors because of their ready accessibility for incorporation into GIS systems (Table 1). For example, PAGE (Pilot Analysis of Global Ecosystems) Coastal Ecosystems Study, conducted by the World Resources Institute, used a variety of GIS layers in an analysis of coastal ecosystems (Burke et al., 2000). It also assembled numerous available GIS datasets and applied them to

Table 1. Sources of compiled global datasets that may be useful in combination with other datasets and remote sensing data for coastal and coral reef management.

	Scale[1]	Data Source, date, update frequency	Availability[2]
Cartographic Origins			
TerrainBase Global Terrain Model (topography and bathymetry)	5' grid	Model built from other models and a variety of data sources, 1994, topography is being updated with SRTM[3] data to produce a product called GTOPO30 (see below).	National Geophysical Data Center. Available online at http://www.ngdc.noaa.gov/seg/fliers/se-1104.shtml
C-MAP Bathymetry Point Data	Various	National Hydrographic Authorities, ongoing updates	Designed for navigation purposes, but includes extensive depth soundings, commercial products with proprietary software
World Vector Shoreline (WVSPlus®)	1:250,000	Navigational charts, ongoing updates	National Geospatial-Intelligence Agency vendors
Coastal zone extent	1:250,000	Based on WVS and Global Maritime Boundaries Database, 2000	Pruett and Cimino. 2000. (unpublished)
Global Self-consistent, Hierarchical, High-resolution Shoreline Database (GSHHS)	1:100,000	WVS supplemented with World Data Bank II (1:3,000,000) processed to remove internal inconsistencies such as erratic points and crossing segments, 1999	NOAA[4] National Geophysical Data Center (NGDC; Wessel and Smith 1996). Available online at http://www.ngdc.noaa.gov/mgg/shorelines/gshhs.html
Global Maritime Boundaries Database	1:250,000	General Dynamics Advanced Information Systems compilation of text data sources with WVS, 2000, future updates	Veridian (Cimino et al., 2000).
Coastal Typology Dataset	1' grid	Geographically indexed data set compiled by LOICZ5 to allow grouping of the worlds coastal zones into clusters based on natural socioeconomic features; 1998	LOICZ5 International Project Office, International Geosphere-Biosphere Programme. Available online at http://www.old.nioz.nl/loicz/data.htm
Gridded Population of the World	2.5' grid (5 km)	Compilations of national population censuses; 1990, 1995	Center for International Earth Science Information Network (CIESIN et al. 2000). Available online at http://sedac.ciesin.org/plue/gpw

Table 1, cont'd.

	Scale	Data Source, date, update frequency	Availability
Global Coral Reef Distribution	1:200,000[6]	Compilation of maps, some updates through 2000	UNEP-WCMC[7] (Spalding et al., 2001)
Global Mangrove Distribution		Compilation of maps, 1997	UNEP-WCMC (Spalding et al., 1997)
Global Seagrass Distribution		Compilation of maps, 2002	UNEP-WCMC (Green and Short 2003)
Global Wetland Distribution		Compilation of maps, 1998	UNEP-WCMC
World Database on Protected Areas		Boundary polygons for about half of the 677 coral reef marine protected areas (MPAs) in the United Nations List of Protected Areas (Green et al. in review), center points, areas, and World Conservation Union (IUCN) management attributes; 1993, 1997, 2003.	World Database on Protected Areas Consortium, UNEP-WCMC (WCMC and IUCN 1998). Text data but not polygons available online, see also Kelleher et al. (1993)
Terrestrial Ecoregions of the World		New classification of smaller terrestrial ecoregion units within biogeographic regions and terrestrial biomes; 2000.	World Wildlife Fund (Olson et al., 2001). Available for download at http://www.worldwildlife.org/ecoregions

Table 1, cont'd.

	Scale	Data Source, date, update frequency	Availability
Derived from Remote Sensing			
Global Land Cover Characteristics Database (IGBP DISCover)	1 km grid	AVHRR[8] satellite assessment of land cover, including land cover near coasts; 1992-1993 data	U.S. Geological Survey et al., (Loveland et al., 2000). Available online at http://edcdaac.usgs.gov/glcc/glcc.asp
MODIS Land Cover Product	1 km	MODIS[9] satellite assessment, quarterly	(Friedl et al., 2002). Available online through http://geography.bu.edu/landcover/userguidelc/
LandScan	30 grid (1 km)	National population censuses, U.S. Bureau of the Census P-95 data, Global Demography Project census data, roads, slope, land cover, Nighttime Lights of the World, WVS coastlines; 2000, 2001, 2002, annual updates	Oak Ridge National Laboratory (Dobson 2000). Available online at http://www.ornl.gov/sci/gist/landscan/index.html
SeaWiFS Global Shallow Bathymetry	1 km grid	Sea-viewing Wide Field-of-View Sensor (SeaWiFS), gives bathymetry from 1-100 m in depth, 2002.	SeaWiFS project and NOAA (Stumpf et al., 2003b). Available online at http://seawifs.gsfc.nasa.gov/cgi/reefs.pl and ftp://samoa.gsfc.nasa.gov/pub/seawifs_bathymetry/
Sea Surface Temperature Operational Products for Coral Bleaching Prediction: (1) Anomalies, (2) Coral Bleaching Hotspots, (3) Degree Heating Weeks	(1) 36 km (2) 50 km (3) 50 km	Advanced Very High Resolution Radiometer (AVHRR) data; 1996-present, monthly	NOAA's Coral Reef Watch (Strong et al., 2000). Available online at http://orbit- net.nesdis.noaa.gov/orad/ coral_bleaching_ index.html; higher resolution (9 km) "Pathfinder" temperatures and climatologies also available (Kilpatrick et al., 2001). Available online at http://podaac.jpl.nasa.gov/sst/
SRTM Digital Topographic data	90 m grid	Shuttle Radar Topography Mission (SRTM), 30 m resolution for U.S., 90 m globally, 2000.	Jet Propulsion Laboratory (Werner, 2001), distributed by U.S. Geological Survey. Available online at http://www2.jpl.nasa.gov/srtm/cbanddataproducts.html

Table 1, cont'd.

	Scale	Data Source, date, update frequency	Availability
Geocover-LC	30 m	Landsat-derived global land cover; 1990 or 2000	EarthSat Corporation
Millennium Coral Reef Maps	30 m	Landsat satellite data and a global geomorphological classification; 1999-2002, planned release of dataset in 2004	Institute for Marine Remote Sensing, University of South Florida. Upon release Shapefiles and GeoTIFFs of the geomorphological maps will be available online at http://imars.marine.usf.edu/corals/index.html. The Landsat data is released in an open access archive at http://seawifs.gsfc.nasa.gov/cgi/landsat.pl. Both Landsat data and Shapefiles of maps can also be accessed via ReefBase at http://www.reefbase.org.

[1] No attempt has been made to standardize units defined for different product map projections; scale is reported in the units defined in the metadata for each data source.
[2] For data that is for sale, sufficient information is given to locate the necessary vendor using Web search tools. If data is available for download at no charge, the URL is given for data access. References indicate books, reports or journal articles discussing the production of the data set.
[3] Shuttle Radar Topography Mission.
[4] National Oceanographic and Atmospheric Administration.
[5] Land-Ocean Interactions in the Coastal Zone.
[6] Source data range from 1:12,500 to 1:10,000,000, but typically were 1:200,000, with varying positional accuracy.
[7] United Nations Environment Programme-World Conservation Monitoring Centre.
[8] Advanced Very-High Resolution Radiometer.
[9] MODerate resolution Imaging Spectroradiometer.

characterizing coastal features and habitats. Their list of incorporated datasets represents a good summary of the GIS-based data layers that can be useful as sources of input data for coastal studies. The published and accessible data are included in Table 1. Although the degree of data validation in the different datasets varies from strict to unknown, together they meet the immediate need for supplemental information in support of a variety of management analyses. Some elements of the most widely employed datasets are summarized below

2.1 WORLD VECTOR SHORELINE DATA

The World Vector Shoreline (WVSPlus®) is a vector-based digital representation of shorelines, maritime boundaries, international boundaries, country areas, and selected depth information produced for the U.S. National Imagery and Mapping Agency (NIMA 1999, now the National Geospatial-Intelligence Agency). The base data is at 1:250,000 scale, and products are also available at 1:1,000,000 to 1:120,000,000 scales. The coastal shoreline is derived from several NIMA products including Digital Landmass Blanking data, with information supplemented form Operational Navigation Charts, Tactical Pilotage Charts, and Joint Operations Graphic products. Other digital map products produced by NIMA and commonly used as input data are the Digital Chart of the World (DCW®) and its updated version Vector Map (Vmap) Level 0 at 1:1,000,000, and Digital Nautical Charts (DNC®) designed to show features needed for safe marine navigation (e.g. Figure 1, A). All projections, datums, and accuracy are documented. Data are distributed commercially by government and private vendors, and derived products at reduced resolutions are also available as part of some open source data sets. The PAGE Coastal Ecosystems Study (Burke et al., 2000) developed new standardized global statistics on extent of coastal ecosystems, such as an estimate of coastline length by country using World Vector Shoreline data.

2.2 WCMC MAPS OF REEFS, MANGROVES, AND MPAS

The United Nations Environment Programme-World Conservation Monitoring Centre's (UNEP-WCMC) marine program has carried out a number of projects to synthesize cartographic information and produce global maps of marine and coastal resources. By combining cartographic data (including some maps from remote sensing studies) from a variety of sources, WCMC succeeded in providing the first global maps of mangroves (Spalding et al., 1997), coral reefs (most recently in Spalding et al., 2001), seagrasses (Green and Short, 2003) and MPAs (Kelleher et al., 1993 and online updates). These represent the first attempts to create global summary data and are very valuable references. However, because they were compiled from a variety of cartographic products with different purposes and spatial scales, the maps are uneven in levels of detail, include positional and other errors, and have undetermined accuracies (e.g. Green et al., in review). In the past, this dataset was the best available global compilation, however, higher resolution global products currently being produced (see 0) will soon allow a complete revision with remote sensing data as the source. The baseline maps are in vector format and maps are viewable using online search tools at websites associated with UNEP-WCMC (http://www.unep-wcmc.org) and ReefBase (Figure 1 A; McManus and Noordeloos, 1998; http://www.reefbase.org).

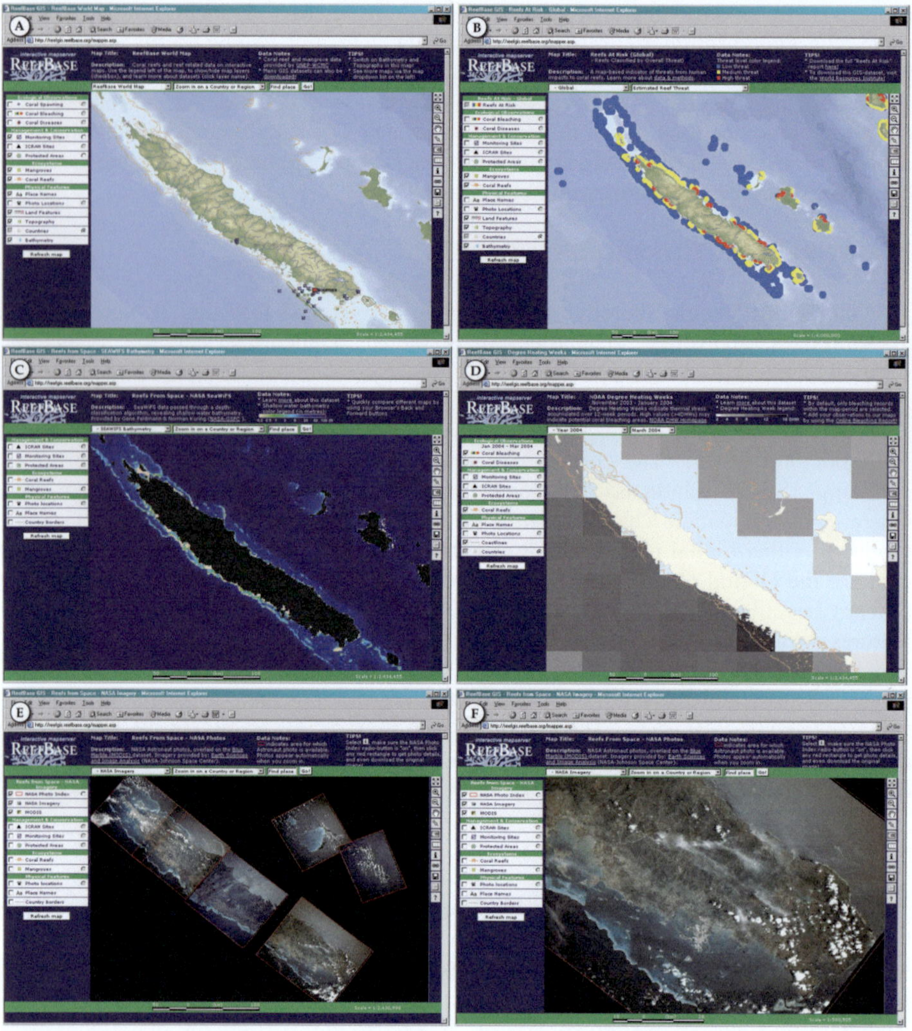

Figure 1. Example of some of the datasets available for interactive viewing in ReefBase, shown centered on New Caledonia for comparison. A. Basic global GIS data including land features from Vmap Level 0, Topography from 1 km base elevation (GLOBE), bathymetry from National Geophysical Data Center Terrain Base Global DTM Version 1.0, and Coral Reefs and Mangroves from the UNEP-WCMC databases. B. Global Reefs at Risk threat levels. C. SeaWiFS 1 km shallow water bathymetry. D. Degree Heating Weeks for March 2004 from NOAA. This image also shows the many georeferencing problems encountered in combining different data sources and projections. The blue mask appears to be shifted compared to island positions because of registration errors between the two data sources. E. General remote sensing image data (in this case, georeferenced astronaut photography). F. Example of the flexibility in focusing on higher resolution remote sensing data, as the user begins to zoom in on astronaut photography.

2.3 WORLD WILDLIFE FUND GLOBAL ECOREGIONS

Conservation planning requires information on the complex distribution of ecological communities for the identification of areas of unique biodiversity. Digital datasets classifying biogeographic realms or biomes down to a local ecosystem scale have been developed for terrestrial ecosystems by the World Wildlife Fund (WWF) (Olson et al., 2001) with online descriptions and browsing and GIS shapefiles available for download (http://www.worldwildlife.org/ecoregions). The ecoregions identified include areas of coastal mangroves, but not salt marshes, beaches, or other major coastal habitats. WWF's Global 200 (Olson et al., 2000) is an identified subset of 200 WWF Global Ecoregions classes that are most important for conservation of global biodiversity. This data set includes freshwater and marine habitats in addition to terrestrial ecoregions. However, the much more limited knowledge about marine biodiversity limited the scope of the analysis to five broad habitat types (polar, temperate shelf and seas, temperate upwelling, tropical upwelling, and tropical coral; Olson et al., 2000).

One potential problem with global ecoregion datasets such as the WWF products is that they are by necessity derived from a synthesis of other data sources, and include a significant number of assumptions that must be considered before the dataset is used in an "outside" analysis (one for which it was not designed). Often, once a product is publicly available, it begins to be assimilated into other studies - whether limitations and assumptions are appropriate to subsequent uses or not. Methodological details and documentation of these assumptions are frequently not readily available to outside users. For example, as a biogeographic framework the defined ecoregions of the WWF Global Ecoregions set are more descriptive of habitat for some taxa (plants, insects) than for other others. The use of smooth mapped ecoregions oversimplifies boundaries across ecotones, mosaic habitats and unique embedded habitats (Olson et al., 2001). Beyond the assumptions implicit in biogeographic approaches, additional assumptions are inherent in the protocols used in mapping the regions in the first place. Expert opinions were heavily used in developing the WWF Global 200 and have a large weight in the decisions taken to prioritize areas for conservation, but expert opinion may be biased toward better-known sites.

3. Global Datasets Derived from Satellite Remote Sensing

In summarizing the different remote sensing datasets available with global or nearly global coverage, we group the datasets by their relative spatial resolution. Subjective terms for spatial resolution often differ among investigators depending on whether satellite or airborne platforms are the norm. In this chapter, we try to use a manager's perspective which would consider high spatial resolution studies as those that have fine enough grain to distinguish coastal habitats. Thus, we define high spatial resolution as $\leq$ 10 meter mapping units (pixels), moderate spatial resolution as 30-100 meter mapping units, and low spatial resolution as $\geq$ 500 meter mapping units.

The general use of satellite remote sensing sensors in tropical coastal management was reviewed by Green et al. (1996, 2000). Sensors and techniques for mangrove mapping (Green et al., 1997, 1998) have also been reviewed. The most recent reviews of the many advances in remote sensing of coral reefs are by Mumby et al. (2004), Andréfouët and Riegl (2004), and Andréfouët et al. (2004). Here, we summarize only the most recent general applications in these areas needed to evaluate the potential for data synthesis using different sensors.

3.1 LOW SPATIAL RESOLUTION (≥500 METER) GLOBAL LAND (COASTAL) PRODUCTS

Multiple global or nearly global land cover products are now available. These global products make it possible to use GIS and modeling tools to develop regional and global derived analyses include land cover adjacent to the coastal zone without conducting primary classifications. With all the varied uses of land cover maps, it is important to know specific information about each map that describes the author's purpose, methods, errors uncertainties or biases in the final product, and validation results or accuracy assessments. Land cover products and measurement of land cover change are important components in many analyses of coastal habitats. Change detection can often require that multiple sources of data be combined (e.g. Petit and Lambin, 2001). Newly released global digital elevation models (DEMs) built from Shuttle Radar Topography Mission (SRTM) data can be combined with land cover data to allow improved estimation of vulnerability to erosion or flooding. For example, by using land cover data in the context of the physical and environmental factors of the landscape (slope, soil type, precipitation regime), it is possible to estimate relative erosion rates (Bryant et al., 1998) which can affect coast waters.

3.1.1 IGBP DISCover global landcover

The International Geosphere Biosphere Program (IGBP) began in 1992 to produce the first global, validated land cover data set called DISCover. The completion of this enormous project was preceded by five years of technical planning and oversight by the IGBP-DIS Land Cover Working Group who defined the technical specifications involved in producing a new global land cover database (Loveland et al., 1999). The first global classification was completed in 1997 and the validation of this data set was concluded in 1999. The data set consists of 17 land cover classes derived from 1-km Advanced Very High Resolution Radiometer (AVHRR) imagery (Table 2). The classification process was divided into four major steps: 1) data set preparation and assessment of AVHRR image quality; 2) computation of monthly Normalized Difference Vegetation Indices; 3) unsupervised clustering derived from the K-Means algorithm to construct preliminary greenness classes (Belward et al., 1999); 4) development of a method to interpret land cover regions based on seasonal changes in greenness; and 5) final land cover product generation (Loveland et al., 1999). Major caveats to the use of the dataset are the image quality of the AVHRR data that was used in the analysis and in difficulties with Landsat TM imagery that was used for validation (Husak et al., 1999). Classification accuracy of the different land cover types varied from 60 to 90% (Scepan, 1999). The IGBP DISCover data, or "Global Land Cover Characteristics Database Version 2.0 Based on 1 km AVHRR data (April 1992-March 1993)" is available from EROS Data Center (EDC) Distributed Active Archive Center (DAAC).

Global landcover datasets are most suited to studies that need to include the extent of human modified areas (such as agriculture) compared to natural habitats (such as forest or wetland). Unfortunately for use in coastal management studies, the IGBP classification scheme, does not specifically distinguish coastal wetland types such as mangroves (Table 2).

Table 2. Land cover classes used in global land cover products. Data are aligned to indicate areas that are expected to be classified similarly. IGBP land cover legend information from http://edcdaac.usgs.gov/glcc/globdoc2_0.html#app2, Geocover-LC legend information from http://www.geocover.com/gc_lc/index.html; approximate alignment courtesy of Jennifer Gebelein (unpublished).

Classification Scheme	
IGBP DISCover/MOD12	**Geocover-LC**
1-Evergreen Needleleaf Forest	2-Forest, Evergreen
2-Evergreen Broadleaf Forest	2-Forest, Evergreen
3-Deciduous Needleleaf Forest	1-Forest, Deciduous
4-Deciduous Broadleaf Forest	1-Forest, Deciduous
5-Mixed Forest	
6-Closed Shrublands	3-Shrub/Scrub
7-Open Shrublands	3-Shrub/Scrub
8-Woody Savannas	
9-Savannas	4-Grassland (>10% ground cover)
10-Grasslands	4-Grassland (>10% ground cover)
11-Permanent Wetlands	9-Wetland, Permanent/Herbaceous
12-Croplands	7-Agriculture, General
13-Urban and Built-Up	6-Urban/Built-Up
14-Cropland/Natural Vegetation Mosaic	7-Agriculture, General
15-Snow and Ice	12-Permanent or Nearly Permanent Ice and/or Snow
16-Barren or Sparsely Vegetated	5-Barren/Minimal Vegetation (<10% ground cover)
17-Water Bodies	11-Water
	8-Agriculture, Rice/Paddy
	10-Wetland, Mangrove
99-Interrupted Areas	
100-Missing Data	13-Cloud/Cloud Shadow/No Data

3.1.2 MODIS global landcover

Data from NASA's Earth Observing System (EOS) MODerate resolution Imaging Spectroradiometer (MODIS) sensor is currently being used to produce global landcover maps with 1 km mapping units. The maps are compiled quarterly to incorporate seasonal dynamics, and follow the same 17-class structure used for the IGBP DISCover project (Table 2). The maps are created using a supervised classification approach with a decision tree classifier that determines the probabilities that each pixel belongs to each class (Friedl et al., 2002). A recursive weighting process, called "boosting," develops classification trees that can incorporate prior knowledge in order to improve the global classification map (McIver and Friedl, 2002). Preliminary validation results indicate classification accuracies for the different classes of 60 to 90% (http://geography.bu.edu/landcover/userguidelc/consistent.htm). The "MODIS/Terra Land Cover Type 96-Day L3 Global 1km ISIN Grid" is available from the Land Processes Distributed Active Archive Center at the EROS Data Center.

3.2 LOW SPATIAL RESOLUTION (≥500 METER) GLOBAL OCEAN PRODUCTS

3.2.1 SeaWiFS 1-km shallow bathymetry

The Stumpf et al. (2003a) bathymetry algorithm has been applied to all the SeaWiFS (Sea-viewing Wide Field-of-view Sensor) data for the five-year mission to produce a global composite map with 0.01 degree map resolution (Stumpf et al., 2003b). The high positional accuracy of SeaWiFS made it possible to build a composite map and allowed elimination of clouds and short-term turbidity events to produce a uniform global bathymetry map. The dataset is available through the SeaWiFS Project (http://seawifs.gsfc.nasa.gov/reefs/) (Figure 1) and through ReefBase (Figure 1C; http://www.reefbase.org). In areas of frequent turbidity, the shallowest depths are underestimated, however, the algorithm performed well compared to reference data.

3.2.2 Sea Surface Temperatures

Warming of the oceans due to global climate change has the potential to induce major coral bleaching episodes around the world (Hoegh-Guldberg, 1999), and understanding patterns of anomalous temperatures near coral reefs is important for reef and fisheries management. Of the variety of different sea surface temperature analyses being conducted globally, a few have great potential for synthesis with other remote sensing data in improving understanding of coastal ecosystems. The U.S. National Oceanographic and Atmospheric Administration (NOAA) provides a number of operational (frequently updated) products on global sea surface temperature and anomalies. Weekly and monthly datasets produced from the Polar-orbiting Operational Environmental Satellite (POES) Advanced Very High Resolution Radiometer (AVHRR) combined with buoy observations for field validation (Reynolds, 1988, Reynolds et al., 2002) provide coarse global coverage with 1° mapping units. The AVHRR Pathfinder algorithm allows improved spatial resolution in sea surface temperature determination with 9 km mapping units (Kilpatrick et al., 2001).

Since 1997, NOAA has provided experimental AVHRR maps of sea surface temperature anomalies with 50 km mapping units for the express purpose of evaluating and predicting the incidence of coral bleaching events (Strong et al., 1997). Sea surface temperature maps and estimates of cumulative thermal stress in terms of degree heating weeks have been successful at predicting and monitoring coral bleaching events around the world (e.g. Liu et al., 2003). For the large-scale bleaching events on the Great Barrier Reef in 1998 and 2002, the maximum sea surface temperature occurring over any 3-day period ("max3d") predicted bleaching better than anomaly-based methods (Berkelmans et al., 2004). Higher resolution (9 km mapping units) temperature anomaly products can further refine and improve the ability to predict coral bleaching (Toscano et al., 2002a, b). Products are under continued revision and development zand are available from NOAA's Coral Reef Watch (http://orbit-net.nesdis.noaa.gov/orad/coral_bleaching_index.html). Degree Heating Weeks data are also shared with the ReefBase online GIS server (Figure 1, D; http://www.reefbase.org)).

The Moderate Resolution Imaging Spectroradiometer (MODIS) onboard the Terra and Aqua satellites also makes observations of sea surface temperatures, chlorophyll and many other bio-optical products. The "level 2" daily data has 1 km spatial resolution, and global binned map products ("level 3", including quality control and eliminating most cloud cover) for ocean color and sea surface temperature are available with 4.88 km, 39 km, or 1 degree mapping units compiled over daily, weekly, monthly and yearly intervals (Esias et al., 1998). Products are available from the Goddard Earth Sciences Distributed Active Archive Center (GES DAAC, http://daac.gsfc.nasa.gov/MODIS/) At the time of this writing, ocean color products from the SeaWiFS (Sea-viewing Wide Field-of-view Sensor) and calibration of MODIS ocean color products are in flux, with new products to be designated soon. In addition, the number of MODIS products and distribution protocol and gateways are also under review.

Another area of active ongoing research is how best to combine other data available from ocean satellites, including wind, currents, cloud cover, and solar radiation with sea surface temperature and degree-heating-weeks models to improve prediction of the biological response that causes coral bleaching (e.g. Wooldridge and Done, 2004). Mumby et al. (2004) point out that global map products of ultraviolet and photosynthetically active radiation could be combined with water column attenuation coefficients to estimate the solar radiation received by marine organisms. The potential for integrating sea surface temperature data with data on reef geomorphology, coastal habitats and adjacent land cover remains to be explored (cf., Andréfouët et al., 2004).

3.3 MODERATE SPATIAL RESOLUTION (20-100 METER) GLOBAL LAND (COASTAL) PRODUCTS

Landsat Multispectral Scanner (MSS), Thematic Mapper (TM) and Enhanced Thematic Mapper (ETM+) have been global workhorses for moderate resolution (30 m) multispectral remote sensing. There are currently several free or low-cost sources of archival data that make it easy to add Landsat information into management analyses (especially the University of Maryland's Global Land Cover Facility, http://glcf.umiacs.umd.edu). NASA's "Geocover" set of global orthorectified Landsat data provides 1980s, 1990s and 2000, orthorectified to facilitate comparisons data circa of change over time (Tucker et al., 2004). Other moderate resolution sensors such as SPOT (Système pour l'Observation de la Terre), HRV (High Resolution Visible), and ASTER (Advanced Spaceborne Thermal Emission and Reflection Radiometer), are also available.

Most multispectral optical sensors use bands selected to respond to vegetation signals in the red and near infrared, so multispectral data are well-suited to the detection of wetland vegetation and mangroves. Spatial scale is a key factor in using different data sources (e.g. Ramsey and Laine, 1997). For example, marsh vegetation becomes increasingly sparse as the habitat degrades. Subpixel analysis of the amount of water, soil and vegetation can help to monitor "health" of a marsh (Kearney et al., 1995). The detection of different levels of habitat loss or degradation can be dependent on the scale of the source data. In temperate marshes, seasonal changes in vegetation (senescence and regrowth) are also important considerations. The more complex and heterogeneous the marsh, the more challenging it can be to get good classifications of wetland vegetation types and accurate estimates of change (Ramsey and Laine, 1997). As a class of coastal vegetation, coastal mangroves exhibit a strong multispectral signal and are well-suited to mapping from most multispectral satellite sensors, but discrimination of mangrove types can also be more challenging (Green et al., 1998).

Although numerous regional studies have mapping data that could be used for synthesis, there is only one nearly global classified landcover product at moderate resolution, the Geocover-LC product produced by Earthsat Corporation.

3.3.1 Geocover-LC

The Earth Satellite Corporation (EarthSat) contracted with the NASA in 1998 to generate an orthorectified satellite image base (Tucker et al., 2004), and then contracted with the National Imagery and Mapping Agency (NIMA) to produce a land cover map of the earth's landmasses. The GeoCover-LC product (Earth Satellite Corporation 2003) is the first moderate resolution land cover map of the world using Landsat Enhanced Thematic Mapper (ETM+) and Thematic Mapper (TM) data. EarthSat used a proprietary process to generate orthorectified tiles of ETM+ data for the Earth's landmasses with a 1.4 ha minimum mapping unit (accuracy equivalent to 1:200,000-scale maps). The first set of maps was generated utilizing 1990 imagery and covers North America and the Caribbean, Africa, Europe and east Asia. The second using 2000 data covers non-Amazonian South America, central and southeast Asia. Both datasets consist of a 13-class land cover legend. Updated coverages for northern Africa, Europe and east Asia were produced using 2000 data and a patented "Cross Correlation Analysis." This method detects change between the 1990 Landsat 4/5 images and the 2000 Landsat 7 images. The full Geocover classification is then reconstructed from the older data and the changed data for the update. Although the coverage was originally envisioned to be global, land cover has not been mapped for Australia, New Zealand, the Amazon, and parts of northern Eurasia. Because it was produced by a commercial entity, Geocover-LC products must be obtained under license.

The Geocover-LC scheme has an uncertain degree of validation, and studies that make use of the data in synthesis applications need to validate its use. For example, some preliminary data indicates that Geocover-LC classes for Caribbean islands may underestimate agricultural land uses, and so the influence of that bias on the use of the data in several ongoing analyses needs to be evaluated (J. Gebelein, L. Burke, and J. Robinson, unpubl data). Unlike the IGBP classification scheme, which does not specifically identify coastal wetland types, the Geocover-LC scheme includes a Wetland/Mangrove class in (Table 2).

3.3.2 Digital elevation models (SRTM)

The Shuttle Radar Topography (SRTM) mission flown in February 2000 collected radar interferometry data over 80% of the world's landcover (Werner, 2001). Data are still being processed and new digital elevation models (DEMs) are gradually being released through 2004 (Rabus et al., 2003). C-band (5.6 cm wavelength) synthetic aperture radar (SAR) data are processed by the NASA Jet Propulsion Laboratory, and products are distributed by the USGS (U.S. Geological Survey) Eros Data Center. Two near-global products are available for public release: 1-arc second (30 meter) SRTM data postings of the continental United States and 3-arc second (90 meter) international continental datasets (Farr and Kobrick, 2000). Data processing and public release is scheduled to be complete in 2005. Versions of standard topographic data (such as the U.S. Geological Survey GTOPO30) are also being updated based on new SRTM data. X-band interferometric SAR data will be processed by the German Remote Sensing Data Center of the German Aerospace Center (DLR). X-band has nearly twice the accuracy in measuring elevations, but a narrower swath width leaves gaps in coverage (Bamler, 1999). Nearly global 30 m data will be available from the DLR (http://www.eoweb.dlr.de) at cost of archiving and distribution (Werner, 2001).

3.4 MODERATE SPATIAL RESOLUTION (30 METER) GLOBAL OCEAN PRODUCTS

For coral reefs, multispectral sensors have been very successful in identifying geomorphological zones (forereef, reef crest, reef flat, lagoon, etc.) and coral reef habitat types for moderate complexity mapping (4-8 habitats) with 60-75% classification accuracy. Classification accuracy declines to a little over 53% for more than 13 classes (Andréfouët et al., 2003a, b). Accuracy of habitat classifications with Landsat data is systematically lower than with higher spatial resolution images, such as Ikonos (4m, Space Imaging, Inc.) data (Andréfouët et al., 2003a; Capolsini et al., 2003).

The proper spectral resolution for accurately distinguishing such important ecological end-members as coral and algae is not available on any currently orbiting satellite sensor (Hochberg and Atkinson, 2003). Thus, detailed assessments of the status of coral reef communities currently requires imaging spectroscopy from airborne platforms and cannot be effectively conducted over regional and larger scales (Mumby et al., 1997).

3.4.1 Millennium Coral Reefs - Global reef geomorphology maps from Landsat-7

By using a consistent dataset of moderate-resolution (30 meter) multispectral Landsat-7 images acquired between 1999 and 2003 (Gasch et al., 2000), a team led by the Institute for Marine Remote Sensing at the University of South Florida is developing the first high-resolution global map of shallow coral reef ecosystems. This effort is summarized at http://eol.jsc.nasa.gov/reefs. In addition to using a consistent data set processed with consistent methods, the project includes an unprecedented standardization of the geomorphological description of reefs around the world to build a globally consistent and rich classification scheme made of several hundred classes based on geomorphology, exposure, depth and continental/oceanic attributes. At present, the maps are being constructed scene-by-scene, with mosaics produced for selected locations (Figure 2).Completion of the mapping is expected in 2004, with product release to follow. More information is available at http://imars.marine.usf.edu/corals/index.html. The global archive of the underlying Landsat data was achieved in early 2004 and has been made available to the research community, see http://seawifs.gsfc.nasa.gov/cgi/landsat.pl.

3.5 HIGH SPATIAL RESOLUTION (≤ 20 METER) DATASETS AVAILABLE FOR MULTIPLE REGIONS

As studies move from the global to regional and local levels, higher spatial resolution datasets become important. It also becomes possible to purchase small amounts of satellite data or aerial remote sensing data. Multispectral satellite sensors remain key data sources for regional and local scale studies, including Landsat, ASTER (Advanced Spaceborne Thermal Emission and Reflection Radiometer), digital astronaut photography, IKONOS, and Quickbird. For coastal areas and reefs close enough to major population centers to support aircraft deployment, airborne acquisition of photographs, multispectral data, hyperspectral data, and LIDAR (light detection and ranging) provide additional data sources that can be combined to provide extensive information about coastal environments. The cost effectiveness of studies using aerial acquisition depends on both the costs of data acquisition and the cost of data processing. For example, although LIDAR data (see Chapter 7) can provide

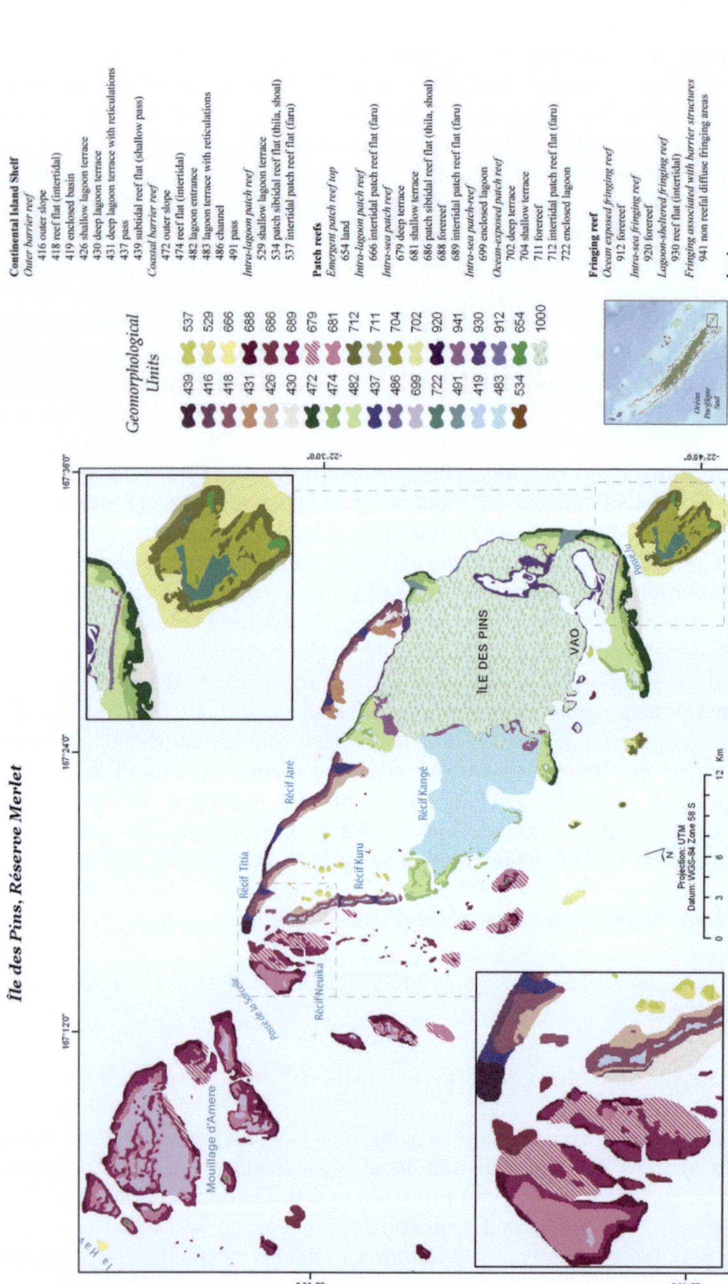

Figure 2. Example of Millennium coral reef map for Île des Pins, New Caledonia (Andréfouët and Torres-Pullizza, 2004), complete global products are scheduled for release at the end of 2004, including their integration in ReefBase.

outstanding mapping accuracy, it requires extensive post-processing which can make it prohibitively expensive for many users.

A discussion of the types of data sources available is beyond the scope of this chapter but is presented in Chapter 11. Here we focus on a few examples of local and regional projects that represent the types of activities being conducted for high spatial resolution coastal studies. The sensors used in these studies cover a wide range of geographic areas, include high spatial resolution information, and are available for public distribution and incorporation into further synthesis studies.

3.5.1 U.S. Coral Reef Taskforce mapping

The *Coral Reef Mapping Implementation Plan* of the U.S. Coral Reef Taskforce (1999) focused on both short- and long-term (>5 years) mapping needs for all reefs in U.S. states and territories. The working group pointed out the need for cost effective methods of mapping the baseline or current status of reefs, as well as the need for routine updates. Mapping activities to date have emphasized use of color aerial photography and IKONOS, with some use of airborne hyperspectral data in Hawaii (Table 3).

The USGS is a participating partner in the Mapping and Information Synthesis Working Group of the U.S. Coral Reef Taskforce (Rohmann, 2003). USGS activities have included work in the Hawaiian Islands, using an approach aimed at examining factors that influence coral reef health. In these studies, satellite sensors in combination with on-land digital camera stations are used to detect suspended sediments due to land run-off events. Together with aerial imaging, satellite remote sensing data are being used to map and study spatial distribution, transport patterns, and amount of sediment introduced onto the reef and the potential impact to the reefs. Airborne SHOALS (Scanning Hydrographic Operational Airborne LIDAR Survey) and digital cameras have also been used in the localized mapping studies. The addition of aerial acquisitions provides the opportunity to acquire several types of data simultaneously for later synthesis. For example, USGS scientists mapped reefs on southern Moloka'i, Hawaii using a combination of aerial photomosaics and bathymetry derived from the SHOALS instrument (Storlazzi et al., 2003).

3.5.2 Regional studies sponsored by The Nature Conservancy

The Nature Conservancy is collaborating with the International Institute of Tropical Forestry, U.S. Forest Service, and U.S. Geological Survey, to produce vegetation/land cover maps for all the islands of the Caribbean under the Caribbean Vegetation and Landcover Mapping Initiative. While the project is still in its development phases, the collection of existing vegetation maps is complete, forming the Caribbean Vegetation Atlas. A consistent classification system suited to Caribbean vegetation has been completed and mapping of the islands based on Landsat data is ongoing. The classification system and mapping efforts will produce updated vegetationmaps. At completion, all products are planned to be online (http://edcintl.cr.usgs.gov/tnc/project/project.html).

The Nature Conservancy has also supported a number of local resource management studies that have included assembling and interpreting data on coral reefs, coastal interfaces, and land cover, in efforts to improve coastal management. Targets have included the Bahamian Archipelago, (http://islands.bio.miami.edu/index.html), the Meso-American Reef system (Mexico, Belize, Guatemala, and Honduras), and Palmyra Atoll.

Table 3. Mapping activities under the U.S. Coral Reef Task Force Mapping Implementation Plan (1999) and status of the work (see also Rohmann, 2003). New mapping efforts will focus on initiating shallow-water mapping projects in the U.S. Freely Associated States, updating shallow-water maps of southern Florida, and developing maps of deeper water (> 30 m) habitats in the Northwestern Hawaiian Islands.

U.S. Territory	Data Source	Status
Southern Florida (including Florida Keys)	Color aerial photography; remap using a suite of technologies	30% in 1998, initiate 100% remap in 2004 to support specific coastal and fisheries management needs
U.S. Virgin Islands and Puerto Rico	Color aerial photography	Complete 2002
Hawaii, main islands	Color aerial photography, airborne hyperspectral, IKONOS	60% complete in 2003; contract established to complete remaining 40% by 2005
Northwestern Hawaiian Islands (including Midway Atoll)	IKONOS	Draft 2003; ongoing effort to collect field validation data
American Samoa	IKONOS	Draft 2003; Final expected in 2004
Comm. Northern Mariana Is.	IKONOS	Draft 2004; Final expected in 2004
Guam	IKONOS	Draft 2004; Final expected in 2004
Freely Associated Islands of Micronesia[1]	IKONOS; Landsat	Initiate mapping effort in 2004
Other U.S. Pacific Remote Islands[2]	IKONOS	pending

[1] The Republic of the Marshall Islands, the Federated States of Micronesia and the Republic of Palau.
[2] Wake, Johnston, and Palmyra Atolls, Kingman Reef, and Jarvis, Howland, and Baker Islands.

4. Analytical Challenges for Data Synthesis

Synthesis of different data types can occur at several different levels. The simplest is the assembly of different geographically referenced data layers into a common geographic information system (GIS) interface for further query or modeling. The typical challenges of any GIS project are encountered: gaining access to the needed data, getting data into a single software package in a common projection, and dealing with incongruence among source data once they have all been incorporated.

One critical challenge to the synthesis of different data types is the scale of the data. Frequently, after identifying sets of data that can be combined for analysis, one of those data sets will be at a much coarser spatial resolution than would be ideal. The one coarse dataset can limit the spatial scale of the synthesized data product, making it less relevant for addressing management questions.

Assumptions that went into methods to derive source data from different data types are often not carefully considered in studies that combine the best available data. The constraints of data availability and resources for analysis often require a "making due" approach to data sources. However we emphasize that, whenever possible, good practice includes evaluating the effect of any assumptions underlying source data on the results of the synthesis product.

Data synthesis can be more complex than combining data layers in a GIS model. New software tools for object oriented classification, such as eCognition, increase the accessibility of 2nd generation remote sensing techniques where images are divided into uniform regions at various scales prior to classification (Benz et al., 2004). Applications of object-oriented approaches to coastal management are now being emphasized, improving basic remote sensing classifications in local-scale studies. In one study, Kaya et al. (2002) used object-oriented classifications of multitemporal Radarsat-1 data to classify populated areas and wetlands, and then used the data in a GIS analysis of malaria risk in coastal Kenya. Beyond multitemporal data, object-oriented approaches can also be used to combine several datasets. For example, data layers from one map product could be used to divide a remotely sensed image into segments for subsequent classifications. The ability to use segmentation to separately evaluate areas with different inherent spatial scales will allow new approaches to data synthesis including remote sensing data (Schiewe et al., 2001).

5. Data Distribution and Impediments to Distribution

A key element in the global and regional sources of data discussed in this chapter is accessibility. Both technical and social issues can influence the accessibility of data for synthesis. Technical issues include issues of data volume, security, format, projection, Internet map server (IMS) technologies, and accessibility and quality of metadata. Social issues relate to data ownership, credit, and stake in synthesis products.

For regional and global studies, the size of datasets can be an issue impeding the distribution and use of data, especially for high spatial resolution datasets. Errors in data conversions or poor methodological choices can also increase the size of datasets, slowing subsequent analyses. For example, we have encountered enormous ESRI Shapefiles (Environmental Systems Research Institute, Inc.) where every pixel in the raster data had been converted into an individual polygon. The size of datasets is a particularly important issue for distribution of data online in IMS applications (including Arc/IMS, Demis Map Server, and open source GIS). IMS technology is often slow even in areas with high Internet bandwidth—but slow data transmission speeds can make online datasets completely inaccessible in areas with only dial-up access to the Internet (areas which may be most in need of access to shared data). Large global datasets also can have significant costs for data storage and distribution that can impact the ability to distribute data at no cost over the long term. Products distributed via DVDs or CDs also have cumulative costs of distribution.

Data format and projection can also be issues. Both proprietary and collaborative formats have key roles in the exchange of map information, with ESRI (1998) Shapefiles leading for vector data and open GeoTIFF standards (http://remotesensing.org/geotiff/geotiff.html) for raster data. A proprietary compression standard, MrSID (Multi-resolution Seamless Image Database) allows selective decompression and use of extremely large remote sensing images over networks. At present, a number of software packages can view MrSID files, but licensing to create the files is priced based on the amount of data encoded (http://www.lizardtech.com). MrSID has been used as the distribution format for orthorectified Landsat mosaics produced by Earthsat for NASA (Tucker et al.,2004), which are of benefit as base remote sensing information for hundreds of applications.

As Extensible Markup Language (XML) standards are developed for GIS applications (led by the OpenGIS Consortium, http://www.opengis.org) it will be easier for users to analyze and display multiple geospatial data types from multiple sources as maps (Schutzberg, 2000).

Although desktop GIS software is generally able to convert data among formats and projections, IMS technologies are less flexible in combining data in different projections, and work best with geographic projections. For integrating data, it can require intensive effort to convert data in a common projection, and to deal with artifacts from re-projection before different datasets can be displayed together seamlessly.

Good documentation of metadata, both in the source products and in derived products, is important for insuring that different data are used appropriately. Without documentation of methods of production, it is difficult to evaluate how the incorporated assumptions affect the quality of the product and its application to a particular problem that may not have been envisioned by the original developer. In particular, issues of spatial scale and assumptions underlying class definitions are important in making wise use of shared data.

A key element for being able to combine multiple datasets is the accessibility of data and data products from a variety of sources. To protect data ownership, many products are now available for viewing on the Web using (IMS) Internet map server technology; however, many fewer products are available for download and incorporation into new analyses. Collaborative agreements may need to be negotiated with owners of data products before the data can be used independently, and depending on the needs of the original producer may be focused on scientific credit, control of data products, or commercial interests. Both commercial and noncommercial data sources may be seeking to recoup the costs of archive maintenance and data distribution, and this can limit the accessibility of good global datasets. For example, commercial products such as Geocover-LC are affordable for use in local studies, but purchasing hundreds of scenes would be cost-prohibitive for regional or global studies.

5.1 DATA SYNTHESIS AND DISTRIBUTION EXAMPLE: REEFS AT RISK AND REEFBASE

Led by the World Resources Institute, the Reefs at Risk in Southeast Asia project had three primary goals: 1) to raise awareness about threats to coral reefs in Southeast Asia and the linkages between human activity and coral reef condition; 2) to develop a standardized indicator of likely threats to coral reefs from human activities; and 3) to promote sharing and improvement of information about coral reefs through data integration and distribution. Reefs at Risk in Southeast Asia (Burke et al., 2002) was the first regional study to follow the global Reefs at Risk Assessment (Bryant et al., 1998), and will be followed by other regional studies using the same approach.

In the Reefs at Risk in Southeast Asia project, information about coral reef status was quite limited for most locations in the region. After assembling the best available information on coral reef status and threats to coral reefs for the region, a major component of the project involved modeling threats to coral reefs to produce a map-based indicator that identified areas where coral reef degradation might be expected to occur (or had already occurred) given current levels of human activity. The indicator also considered mitigating factors such as effective management and the natural vulnerability of a location to pollution and sediment.

Using a Geographic Information System (GIS) and over 25 input data layers, the analysis estimated pressure on coral reefs due to threats from coastal development, marine-based pollution, sedimentation from inland sources, overfishing, and destructive fishing (Figure 3). Data sets that served as input to the threat models included: land cover type, elevation and slope, precipitation, and soil type (for the analysis of sedimentation); population density and coral reef area (for the analysis of overfishing pressure); cities, settlements, population growth, mines, and tourist centers (for the analysis of coastal development); and ports, oil wells and rigs, and shipping lanes (for the analysis of marine-based pollution).

During the project, collaborators expanded and improved the data sets on coral reef locations, coral condition, coral bleaching, marine protected areas, management effectiveness of protected areas, tourism pressures, and destructive fishing. To assemble the needed data, collaborators included eleven national research institutions and universities, three non-governmental organizations in Southeast Asia, and eleven international collaborators.

World Resources Institute has also just completed a Reefs at Risk regional analysis for the Caribbean (Burke and Maidens, 2004; see Figure 1B for a sample product). Although this project built on the foundation of previous datasets and collaborations, several new challenges had to be addressed, including accessing higher spatial resolution data for islands in the Lesser Antilles, and acquiring copies of updated coral reef maps prior to their release by the Millennium Coral Reef Mapping project. New partners working in each region were sought out to verify and validate mapping data and derived products. For example, the Reefs at Risk team held a threat assessment workshop to engage researchers, regional non-governmental organizations, and resource managers in the effort, and to incorporate their recommendations into the product.

By assembling a variety of data sets together and acting as a catalyst to draw information from numerous partners, the Reefs at Risk project is able to synthesize disparate data and produce assessments of the risk of human activities to coral reefs. These assessments are useful to managers at scales from local to global, and are unavailable through any other process. Specific actions are recommended in the final reports. For example, for the Caribbean recommendations were made for ways to create public interest in protecting reefs, for improving local and national expertise for managing coral reef ecosystems, for improving management approaches, and for appropriate international conservation action (Burke and Maidens, 2004).

Through collaboration with the ReefBase project (McManus and Noordeloos, 1998; Robinson et al., 2000) the Reefs at Risk products are available in an online GIS (Figure 1; http://www.reefbase.org) and can be integrated with a variety of other coral reef datasets previously discussed. Some of these include NOAA coral hotspots maps and the UNEP-WCMC data on protected areas, mangroves, and coral reefs, SeaWiFS shallow water bathymetry, and global coastline data. In addition to remote sensing and cartographic data, ReefBase also hosts *in situ* data collected by the Global Coral Reef Monitoring Network (http://www.gcrmn.org/index.html). ReefBase will also serve as a repository for the Millennium Global Coral Reef Maps when they are complete, and will link to sources of download of remote sensing imagery from Landsat, human spaceflight photography, and other shared data. Thus the ReefBase project provides interactive access to a variety of data products, as well as a guide for accessing source data when needed.

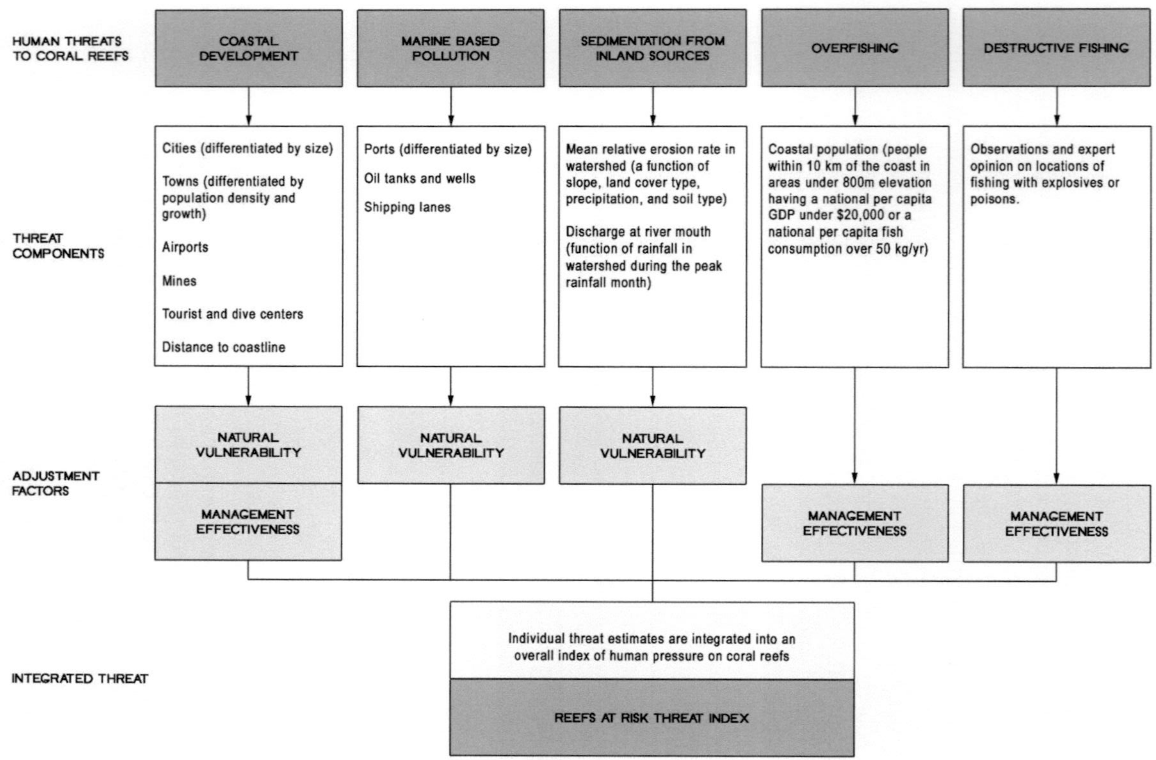

Figure 3. Overview of the Reefs at Risk Threat model indicating the information used to address the different themes. For more information, see Burke et al. (2002).

6. Conclusions

There is a great and ongoing need for synthesis of complementary datasets to answer fundamental questions in coastal ecosystem management. These questions cannot be answered using a single data source or field-based study. In particular, those seeking to link data from land to sea have a strong need to use a variety of data sources to understand the processes of change.

Data synthesis is becoming more tractable due to the availability of GIS software that can readily combine raster and vector data, and use newer approaches such as object-oriented classification. Although there are numerous hurdles to acquiring and adapting available global and regional datasets to synthesis studies, the number of datasets and their accessibility is improving. Studies incorporating data synthesis methods have the potential to transform our understanding of the function of coastal and reef ecosystems, and the risks created by human activities. Synthesis of data types is the mechanism for implementing analyses in support of an integrated coastal management model (Cicin-Sain, 1993; Vallega, 1993). By providing analysis that is more than the sum of the parts, it can lead to improved focus for management activities and future research.

7. Acknowledgements

Support was provided by NASA through "Using Landsat 7 Data in a GIS-based Revision of ReefBase (A Global Database on Coral Reefs and Their Resources): Distributing Information on Land Cover and Shallow Reefs to Resource Managers," NASA Office of Earth Sciences, Earth Science Enterprise, Carbon Applications. 2001-2004. and "A Millennium Global Scale Coral Reef Ecosystem Assessment using High Resolution Remote Sensing Data". NASA Office of Earth Sciences, Earth Science Enterprise, Interdisciplinary Science. 2001-2004. We thank all the partners and collaborators on these projects, especially G. Feldman, N. Kuring, J. Maidens, M. Noordeloos, and R. Stumpf. We appreciate information on NOAA Coral Reef Taskforce activities provided by S. Rohmann, and background information on land cover datasets shared by J. Gebelein. A. Spraggins, F. Muller-Karger, D. Torres, and C. Kranenburg are all critical players in producing and sharing the Millenium Coral Reef maps that are at the center of the collaborations that led to this review. Last, but certainly not least, we thank Jennifer Rhatigan, Cindy Evans, and one anonymous reviewer for comments that improved the work.

8. References

Andréfouët S., E.J. Hochberg,, C. Chevillon, F.E. Muller-Karger, J.C. Brock, and C. Hu. 2004. Multi-scale integration of remote sensing tools to understand physical and biological processes in coastal environments: examples on coral reefs. In: Remote Sensing of Coastal Aquatic Environments, R.L.Miller, C.E. Del Castillo, and B.A. McKee (Editors). Kluwer Academic Publishers, in press.

Andréfouët, S., P. Kramer, D. Torres-Pulliza, K.E. Joyce, E.J. Hochberg, R. Garza-Perez, P.J. Mumby, B. Riegl, H. Yamano, W.H. White, M. Zubia, J.C. Brock, S.R. Phinn, A. Naseer, B.G. Hatcher, and F.E. Muller-Karger. 2003a. Multi-site evaluation of IKONOS data for classification of tropical coral reef environments. Remote Sensing of Environment, 88:128-143.

Andréfouët, S., and B. Riegl. 2004. Remote sensing: a key-tool for interdisciplinary assessment of coral reef processes. Coral Reefs, in press.

Andréfouët, S., J.A. Robinson, C. Hu, B. Salvat, C. Payri, and F.E. Muller-Karger. 2003b. Influence of the spatial resolution of SeaWiFS, Landsat 7, SPOT and International Space Station data on landscape parameters of Pacific Ocean atolls. Canadian Journal of Remote Sensing, 29:210-218.

Andréfouët S. and D. Torres-Pullizza. 2004. Atlas des récifs coralliens de Nouvelle-Calédonie, IFRECOR Nouvelle-Calédonie, IRD, Nouméa.

Bamler, R. 1999. The SRTM Mission – a world-wide 30 m resolution DEM from SAR interferometry in 11 days. Photogrammetrische Woche, Universitaet Stuttgart [Universitaet Stuttgard, Photogrammetric Week] 1999(47):145-154.

Belward, A., J. Estes, and K. Kline. 1999. The IGBP-DIS Global 1-km Land-Cover Data Set DISCover: A Project Overview. Photogrammetric Engineering and Remote Sensing, 65(9):1013-1017.

Berkelmans, R., G. De'ath, S. Kininmonth, and W. Skirving. 2004. A comparison of the 1998 and 2002 coral bleaching events on the Great Barrier Reef: spatial correlation, patterns and predictions. Coral Reefs, in press.

Benz, U.C., P. Hoffmann, G. Willhauck, I. Lingenfelder, and M. Heynen. 2004. Multi-resolution, object-oriented fuzzy analysis of remote sensing data for GIS-ready information. ISPRS Journal of Photogrammetry and Remote Sensing, 58:239-258.

Bookman, C.A., T.J. Culliton, and M.A. Warren. 1999. Trends in U. S. Coastal Regions, 1970–1998. U.S. Department of Commerce, National Oceanic and Atmospheric Administration, National Ocean Service, Special Projects Office, Silver Spring, Maryland.

Bryant, D., L. Burke, J. McManus, and M. Spalding, 1998. Reefs at Risk: A Map-based Indicator of Threats to the World's Coral Reefs. World Resources Institute, International Center for Living Aquatic Resources Management, World Conservation Monitoring Centre, and United Nations Environment Programme, Washington D.C.

Burke, L., Y. Kura, K. Kassem, C. Revenga, M. Spalding, and D. McAllister. 2000. Pilot analysis of global ecosystems (PAGE): Coastal ecosystems. World Resources Institute, Washington, D.C.

Burke, L. and J. Maidens. 2004. Reefs at Risk in the Caribbean. World Resources Institute, Washington D.C.

Burke, L., E. Selig, and M. Spalding, 2002. Reefs at Risk in Southeast Asia. World Resources Institute, Washington D.C.

Capolsini, P., S. Andréfouët, C. Rion, and C. Payri. 2003. A comparison of Landsat ETM+, SPOT HRV, Ikonos, ASTER and airborne MASTER data for coral reef habitat mapping in South Pacific islands. Canadian Journal of Remote Sensing 29(2):187-200.

CIESIN (Center for International Earth Science Information Network), International Food Policy Research Institute, and World Resources Institute. 2000. Gridded Population of the World, Version 2. Columbia University, Palisades, New York.

Cicin-Sain, B. 1993. Sustainable development and integrated coastal management. Ocean & Coastal Management, 21:11-43.

Cimino, J., L. Pruett, and H. Palmer. 2000. Management of Global Maritime Limits and Boundary Using Geographical Information Systems. Integrated Coastal Zone Management, ICG Publishing Ltd [Henley Media Group] Spring 2000:91-97.

Dahdouh-Geubas, F. 2002. The use of remote sensing and GIS in the sustainable management of tropical coastal ecosystems. Environment, Development, and Sustainability, 4:93-112.

Dobson, J.E., E.A. Bright, P.R. Coleman, R.C. Durfee, and B.A. Worley, 2000. A Global Poulation Database for Estimating Population at Risk. Photogrammetric Engineering & Remote Sensing 66(7):849-857.

Duggin, M.J., and C.J. Robinove, 1990. Assumptions Implicit in Remote Sensing Data Acquisition and Analysis, International Journal of Remote Sensing 11(10):1669-1694.

Esaias, W.E., M.R. Abbott, I. Barton, O.B. Brown, J.W. Campbell, K.L. Carder, D.K. Clark, R.L. Evans, F.E. Hoge, H.R. Gordon, W.P. Balch, R. Letelier, P.J. Minnett. 1998. Overview of MODIS Capabilities for Ocean Science Observations. IEEE Transactions on Geoscience and Remote Sensing 36:1250-1265.

ESRI. 1998. ESRI Shapefile technical description. ESRI Corporation, Redlands, California. Available online at http://www.esri.com/library/whitepapers/pdfs/shapefile.pdf

Farnsworth, E.J. and A.M. Ellison. 1997. The global conservation status of mangroves. Ambio 26:328-334.

Farr, T.G. and M. Kobrick. 2000. Shuttle Radar Topography Mission produces a wealth of data, Eos, Transactions of the American Geophysical Union 81:583-585.

Friedl, M.A., D.K. McIver, J.C.F. Hodges, X. Zhang, D. Muchoney, A.H. Strahler, C.E. Woodcock, S. Gopal, A. Schnieder, A. Cooper, A. Baccini, F. Gao, and C. Schaaf. 2002. Global land cover from MODIS: Algorithms and early results, Remote Sensing of Environment, 83:135-148.

Gasch, J., T. Arvidson, S.N. Goward, S. Andréfouët, C. Hu, and F.E. Müller-Karger. 2000. An Assessment of Landsat 7/ETM+ Coverage of Coral Reefs Worldwide. In: Proceedings of the International Geoscience and Remote Sensing Symposium (IGARSS 2000). IEEE, Honolulu, Hawaii.

Green, E.P., C.D. Clark, P.J. Mumby, A.J. Edwards, and A.C. Ellis. 1998. Remote sensing techniques for mangrove mapping. International Journal of Remote Sensing, 19:935-956.

Green, E.P., P.J. Mumby, A.J. Edwards, C.D. Clark, and A.C. Ellis. 1997. Estimating leaf area index of mangroves from satellite data. Aquatic Botany, 58:11-19.

Green, E.P., P.J. Mumby, A.J. Edwards, and C.D. Clark. 1996. A review of remote sensing for the assessment and management of tropical coastal resources. Coastal Management, 24:1-40.

Green, E.P., P.J. Mumby, A.J. Edwards, C.D. Clark. 2000. Remote sensing handbook for tropical coastal management. Coastal Management Sourcebooks 3, UNESCO, Paris.

Green, E.P. and F.T. Short. 2003. World atlas of seagrasses. University of California Press, Berkeley.

Green, E., R. Wood, R.P. Stumpf, G.C. Feldman, N. Kuring, B. Franz, A. Holt, C. Ravillious, J. Oliver, and J. A. Robinson. All our eggs in one basket: the present state of tropical marine biodiversity conservation. In review, *Coral Reefs*.

Hinrichsen, D. Coasts in crisis. Issues in Science and Technology 12(4):39-47.

Hochberg, E.J. and M.J. Atkinson. 2003. Capabilities of remote sensors to classify coral, algae, and sand as pure and mixed spectra. Remote Sensing of Environment, 85:174-189.

Hodgson, G. 1999. A global assessment of human effects on coral reefs. Marine Pollution Bulletin, 38:345-355.

Hoegh-Guldberg, O. 1999. Climate change, coral bleaching and the future of the world's coral reefs. Marine and Freshwater Research, 50:839-866.

Husak, G.J., B. Hadley, and K. McGwire. 1999. The IGBP-DIS Global 1-km Land-Cover Data Set DISCover: a project overview, Photogrammetric Engineering and Remote Sensing, 65(9):1033-1039.

Kaya, S., T.J. Pultz, C.M. Mbogo, J.C, Beier, and E. Mushinzimana. 2002. The use of radar remote sensing for identifying environmental factors associated with malaria risk in Coastal Kenya. In*:* Proceedings of the International Geoscience and Remote Sensing Symposium (IGARSS 2002). IEEE, Toronto.

Kearney, M.S., A.S. Rogers, J.T. Townshend, W.L. Lawrence, K. Dorn, K. Eldred, D. Stutzer, F. Lindsay, E. Rizzo. 1995. Developing a model for determining coastal marsh "health." Proceedings of the Third Thematic Converence on Remote Sensing for Marine and Coastal Environments, Seattle, Washington II:527-537.

Kelleher, G., C. Bleakley, and S. Wells. 1995. A Global Representative System of Marine Protected Areas, Volume 1. World Bank, Washington, D.C.

Kennish, M.J. 2000. Anthropogenic impacts and the National Estuary Program. In*:* Estuary Restoration and Maintenance. CRC Press, Boca Raton, Florida.

K.A. Kilpatrick, G. Podesta, and R. Evans. 2001. Overview of the NOAA/NASA Advanced Very High Resolution Radiometer Pathfinder algorithm for sea surface temperature and associated matchup database. Journal of Geophysical Research, 106:9179-9197.

Lindén, O. 1990. Human impact on tropical coastal zones. Nature and Resources, 26:3-11.

Liu, G., W. Skirving, and A.E. Strong. 2003. Remote sensing of sea surface temperatures during 2002 Barrier Reef coral bleaching. Eos, Transactions of the American Geophysical Union 84(15):137-144.

Loveland, T., Z. Zhu, D. Ohlen, J. Brown, B. Reed, and L. Yang. 1999. An Analysis of the IGBP Global Land-Cover characterization process, Photogrammetric Engineering and Remote Sensing 65(9): 1021-1032.

Loveland, T.R., B.C. Reed, J.F. Brown, D.O. Ohlen, Z. Zhu, L. Yang, and J. Merchant. 2000. Development of a Global Land Cover Characteristics Database and IGBP DISCover from 1-km AVHRR data. International Journal of Remote Sensing, 21(6-7):1303–1330.

Mackinnon, J. 1997. Protected Areas systems review of the Indo-Malayan Realm. Asian Bureau for Conservation, Canterbury, UK.

McIver, D.K. and M. A. Friedl. 2002. Using prior probabilities in decision-tree classification of remotely sensed data, Remote Sensing of Environment, 81:253-261.

McManus, J.W. 1997. Tropical Marine Fisheries and the Future of Coral Reefs: A Brief Review with Emphasis on Southeast Asia. Coral Reefs 16(Suppl.):S121-S127. *Also* Proceedings of the Eighth International Coral Reefs Symposium 1:229-234.

McManus, J. W. and M. Noordeloos. 1998. Toward a Global Inventory of Coral Reefs (GICOR): remote sensing, international cooperation and *ReefBase* Proceedings of the Fifth International Conference on Remote Sensing of the Marine Environment, I:83-89.

Mumby, P.J., E.P. Green, A.J. Edwards, and C.D. Clark. 1997. Coral reef habitat-mapping: how much detail can remote sensing provide? Marine Biology, 130:193-202.

Mumby, P.J., E.P. Green, A.J. Edwards, and C.D. Clark. 1999. The cost effectiveness of remote sensing for tropical coastal resources assessment and management. Journal of Environmental Management, 55: 157-166.

Mumby, P., W. Skirving, A. Strong, J. Hardy, E. LeDrew, E. Hochberg, R. Stumpf, and L. David. 2004. Remote sensing of coral reefs and their physical environment. Marine Pollution Bulletin, 48(3-4): 219-228.

National Research Council, Mapping Science Committee, Committee on National Needs for Coastal Mapping and Charting. 2004. A geospatial framework for the coastal zone: national needs for coastal mapping and charting. National Academies Press, Washington, D.C.

Nicholls, R.J., E.M.J. Hoosemans, and M. Marchand. 1999. Increasing flood risk and wetland losses due to global sea-level rise: regional and global analyses. Global Environmental Change, 9:S69-S87.

NIMA (National Imagery and Mapping Agency). 1999. Performance Specification: World Vectors Shoreline (WVSPLUS®), MIL-PRF-0089012A. NIMA, Washington, D.C..

Olson, D.M., E. Dinerstein, R. Abell, T.F. Allnutt, C. Carpenter, L. McClenachan, J. D'Amico, P. Hurley, K. Kassem, H. Strand, M. Taye and M. Thieme. 2000. The Global 200: a representation approach to conserving the Earth's distinctive ecoregions. World Wildlife Fund, Washington, D.C.

Olson, D.M., E. Dinerstein, E.D. Wikramanayake, N.D. Burgess, G.V.N. Powell, E.C. Underwood, J.A. D'Amico, I. Itoua, H.E. Strand, J.C. Morrison, C.J. Loucks, T.F. Allnutt, T.H. Ricketts, Y. Kura, J.F. Lamoreux, W.W. Wettengel, P. Hedao, and K.R. Kassem. 2001. Terrestrial ecoregions of the world: a new map of life on Earth. Bioscience, 51(11):933-938.

Petit, C. and E.F. Lambin. 2001. Integration of multi-source remote sensing data for land cover change detection. International Journal of Geographical Information Science, 15(8):785-803.

Rabus, B., M. Eineder, A. Roth, and R. Bamler. 2003. The shuttle radar topography mission- a new class of digital elevation models acquired by spaceborne radar. ISPRS Journal of Photogrammetry and Remote Sensing, 57:241-262.

Ramsey, E.W., III, and S. C. Laine. 1997. Comparison of Landsat Thematic Mapper and high resolution photography to identify change in complex coastal wetlands. Journal of Coastal Research 13:281-292.

Reynolds, R.W., 1988: A real-time global sea surface temperature analysis. Journal of Climate, 1:75-86.

Reynolds, R.W., N.A. Rayner, T.M. Smith, D.C. Stokes, and W. Wang. 2002. An improved in situ and satellite SST analysis for climate. Journal of Climate, 15:1609-1625.

Robinson, J.A., G.C. Feldman, N. Kuring, B. Franz, E. Green, M. Noordeloos, and R.P. Stumpf. 2000. Data fusion in coral reef mapping: working at multiple scales with SeaWiFS and astronaut photography. Proceedings of the Sixth International Conference on Remote Sensing for Marine and Coastal Environments, 2:473-483.

Rohmann, S. 2003. Mapping moderate depth habitats of the U.S. Pacific Islands with emphasis on the Northwestern Hawaiian Islands: an implementation plan, Version 2. NOAA National Ocean Service, Special Projects Office, Silver Spring, Maryland. Available online at http://biogeo.nos.noaa.gov/projects/mapping/Pacific_mod_depth_MIP.pdf

Scepan, J. 1999. Thematic validation of high-resolution global land-cover data sets, Photogrammetric Engineering and Remote Sensing, 65:1051-1060.

Schiewe, I.J., G.L. Tufte, and I.M. Ehlers. 2001. Potential and problems of multi-scale segmentation methods in remote sensing. GIS 06/01:34-39.

Schutzberg, A. 1999. XML, GIS and you. GIS Vision, September 2000. Available online at http://www.gisvisionmag.com/GISVision/vision.php?article=/GISVision/Review/XML.html

Short, F. 1999. "Global Seagrass Survey Results 1997–1998." On-line at http://www.seagrass.unh.edu. Last viewed 30 October 2003.

Spalding, M., F. Blasco, and C. Field. 1997. World mangrove atlas. International Society for Mangrove Ecosystems, Okinawa, Japan.

Spalding, M., C. Ravilious, and E.P. Green. 2001. World atlas of coral reefs. University of California Press, Berkeley.

Storlazzi, C., J. Logan, and M. Field. 2003. Quantitative morphology of a fringing reef tract from high resolution laser bathymetry: Southern Molokai, Hawaii. Geological Society of America Bulletin 115(11):1344-1355.

Strong, A.E., C.B. Barrientos, C. Duda, and J. Sapper. 1997. Improved satellite techniques for monitoring coral reef bleaching. Proceeding of the Eighth International Coral Reef Symposium, Panama City, Panama, pp. 1495-1498.

Strong, A.E., E. Kearns, and K.K. Gjovig. 2000. Sea surface temperature signals from satellites - an update. Geophysical Research Letters, 27(11):1667-1670.

Stumpf, R.P., K. Holderied, and M. Sinclair. 2003a. Determination of water depth with high-resolution satellite imagery over variable bottom types. Limnology and Oceanography, 48:547-556.

Stumpf, R.P., K. Holderied, J.A. Robinson, G. Feldman, and N. Kuring. 2003b. Mapping water depths in clear water from space. Coastal Zone 03, Baltimore, Maryland, 13-17 July 2003. Available online at http://eol.jsc.nasa.gov/newsletter/CoastalZone/default.htm.

WCMC (World Conservation Monitoring Centre) and IUCN World Comission on Protected Areas. 1998. 1997 United Nations List of Protected Areas. IUCN, Gland, Switzerland.

Toscano, M.A., K.S. Casey, and J. Shannon. 2002a. Use of high resolution Pathfinder SST data to document coral reef bleaching. Proceedings of the Seventh International Conference on Remote Sensing of Marine and Coastal Environments, Miami, Florida.

Toscano, M.A., G. Liu, I.C. Guch, K.S.Casey, A.E. Strong, and J.E. Meyer. 2002b. Improved prediction of coral bleaching using high-resolution HotSpot anomaly mapping. Proceedings of the Ninth International Coral Reef Symposium, Bali, Indonesia, pp. 1143-1148.

Tucker, C.J., D.M. Grant, and J.D. Dykstra. 2004. NASA's global orthorectified Landsat data set. Photogrammetric Engineering and Remote Sensing, 70:313-322.

U.S. Coral Reef Task Force, Mapping and Information Synthesis Working Group. 1999. Coral Reef Mapping Implementation Plan (2nd Draft). NOAA, NASA and USGS, Washington, DC. Available online at http://coralreef.gov/MIP.pdf .

Vallega, A. 1993. A conceptual approach to integrated coastal management. Ocean & Coastal Management, 21:149-162.

Werner, M. 2001. Status of the SRTM data processing: when will the world-wide 30m DTM data be available? Photogrammetrische Woche, Universitaet Stuttgart [Universitaet Stuttgard, Photogrammetric Week] 2001:159-169.

Wessel, P., and W.H.F. Smith. 1996. A global self-consistent, hierarchical, high-resolution shoreline database. Journal of Geophysical Research, 101(B4):8741-8743.

Wooldridge, S., and T. Done. 2004. Learning to predict large-scale coral bleaching from past events: A Bayesian approach using remotely sensed data, in-situ data, and environmental proxies. Coral Reefs, in press.

Chapter 13

RECOMMENDATIONS FOR SCIENTISTS AND MANAGERS FOR APPLICATION OF REMOTE SENSING TO COASTAL WATERS

ELLSWORTH F. LEDREW[1] AND LAURIE L. RICHARDSON[2]
[1] *Department of Geography, University of Waterloo, Waterloo Ontario, N2L 3G1, Canada*
[2]*Department of Biological Sciences, Florida International University, Miami, Florida 33199 USA*

1. Introduction

The science and engineering capabilities of remote sensing have evolved considerably over the slightly more than three decades since the 1972 launch of the then-called ERTS, now known as Landsat. In the early years, the expertise, resources, and rapidly expanding knowledge base necessary to 'decipher' remote sensing imagery was not readily obvious to either the scientist or the manager. In our experience, in these first years scientists and manangers assumed that ordering costly satellite data involved simply noting which 'product' was desired – for example, a fully atmospherically corrected and georegistered, processed, and classified map that indicated the spatial distribution of a desired surface cover. Of course, this scenario is only now becoming a reality.

This lack of connectivity between the scientist or manager and production of an expertly produced remote sensing product was particularly evident through the first two decades of the widespread availability of satellite data. This resulted in a backlash. When the technology, including user friendly image processing software and atmospheric correction packages, was eventually ready, it was then difficult to get the prospective users of this tool to explore the true potential.

We are now at a very interesting crossroads where the capabilities of the tool are very promising indeed. At the same time, environmental problems confronting the users (both scientists and managers) require the type of data that can be derived from the synoptic and repetitive coverage of the earth's surface and atmosphere by satellite based instrumentation. We now have numerous case studies for which remotely sensed data are an *operational* component of the analysis of environmental issues at regional, and even local, scales. Additionally, there is now a clear understanding of the limitations and opportunities that remote sensing affords and a move away from 'overselling' the product. As an added bonus, the cost of remote sensing data has moved from prohibitive to reasonable.

The growing acceptance of remote sensing as critical factor in environmental analysis is reflected in the ongoing development of the Global Earth Observations System of Systems (GEOSS). GEOSS evolved out of the World Summit on Sustainable Development in Johannesburg, 2002. That Summit highlighted the urgent need for coordinated observations relating to the state of the Earth. At the invitation of the United States, thirty-three nations and the European Commission joined together at the first Earth Observation Summit on 31 July 2003 in Washington DC (http://www.earthobservationsummit.gov/index.html) to adopt a Declaration that called for action in strengthening global cooperation on Earth observations. The Washington

L.L. Richardson and E.F. LeDrew (eds.), Remote Sensing of Aquatic Coastal Ecosystem Processes: Science and Management Applications, 307-315.

Summit Declaration establishes the objective *"to monitor continuously the state of the Earth, to increase understanding of dynamic Earth processes, to enhance prediction of the Earth system, and to further implement our international environmental treaty obligations",* and thus the need for *"timely, quality, long-term, global information as a basis for sound decision making*" (ibid). Thus, the purpose of the Summit was to promote the development of a *comprehensive, coordinated, and sustained* Earth observation 'system of systems' among governments and the international community. The objective is to understand and address global environmental and economic challenges and benefits of an international global observation system. A process is in place to develop a conceptual framework and implementation plan for building this comprehensive, coordinated, and sustained Earth observation system of systems.

Of particular interest for readers of this book is the creation of a Users Interface Working Group acknowledging the operational nature of the System of Systems. A primary rationale is "…GEOSS benefits will be realized by a broad range of user communities, including managers and policy makers in the targeted societal benefit areas, scientific researchers and engineers, civil society, governmental and non-governmental organizations and international bodies" (*User Interface Proposal,GEOSS working document, April 2005*). GEOSS is poised to be the source for operational multiplatform data that can be used in day-to-day environmental analysis. There is great promise for confident use of remote sensing in a variety of environmental management strategies.

The rationale for this book, which began at about the time of the Johannesburg summit, was our belief that remote sensing has matured to the point where current image data are of sufficient spatial and spectral resolution to allow identification of remote sensing 'fingerprints' of importance to coastal ecosystems. Detection of these fingerprints in turn allows detection of change or variability in such ecosystems at a level that has not been previously possible. We believe that remote sensing instrumentation, technology, and image analysis can be used to conduct quantitative science, and can be used to interpret an existing historical archive of image data that, in some cases, extends back in time over 30 years (e.g. Landsat imagery). Even if such earlier data are not of sufficient spectral or spatial resolution to allow identification of 'fingerprints', data fusion techniques can link the two (current and past) data sets to provide valuable information on the nature, rate, and location of change. When these data are combined with data sets such as sedimentation rates, deforestation extent and timing, extent of pollution episodes, etc., we envision a powerful integrated data tool that can be used to derive meaningful policy regarding coastal ecosystems.

The coastal ecosystem is affected by natural and anthropogenic processes that can cause perturbations at a variety of scales in both time and in space. We need to demonstrate the realistic value of the use of remotely sensed data, in particular when used in combination with traditional *in situ* data, to understand these perturbations at the process level as well as to track such changes over time. We see this approach as crucial for providing credible information for development and validation of management policy. While similar studies have been completed and are widely used for remote sensing applications in land based vegetation systems and blue water oceanography, the remote sensing of coastal ecosystems has lagged behind. This is, in large part, due to the fact that coastal aquatic systems present notable challenges, for example highly complex and interacting spectral signatures, as should be evident from reading this book. However, research in this area has been very active over the past decade and is yielding important new findings and success stories as we write.

In this book we have examined the role of remote sensing as applied to the dynamics and variability of the coastal ecosystem. We have included an assessment of the current status of remote sensing instrumentation and analytical approaches available for measurement of coastal "fingerprints", supported by real case studies with proven results. The potential and role of integrated data systems (current remote sensing capabilities supplemented with archived data and parallel data sampling) for use in policy development and implementation are also addressed and supported with specific examples of successful implementations of this approach. Finally the individual authors of the chapters in this book have surveyed the plans for development of satellite-borne sensors specifically designed for use in the coastal zone, and assessed the opportunities for improving our understanding of coastal ecosystem processes and effectively managing coastal habitats using satellite data.

2. Science Applications

To summarize, for those potential readers who may still be hesitant to tackle the technologically sophisticated field of remote sensing, we have included within the 'Science Applications' section of this book four papers that exemplify the state-of-the-art in integrating remote sensing and science. These are:

- the development of a near real-time index ("Ocean hot spots") that maps the potential for coral bleaching on a global basis (Skirving et al.)
- a thorough analysis of estimating chlorophyll a concentrations in coastal waters where the nature of the signal interactions are considerably more complex than in Case 1 or 'blue' waters (Schalles)
- the theory and practical application of a modeling tool for spectral analysis in coastal and deep waters (Gege and Albert), with the program available at http://extras.springer.com
- an examination of the actual biological processes of coral reefs and the potential for analysis and scaling up with remote sensing (Brock et al.).

The global 'bleaching warning' products described by Skirving et al. (Chapter 2) have been validated in the 'real-world' experiment by providing notice of the 1998 bleaching on the Great Barrier Reef before the local management was aware of it. This has led to a formal alliance between NOAA of the United States, who developed the products, and The Australian Institute for Marine Science (AIMS) and the Great Barrier Reef Marine Park Authority (GBRMPA) to further refine the indices and build them into the ongoing management of the Reef. These products are now available on-line through the Coral Reef Watch and include the current Degree Heating Weeks (DHW) and maximum DHW since the start of the historical analysis in 1985. Managers are alerted by a warning symbol when the 'Hot Spot' value is greater than zero. The symbol grows and size and flash intensity as the severity of the heat stress increases over the season. Current work in progress includes 1 to 4 week forecasts and improved products for climate modeling. This surely is a clear demonstration of the value of satellite based data for day-to-day resource management.

Schalles (Chapter 3) reviewed the considerable knowledge of radiative transfer in Case 1, or 'blue' waters. The bio-optics are largely based upon the interaction of radiation with living phytoplankton cells, organic tripton or detritus from the death and decay of phytoplankton cells, and dissolved organic matter produced by phytoplankton metabolism and decay of organic tripton. Case 2 or shallow coastal waters have, in addition to these constituents, the material suspended by turbulent interaction with the sea floor, inorganic and organic tripton originating from the land surface, colored dissolved organic matter (CDOM), and anthropogenic particulate and dissolved materials. Clearly the physics of radiant interaction become significantly more complex and the effect will reflect the diurnal and seasonal variability of processes along the dynamic transition between land and water. Schalles provided a detailed and extensive explanation of the optical issues, reviewed a wide range of relevant observational spectra, and provided detailed explanations for specific spectral responses. He concludes that spectral remote sensing of coastal waters is an operational, albeit complicated, procedure. Such information is of direct benefit to, for example, aquaculture operations. An important observation is that this type of analysis demands hyperspectral imaging with the spectral resolution that is necessary to detect subtle anomalies in specific bands. There are several initiatives under way for satellite based hyperspectral imaging by Canada (HERO Satellite) and others.

Gage and Albert (Chapter 4) focused on the numerical modeling of the radiation stream through the water column. They developed a sensor-independent software tool that generates spectra for particular water column characteristics, or can analyze spectra that have been observed for specific water columns. This "Water colour Simulator" or WASI is explained in depth and the program itself and documentation are available at http://extras.springer.com. Thirty-three equations that describe analysis of eight different spectrum types (e.g. Absorption, attenuation, specular reflectance, etc.) are provided in the chapter. The analyst can trace through the process for each spectrum calculation. The implementation method is discussed, as is the error assessment. The interface is 'user-friendly' and there are clear illustrations for a variety of practical issues. This software tool will be a decided asset for those wishing to experiment with the properties of radiance interaction for their particular site, and may very well serve as a basis for the decision to use or not use remote sensing within a specific science or management application. Specifically, the WASI program allows the scientist to evaluate what can be realistically mapped or measured with remote sensing as well as determine the contributions to the sensor signal by specific *in situ* properties.

Brock, Yates and Halley (Chapter 5) provided a comprehensive conceptual framework for incorporating remote sensing information into estimation of the community metabolism of reef systems. The authors note that the transition of benthic communities from coral-rich to micro- or macroalgal domination, which involve changes in the community metabolism, are clearly associated with shifts in the system boundary conditions. Models of coral reef ecosystems that include parameterized process functions that are scaled in both space and time by remote sensing information can be an important component of management decisions that respond to changing stressors. A comprehensive field program that addresses scaling of reef metabolism using remote sensing is detailed as an illustration of this conceptual framework. The net heterotrophic state of the reef under study is considered to be the result of organic detritus of land origin, followed by *in situ* remineralization. This dictates that water column nutrient concentrations be assessed at the benthic boundary layer to be of greatest value in management of this reef system.

3. Monitoring Applications

In the 'Monitoring Applications' section of this book are three Chapters that focus on assessing temporal change within different aspects of coastal ecology. These are:

- a detailed presentation of an *in situ* instrumentation platform for remote identification coral reef stress (Hendee et al., Chapter 6)
- illustration of the use of airborne lidar for shore DEM generation for predictive modeling of flood risk (Webster and Forbes, Chapter 7)
- assessment of the historical archives of image data relevant to coastal ecosystem change and variability (Gebelein, Chapter 8)

Chapter 6 (which can be considered as a complement to Skirving et al., Chapter 2) discusses the CREWS (Coral Reef Early Warning System) network. This network provides continuous monitoring of wind speed and gusts, direction, air and sea temperature, salinity, PAR and discrete or broadband ultraviolet radiation, along with video records of the CREWS array itself when within line of sight of a land observing post, and underwater video of the substrate. Ensemble anomalies can be determined through ongoing stochastic analysis of the data. By observation via video of the state of the substrate, these data anomalies may be confirmed as cases of coral bleaching. The authors provided a comprehensive list of issues related to permission approval, the planning sequence for installation, site criteria, and ongoing maintenance. Data validation and expert system analysis to aid interpretation are explained. Of particular interest is the provision at some CREWS sites of measurements of coral fluorescence efficiency. The next generation of pulse amplitude modulation (PAM) fluorometry has been incorporated into the CREWS architecture. A reduction in coral fluorescent yield has been associated with coral response to stress, such as a temperature increase. This instrumentation requires some biofouling maintenance, a procedure that must be built into the implementation plan. Although these instruments can be expensive to build, install and maintain (over $100K US) they will provide the definitive *in situ* temporal record to assess the impact of natural environmental stress. Their deployment should be extensive to provide a comprehensive regional and intraregional assessment of coral stress. We will then have a spatially coherent and temporally consistent archive from which to examine coral ecosystem health and consequently provide a sound basis for reef management.

Webster and Forbes (Chapter 7) highlighted the management issues associated with coastal floods that may result from the increased instances of storm surge as well as the sea level rise expected with global warming. The authors point out that the risk of storm surge in low-lying coastal areas can affect 10 million people today. This will increase to 50 to 80 million people by the 2080s, given the IPCC (Intergovernmental Panel on Climate Change) projections of the impact of climate change, and depending upon adaptive strategies and rate of population increase. Remote Sensing information from radar satellites such as RADARSAT and ENVISAT may provide accurate spatial dimensions of flooding and are useful for planning responses to flood, and airborne lidar can be useful in mapping heights and creating a Digital Terrain Model (DEM) with sufficient accuracy to map potential flood impact areas for specific flood and surge projections. The authors provide an in-depth analysis of a lidar mapping campaign as well as error assessment in creating DEMs for such flood and/or surge scenarios. Application is tested against a recent storm surge and more severe scenarios are assessed. They note that "…flood risk maps and information products have made it to the hands of the coastal resource managers, who have to deal with these risks on an annual basis, and have been incorporated into their GIS system." An evolving technology has been made operational for coastal management.

Gebelein (Chapter 8) draws attention to the vulnerability of the transitional ecosystems along coasts where human impact may be the most profound. The clear contribution of remote sensing is the digital and coherent record of change since the launch of first resource satellite, ERTS (Earth Resources Technology Satellite, renamed Landsat) in 1972. The challenge is knowing where the required imagery is archived (if at all!), how to derive meaningful information with clear confidence boundaries, and how to identify and assess the change using the satellite information combined with other image or non-image data. Looking at the Global Land Cover Facility at the University of Maryland as a case study, the author examines applications using a variety of high volume data sets. Several benchmark studies involving both image and derived information types are cited. URL's for many data archives are listed with a summary of the relevant data products. Geographic Information System (GIS) data from the World Resources Institute are reviewed as an important source for assessing environmental change and variability. As an example, the 2004 Reefs at Risk in the Caribbean (Burke et al., 2004) includes "…reef locations, threats from anthropogenic activities, river mouths, bathymetry, ports, oil/gas wells, airports marine protected areas, population densities, soil type, dive centers, and watershed boundaries." Amalgamation of such diverse but geocoded data sets is now a major tool for coastal management decisions. However, the author notes that there are inherent error sources due to geometric rectification procedures, radiometery of different sensor systems, stability of space versus aerial platforms and atmospheric attenuation. An ongoing research issue is the attenuation of the underwater signal through the water column, both downward and upward, and the impact on the spectral information of submerged targets. The author's balanced assessment of data integration issues is critical reading before application of such information to management problems.

4. Management Applications

The last third of the book is devoted to management applications. In this section:

- the spaceborne systems that can address coastal management issues are examined (Whitehouse and Hutt , Chapter 9)
- the practical integration of remote sensing data into existing field programs is explored (Phinn et al., Chapter 10)
- problems and solutions concerning the integration of remote sensing and management of tropical coral reefs in developing nations are presented (Newman et al., Chapter 11)
- the management-level concerns of data synthesis are discussed (Robinson et al., Chapter 12)

Whitehouse and Hutt (Chapter 9) note that "…spaceborne sensors have helped us to realize that the aquatic environment is not as homogeneous as once believed." This has to be balanced against the reality of effective revisit time, reception and processing time, and ordering time when planning a response to a marine disaster, such as an oil spill. The authors provide a series of tables with each satellite available for coastal management identified with relevant characteristics (revisit time etc.). For each of a variety of applications, the system is assessed in terms of whether it is optimal for the project or sub-optimal but merits consideration. In their assessment of spaceborne versus airborne or *in situ* observations, the authors write that they should not be viewed as redundant or competitive. Rather, all systems have their limitations, which are countered by relevant data from another system or systems. One must be aware of the issues surrounding each data source to be able to bring together an ensemble of data to address the management problem. We have progressed beyond the hype and overselling of the 1980s and 1990s and now have a good understanding of what each satellite service can deliver. The assessments in the Tables of this chapter will be a much needed resource for the coastal manager, and web sites that provide updates are given in the Conclusions section.

The primary message that Phinn et al. (Chapter 10) contribute is that coastal management must include three components that, in reality, are a continuum. This continuum includes initial baseline mapping and inventory (a starting snapshot in time), monitoring (the mapped and measured changes through time) and modeling (understanding how the system works and what will be the results of changing stressors through time). The authors detail the conceptual basis for their approach to linking environmental indicators to remote sensing data, and present the derived information to policy makers and stakeholders of various types. An increasingly important component of management decisions is the impact and cause of environmental change, a problem that temporal assemblages of image data are well suited to tackle. There are, however, several technical issues to be addressed, such as precision in registering one image to another, normalization of radiometric scales, correction of atmospheric attenuation, and the procedure used to detect and identify change. Furthermore, change detection involves greater complexity in validation procedures. In some applications when archive images are used, the confirmation at the earlier date may not be possible. The recent challenge to the analyst is the inclusion of increased spatial and spectral information of the new generation of satellites. These systems promise more information but this must be balanced against the validity of comparison with earlier data. The authors provide a balanced assessment of the real validity of incorporating remote sensing into a management cycle and highlight the following research objectives to improve the value of remote sensing: "1) the identification and development of algorithms (and related spectral resolution) to relate reef bio-optical properties with relevant biophysical controls; 2) further development of techniques to remove the attenuating effects of the overlying water column; 3) greater incorporation of biogeochemical cycles (e.g. climatic and oceanographic data) with remote sensing data to understand the processes that influence the biology of reefs and their subsequent bio-optical properties; and 4) evaluation and increased utilization of a greater range of image data sources" (Phinn et al., Chapter 10).

The authors provide an example of the use of remote sensing of an intrusive algal bloom in a regional management application in southeast Queensland, Australia. There is particular attention paid to the details of the technical process and involvement of, and communication with, the local community. As such, it is a benchmark case study for coastal managers who need to assess the potential of remote sensing for their particular program.

Newman et al. (Chapter 11) provided a complement to the coastal and water surface focus of Whitehouse and Hutt of Chapter 9 and concentrate on the remote sensing and management of tropical coral reefs. Typically these are found in regions of developing economies where the is a conflict between the need to effectively manage the resource, perhaps as a coral reef marine protected area (CRMPA), and project it from pollution, ecotourism and other stressors, and the need of the local people to use the reef as a food resource. A local fisherman may be quite aware of the long-term damage that pipe bombing does, but his first priority is to get food for his family's dinner. The authors note the impact of the 'digital divide' between the local community and aid organizations. A remote sensing product cannot be 'parachuted' in with an expectation for a change in behavior that the 'educated' organization is advocating. An approach based upon 'design with culture' that involves the community from the formative stages is discussed. Effective communication is a critical component. In the instance when a group may not respond to a map because their tradition is to list things, not to spatially orient things, other approaches, such as story telling, may be more effective. The authors note that: "Being attentive to local traditions, knowledge, and beliefs provides significant benefits, as the local people:

- are directly aware of the current status of the marine resource on which they depend
- possess an inherent temporal knowledge of patterns within and changes to the marine environment
- are valuable resources for developing monitoring and managing strategies.

It is also important to be aware of the challenges that accompany local knowledge and the limitations of local knowledge's usefulness, which may include:

- the local knowledge is contained in a verbal medium and is often unverified
- the attitudes of local people towards their resource may not include a conservation outlook (which can discourage proper management plans)
- the local knowledge is exclusive to the area, with minimal overlap between regions within small spatial proximities
- the variability of human and physical events from one shallow coastal water environment to another may prevent the incorporation of an approach that was deemed successful elsewhere." (Newman et al., Chapter 11, this book).

In Chapter 12, Robinson et al., address the issue of data synthesis. With the explosion in Geographical Information Systems (GIS) which include layers of digital maps, data fusion which includes integration of many different data types to distill more meaningful data, and change detection techniques whereby typically temporal sequences of data are co-registered and analyzed, there is a concomitant explosion of ideas and processes that deal with synthesis of different data types. The authors review data sets of both cartographic and remote sensing origin, providing an assessment of availability. The remote sensing information is reviewed at high (< 10 meter pixels), moderate (30 to 100 meter) and low resolution with several examples given with URLs. Several, such as the US Coral Reef Taskforce mapping and the Nature Conservancy, are custom-designed for coastal management applications. The authors assess the analytical challenges for data synthesis, which include problems with misregistration of different data types, various resolutions combined in a single product, and new analysis challenges associated with differences between raster and vector data sets. Development of second generation image analysis tools, such as object-oriented classification, have been able to address some of these issues. Another issue is accessibility to information with problems that range from inappropriate spatial data analysis before archiving and distribution (e.g. every pixel being a polygon) to legal ownership of certain data components now that some image sources are private entities. The chapter concludes with an example of a very successful data integration project, the "Reefs at Risk" activities in both Southeast Asia and, more recently, the Caribbean. These integrated data products are accessible through the ReefBase project which includes these data as well as many others, such as the NOAA coral hotspot information (Skirving et al., Chapter 2, this book). We are finally entering an era when disparate data can be combined to provide new analytical insights into processes of stress on natural coastal ecosystems.

5. Where Are We Heading?

These conclusions and recommendations convey a very optimistic note, but are based on what we have seen in this field in recent years. The tremendous progress that we have observed during the past decade has been the result of moving beyond the concept of producing surface-cover classifications of the image. We are now examining physical processes that are represented through a combination of increasingly sophisticated image data and other georeferenced data sets, and are working with multiple indicators that can be combined to make the 'fingerprint'. This fingerprint can provide insight into what is happening, both in space and in time. For example, we are embarking upon models of community metabolism for coral reef systems that include parameterized process functions scaled in space and time by remote sensing image data.

We are also seeing tangible remote sensing support being used for analysis of the particular problems of developing economies which, typically, have been on the other side of the 'digital divide'. We are not only providing data, or western notions of environmental solutions, but have moved to the point that the management at the community level is being empowered by having visual evidence of what is occurring around them. More importantly, the skills and technical resources to build this information have been successfully integrated into the activities and plans of an increasing number of developing nations. The satellite image is truly becoming a global currency of information.

We believe that we are now at the turning point where remote sensing can be integrated into aquatic coastal ecosystem science and management at the process level. It will be viewed as one necessary data source that is credible and readily analyzed in concert with traditional information to yield a new dimension for the scientist and manager. With modest training, one can produce valuable information on a desk-top computer with data provided on a DVD. It is our hope that this book can help foster communication between scientists, managers, and remote sensing specialists. At this point in time the transition necessary to integrate remote sensing in existing and planned projects has never been easier.

INDEX

T

U

V

W

Z

MIX
Papier aus verantwortungsvollen Quellen
Paper from responsible sources
FSC® C105338

If you have any concerns about our products,
you can contact us on
ProductSafety@springernature.com

In case Publisher is established outside the EU,
the EU authorized representative is:
Springer Nature Customer Service Center GmbH
Europaplatz 3, 69115 Heidelberg, Germany

Printed by Libri Plureos GmbH
in Hamburg, Germany